Genetics and Genomics of High-Altitude Crops

Vijay Gahlaut • Vandana Jaiswal
Editors

Genetics and Genomics of High-Altitude Crops

Springer

Editors
Vijay Gahlaut
University Centre for Research & Development
Chandigarh University
Mohali, Punjab, India

Vandana Jaiswal
Biotechnology Division
Institute of Himalayan Bioresource Technology
Palampur, Himachal Pradesh, India

ISBN 978-981-99-9174-7 ISBN 978-981-99-9175-4 (eBook)
https://doi.org/10.1007/978-981-99-9175-4

This Springer imprint is published by the registered company Springer Nature Singapore Pte Ltd.
The registered company address is: 152 Beach Road, #21-01/04 Gateway East, Singapore 189721, Singapore

Paper in this product is recyclable

To the resilient farmers and scientists who have tirelessly cultivated knowledge and crops at the highest reaches of our planet. Your unwavering dedication to understanding the secrets of high-altitude genetics and genomics has not only enriched our scientific understanding but has also contributed to the sustenance and resilience of communities around the world.
May this book serve as a tribute to your passion, perseverance, and the indomitable human spirit that thrives even in the most challenging of environments. Your commitment to unlocking the genetic code of high-altitude crops has the potential to change the course of agriculture and food security for generations to come.
In honor of your tireless efforts, we dedicate this book, Genetics and Genomics of High-Altitude Crops, with gratitude and admiration.

Foreword

This volume *Genetics and Genomics of High-Altitude Crops*, edited by Dr. Vijay Gahlaut and Dr. Vandana Jaiswal, provides an up-to-date account of knowledge on genetics and genomics of HACs using some selected crops, on which work has already been undertaken. The subject has been an active area of research, but the wealth of information generated on this subject so far was scattered in the form of research articles published in journals and was not available at one place for students, teachers, and research workers. The editors successfully filled this gap and provided all the available information at one place in the form of chapters written by experts in their own areas of research. The HACs, which have largely been used for the study of genetics and genomics include Amaranthus, buckwheat, Chenopodium, and saffron. The research work done on these crops has improved our understanding of how genetics and genomics of HACs have evolved towards adaptation to the rigours of high-altitude environments.

The chapters within this book are organized into the following four sections: (1) the genetics and genomics of HACs, (2) the effects of climate change on HACs, (3) nutritional significance of HACs, and (4) molecular breeding for HACs. The authors of different chapters are experts in these different fields, providing authoritative and up-to-date information. Therefore, the book should prove a valuable resource for researchers and plant biologists, who are either already engaged or plan to initiate study of these HACs.

The editors of this volume deserve appreciation for undertaking the task of bringing knowledge to this important subject at one place. The book should prove useful to those for whom it has been written.

Ch. Charan Singh University, Meerut, India
Murdoch University, Perth, WA, Australia
CIMMYT-BISA, New Delhi, India

P. K. Gupta

Preface

In the realm of agriculture, where the challenges of food security and sustainability loom ever larger, the study of high-altitude crops (HACs) has emerged as a fascinating and crucial field of research. High-altitude environments, with their harsh conditions of extensive UV-B irradiation, extreme low temperatures, and intense hypoxia, impose remarkable challenges on plant life. These environmental constraints, in turn, give rise to genomic variations in HACs, sparking the intricate dance of adaptation to thrive in the face of adversity.

The genetic and genomic exploration of these HACs is at the heart of our book, *Genetics and Genomics of High-Altitude Crops*. This volume brings together a wealth of knowledge and insights garnered from years of dedicated research. It compiles and summarizes the findings of numerous studies that have dissected the genetic and genomic regions of HACs, shedding light on their remarkable capacity to cope with environmental stresses.

Yet, until now, these valuable studies have been dispersed throughout the scientific literature. This book seeks to rectify that by presenting a comprehensive collection of recent advances in the genetics and genomics of HACs. It offers readers a succinct description of the applications and implications of this fascinating field, making it accessible to both the unfamiliar and the seasoned. The genomic analysis of other HACs, such as the Tibetan wheat accessions and Eutrema species, has deepened our understanding of how plants adapt to high-altitude conditions.

The chapters in this book are organized into four sections: genetics and genomics of HACs, the potential effects of climate change on these crops, their nutritional significance, and the promising field of molecular breeding. Experts in the subject have contributed to each chapter, ensuring that the content is both authoritative and up to date. Our hope is that this book will serve as an invaluable resource for researchers and plant biologists, providing them with the most current information in this dynamic field.

The first chapter focuses into the genetics and genomics of pulse crops in mountainous regions. These crops hold immense promise in diversifying food sources and addressing global food security challenges. The chapter explores how genomics can revolutionize pulse crop cultivation in challenging terrains, ultimately contributing to a more food-secure and sustainable future. The second chapter was on kodo millet, a climate-resilient and nutrition-rich crop. Despite its potential, kodo millet has faced limitations due to limited genetic resources and the absence of a genome sequence. This chapter summarizes the current research status and highlights the future potential of integrating genomics and gene editing techniques into crop improvement programs. Chapter 3 ventures into the world of saffron, a high-value industrial herb with a rich history. By understanding the genetic regulation of saffron's bioactive compounds, we can enhance its cultivation and quality, benefiting both traditional and modern medicine. Chapter 4 explores the genetics and genomics of the Fritillaria species, a group of plants with a long history of use in traditional medicine. This chapter offers insights into the bioactive molecules found in Fritillaria and their potential therapeutic applications. It also delves into the genetic factors shaping the genome size of these plants.

The fifth chapter introduces us to *Bunium persicum*, also known as black cumin, a plant with significant economic importance and medicinal value. The chapter discusses the genetic resources and advanced molecular tools needed to improve this crop, ensuring its sustainable consumption and cultivation. The sixth chapter takes us to the world of buckwheat, a versatile pseudocereal crop. This chapter summarizes our current understanding of buckwheat genetics, genomics, and breeding, highlighting its potential for enhanced productivity and adaptability in changing environmental conditions. Chapter 7 explores the genetic resources of Amaranthus, buckwheat, and Chenopodium, collectively known as orphan crops. These underutilized crops offer genetic diversity and nutritional benefits, making them essential for food security. The chapter emphasizes the importance of genetic variability and the need for genomic studies to harness their potential fully. Chapter 8 delves into the genetic resources and breeding approaches for Ferula assa-foetida, a valuable herb with a wide range of applications. This chapter discusses the challenges faced by this species and the potential for genetic improvement through agrotechnology.

Our final chapter explores the future prospects of HAC improvement through genomics. High-altitude regions, with their unique climatic conditions, are vital for crop diversity and food security. The chapter discusses the role of genomics techniques in enhancing crop yield, climate resilience, and nutritional value in challenging landscapes.

As we embark on this journey through the genetics and genomics of HACs, we hope that this book will serve as an illuminating guide for researchers, scientists, and anyone interested in the future of agriculture in these dynamic and demanding environments. The collective knowledge contained within these pages offers a glimpse into the remarkable adaptability and potential of HACs, highlighting the path forward towards a more sustainable and food-secure world.

Mohali, Punjab, India — Vijay Gahlaut
Palampur, Himachal Pradesh, India — Vandana Jaiswal

Contents

1 **On the Road to a Sustainable and Climate-Smart Future: Recent Advancements in Genetics and Genomics of Pulse Crops in the Hills** 1
Kanishka R. C., Moatoshi Jamir, Sakuonuo Theunuo, Basavaraja T., Harendra Verma, and Rahul Chandora

2 **Genetics and Genomics of Kodo Millet (*Paspalum scrobiculatum* L.)** 47
Pooja Shukla, Shivani Shukla, Kajal Pandey, Pooja Choudhary, Ravikesavan Rajasekaran, and Mehanathan Muthamilarasan

3 **Genetic and Molecular Advancements in Saffron (*Crocus sativus* L.)** 65
Vishek Choudhary, Anita Choudhary, Vijay Gahlaut, and Vandana Jaiswal

4 **Insight into the Genetics and Genomics Studies of the *Fritillaria* Species** 89
Vinay Kumar, Shagun Sharma, and Pankaj Kumar

5 **Genetic and Genomic Resources of *Bunium persicum* (Boiss.) Fedtsch** 115
Sapna, Satakshi Sharma, Ramesh Chauhan, and Satbeer Singh

6 **Genetic and Breeding Advancement in Buckwheat: A Pseudocereal of Himalaya** 131
Vishal Kumar, Priya Kumari, Himanshi Gangwar, Vishek Choudhary, Vijay Gahlaut, and Vandana Jaiswal

7 **Advancing Food Security with Genetic Resources of *Amaranthus*, Buckwheat, and *Chenopodium*** 159
Kanishka R. C, Mithilesh Kumar, Gopal Katna, Kirti Rani, Krishan Prakash, Rakesh Kumar Bairwa, and Rahul Chandora

8 Genetic Resources and Breeding Approaches for Improvement of *Ferula assa-foetida* (*Heeng*) 199
Niketa Yadav, Satbeer Singh, Ramesh Chauhan, Sanatsujat Singh, and Ashok Kumar

9 Future Prospects: High-Altitude Crop Improvement via Genomics 217
Vikas Kumar Singh, Ronika Thakur, Jatin Sharma, Ashita Bisht, Kumar Sanu, Arushi Arora, Deepak Bhamare, Neeraj Pal, and Dinesh Kumar Saini

Editors and Contributors

About the Editors

Vijay Gahlaut currently holds the position of Assistant Professor at the University Centre for Research and Development, Chandigarh University, Mohali. His dedicated research is centered around the identification of novel genomic regions and genes responsible for the regulation of abiotic stress in crop plants. Dr. Gahlaut's expertise spans across multiple domains, including crop biotechnology, quantitative genetics, QTL/GWAS (Quantitative Trait Loci/Genome-Wide Association Studies), and epigenetics. Utilizing a diverse array of bioinformatic and genomic methodologies, he has achieved remarkable success in pinpointing numerous critical genomic regions, epialleles, and candidate genes within various crop species. These findings hold significant importance as they contribute to enhancing the resilience of crops to challenging environmental conditions, particularly water scarcity and heat stress. Dr. Gahlaut's impactful research efforts have resulted in the publication of over 50 research papers in esteemed international journals, in addition to several contributions in the form of book chapters. Furthermore, he actively contributes to the scientific community by serving as an Associate Editor and an independent peer reviewer for several renowned journals. Notably, his remarkable achievements have earned him the distinguished honor of being inducted as a Member of the National Academy of Sciences, India (NASI), a recognition that underscores the significant impact of his work in the field of crop genetics and abiotic stress tolerance.

Vandana Jaiswal is currently working as a Scientist at CSIR-IHBT, Palampur, India. Her work focuses on unveiling the genetics of economically important traits and the improvement of Himalayan crops through molecular approaches. Saffron is one of the most important Himalayan crops, which is triploid in nature, with no seed setting reported, and propagated exclusively through corms. Due to its sterile nature, genetic improvement of saffron is very challenging. She has initiated the application of molecular techniques for genetic improvement of saffron. She has published more than 40 research papers and book chapters. She also holds the responsibility of Associate Editor and independent peer reviewer for many journals. She is the recipient of various prestigious awards like the CSIR-Young Scientist Award, DST-INSPIRE Faculty Award, Early Career Research Award, and the Women Scientist Awards from various Indian government agencies.

Contributors

Arushi Arora Department of Plant Breeding and Genetics, Punjab Agricultural University, Ludhiana, Punjab, India

Rakesh Kumar Bairwa ICAR-Directorate of Mushroom Research, Solan, Himachal Pradesh, India

T. Basavaraja ICAR-Indian Institute of Pulses Research (IIPR), Kanpur, Uttar Pradesh, India

Deepak Bhamare School of Agricultural Biotechnology, Punjab Agricultural University, Ludhiana, Punjab, India

Ashita Bisht CSK Himachal Pradesh Krishi Vishwavidyalaya, Highland Agricultural Research and Extension Centre, Kukumseri, Himachal Pradesh, India

Rahul Chandora ICAR-National Bureau of Plant Genetic Resources (NBPGR), Regional Station, Shimla, Himachal Pradesh, India

Ramesh Chauhan Division of Agrotechnology, Council of Scientific and Industrial Research - Institute of Himalayan Bioresource Technology, Palampur, Himachal Pradesh, India

Anita Choudhary Biotechnology Division, CSIR-Institute of Himalayan Bioresource Technology, Palampur, Himachal Pradesh, India

Academy of Scientific and Innovative Research (AcSIR), Ghaziabad, Uttar Pradesh, India

Pooja Choudhary Department of Biotechnology, Jaypee Institute of Information Technology, Noida, Uttar Pradesh, India

Vishek Choudhary Biotechnology Division, CSIR-Institute of Himalayan Bioresource Technology, Palampur, Himachal Pradesh, India

Academy of Scientific and Innovative Research (AcSIR), Ghaziabad, Uttar Pradesh, India

Vijay Gahlaut Biotechnology Division, CSIR-Institute of Himalayan Bioresource Technology, Palampur, Himachal Pradesh, India

Department of Biotechnology and University Center for Research and Development, Chandigarh University, Mohali, Punjab, India

Himanshi Gangwar Biotechnology Division, CSIR-Institute of Himalayan Bioresource Technology, Palampur, Himachal Pradesh, India

Academy of Scientific and Innovative Research (AcSIR), Ghaziabad, Uttar Pradesh, India

Vandana Jaiswal Biotechnology Division, CSIR-Institute of Himalayan Bioresource Technology, Palampur, Himachal Pradesh, India

Academy of Scientific and Innovative Research (AcSIR), Ghaziabad, Uttar Pradesh, India

Moatoshi Jamir ICAR Research Complex for NEH Region, Nagaland Centre, Medziphema, Nagaland, India

R. C. Kanishka ICAR-National Bureau of Plant Genetic Resources (NBPGR), Regional Station, Shimla, Himachal Pradesh, India

Gopal Katna Chaudhary Sarwan Kumar Himachal Pradesh Krishi Vishvavidyalaya (CSK HPKV), Palampur, Himachal Pradesh, India

Ashok Kumar Division of Agrotechnology, Council of Scientific and Industrial Research - Institute of Himalayan Bioresource Technology, Palampur, Himachal Pradesh, India

Mithilesh Kumar Agricultural Research Station, Mandor of Agriculture University, Jodhpur, Rajasthan, India

Pankaj Kumar Department of Biotechnology, Dr. Y.S. Parmar University of Horticulture and Forestry, Solan, Himachal Pradesh, India

Vinay Kumar Department of Biotechnology, Dr. Y.S. Parmar University of Horticulture and Forestry, Solan, Himachal Pradesh, India

Vishal Kumar Biotechnology Division, CSIR-Institute of Himalayan Bioresource Technology, Palampur, Himachal Pradesh, India

Academy of Scientific and Innovative Research (AcSIR), Ghaziabad, Uttar Pradesh, India

Priya Kumari Biotechnology Division, CSIR-Institute of Himalayan Bioresource Technology, Palampur, Himachal Pradesh, India

Academy of Scientific and Innovative Research (AcSIR), Ghaziabad, Uttar Pradesh, India

Mehanathan Muthamilarasan Department of Plant Sciences, School of Life Sciences, University of Hyderabad, Hyderabad, Telangana, India

Neeraj Pal Department of Molecular Biology and Genetic Engineering, College of Basic Science & Humanities, Govind Ballabh Pant University of Agriculture and Technology, Pantnagar, Uttarakhand, India

Kajal Pandey Department of Biotechnology, Jaypee Institute of Information Technology, Noida, Uttar Pradesh, India

Krishan Prakash ICAR-Indian Agricultural Research Institute, Hazaribagh, Jharkhand, India

Ravikesavan Rajasekaran Department of Millets, Centre for Plant Breeding and Genetics, Tamil Nadu Agricultural University, Coimbatore, Tamil Nadu, India

Kirti Rani ICAR-National Bureau of Plant Genetic Resources (NBPGR), Regional Station, Jodhpur, Rajasthan, India

Dinesh Kumar Saini Department of Plant Breeding and Genetics, Punjab Agricultural University, Ludhiana, Punjab, India

Department of Plant and Soil Science, Texas Tech University, Lubbock, TX, USA

Kumar Sanu Department of Genetics and Plant Breeding, CSKHPKV, Palampur, Himachal Pradesh, India

Sapna Division of Agrotechnology, Council of Scientific and Industrial Research - Institute of Himalayan Bioresource Technology, Palampur, Himachal Pradesh, India

Jatin Sharma Department of Plant Breeding and Genetics, Punjab Agricultural University, Ludhiana, Punjab, India

Satakshi Sharma Division of Agrotechnology, Council of Scientific and Industrial Research - Institute of Himalayan Bioresource Technology, Palampur, Himachal Pradesh, India

Shagun Sharma Department of Biotechnology, Dr. Y.S. Parmar University of Horticulture and Forestry, Solan, Himachal Pradesh, India

Pooja Shukla Department of Plant Sciences, School of Life Sciences, University of Hyderabad, Hyderabad, Telangana, India

Shivani Shukla Department of Plant Sciences, School of Life Sciences, University of Hyderabad, Hyderabad, Telangana, India

Sanatsujat Singh Division of Agrotechnology, Council of Scientific and Industrial Research - Institute of Himalayan Bioresource Technology, Palampur, Himachal Pradesh, India

Satbeer Singh Division of Agrotechnology, Council of Scientific and Industrial Research - Institute of Himalayan Bioresource Technology, Palampur, Himachal Pradesh, India

Vikas Kumar Singh Department of Genetics and Plant Breeding, Ch. Charan Singh University, Meerut, Uttar Pradesh, India

Ronika Thakur Department of Genetics and Plant Breeding, CSKHPKV, Palampur, Himachal Pradesh, India

Sakuonuo Theunuo ICAR Research Complex for NEH Region, Nagaland Centre, Medziphema, Nagaland, India

Harendra Verma ICAR Research Complex for NEH Region, Nagaland Centre, Medziphema, Nagaland, India

Niketa Yadav Division of Agrotechnology, Council of Scientific and Industrial Research - Institute of Himalayan Bioresource Technology, Palampur, Himachal Pradesh, India

Abbreviations

2,4-D	2–4 dichlorophenoxyacetic acid
AC	Activated charcoal
AFLP	Amplified fragment length polymorphism
AICSMIP	All India Coordinated Research Project on Small Millets
ALDH	Aldehyde dehydrogenase
AMOVA	Analysis of molecular variance
ARL	ARGOS-LIKE
AS	Ascochyta blight
BAC	Bacterial artificial chromosome
BBBMs	Biotechnology-based breeding methods
BGDB	Buckwheat genome database
bp	Base pairs
Ca	Calcium
CAAS	Chinese Academy of Agricultural Sciences
CBB	Common bacterial blight
CCD2	Carotenoid cleavage dioxygenase 2
CCS	Circular consensus sequencing
CeHAB	Center for high-altitude biology
CGIAR	Consultative Group on International Agricultural Research
ChS	Chalcone synthase
CIM	Composite interval mapping
cM	Centimorgan
CPA	Cyclopiazonic acid
CRISPR	Clustered regularly interspaced short palindromic repeats
CRISPR/Cas	Clustered regularly interspaced short palindromic repeats/CRISPR-associated protein
CrtISO	Carotenoid isomerase
CuLCrV	Cucurbit leaf crumple virus
CWR	Crop wild relatives
DArT	Diversity arrays technology markers
DH	Doubled haploid
DODA	l-DOPA 4,5-dioxygenase
DPPH	2,2-diphenyl-1-picrylhydrazyl
EMT	Epithelial-mesenchymal transition

EPRV	Endogenous pararetrovirus
EST	Expressed sequence tags
FAO	Food and Agriculture Organization
Fe	Iron
FISH	Fluorescent in situ hybridization
FPKM	Fragments per kilobase million
FRAP	Ferric reducing antioxidant power
FW	Fusarium wilt
GAB	Genomics-assisted breeding
Gb	Gigabase
GBS	Genotyping by sequencing
GBSS	Granule-bound starch synthase
GCV	Genotypic coefficient of variation
GEBV	Genomic estimated breeding value
GHI	Global hunger index
GI	Geographical indication
GM	Genetically modified
GS	Genomic selection
GWAS	Genome-wide association studies
HDAC	Histone deacetylase
IBA	Indole-3-butyric acid
IBPGR	International Board for Plant Genetic Resources
ICAR	Indian Council of Agricultural Research
ICARDA	International Centre for Agricultural Research in Dry Areas
ICRISAT	International Crop Research Institute for the Semi-Arid Tropics
IHR	Indian Himalayan Region
IITA	International Institute of Tropical Agriculture
InDel	Insertion/deletion
iPBS	Inter-primer binding site
IRAP	Inter-retroelement amplified polymorphism
ISM	Indian System of Medicine
ISSR	Inter-simple sequence repeat
ITS	Internal transcribed spacer
IYM	International Year of Millets
LTS	Long-term storage
LYC-B	Lycopene β-cyclase
MAB	Marker-assisted breeding
MAGIC	Multiple advanced generation intercross
MARS	Marker-assisted recurrent selection
MAS	Marker-assisted selection
MI	Marker index
MS-AFLP	Methyl sensitive amplified fragment length polymorphism
MTA	Marker-trait associations
MTS	Mid-term storage
MUFA	Monounsaturated fatty acid

NAA	Naphthalene acetic acid
NADP	Nicotinamide adenine dinucleotide phosphate
NAM	Nested association mapping
NBPGR	National Bureau of Plant Genetic Resources
NGS	Next-generation sequencing
NILs	Near isogenic lines
NIR	Near-infrared
NJ	Neighbor-joining
NJA	Neighbor-joining analysis
PAL	Phenylalanine ammonia lyase
PCA	Principal component analysis
PCoA	Principal coordinate analysis
PCR	Polymerase chain reaction
PCV	Phenotypic coefficient of variation
PDS	Phytoene desaturase
PEBP	Phosphatidyl ethanolamine-binding protein
PGRs	Plant growth regulators
PIC	Polymorphic information content
PREs	Polyphenol-rich extracts
PS	Photoperiod sensitivity
PTC	Plant tissue culture
PUFA	Polyunsaturated fatty acids
PVE	Phenotypic variance
QTL	Quantitative trait loci
RAD-seq	Restriction-site associated DNA sequencing
RAPD	Random amplified polymorphic DNA
RDI	Recommended dietary intake
RFLP	Restriction fragment length polymorphism
RILs	Recombinant inbred lines
ROS	Reactive oxygen species
RPKM	Reads per kilobase per million
RRC	Rodale Research Center
SAE	Saffron aqueous extract
SAGE	Serial analysis of gene expression
SCAR	Sequence characterized amplified region
SCN	Soybean cyst nematode
SCoT	Start codon targeted
SDG	Sustainable development goal
Se	Selenium
SEMs	Structural equation models
SiGMFV	SIDA golden mosaic Florida virus
SM	Secondary metabolites
SMD	Sterility mosaic disease
SMRT	Single molecule real time
SNP	Single nucleotide polymorphism

SRAP	Sequence-related amplified polymorphism
SSR	Simple sequence repeat
STS	Sequence-tagged sites
TALENs	Transcription activator-like effector nucleases
TE	Transposable elements
TNAU	Tamil Nadu Agricultural University
TPKM	Transcripts per kilobase million
TraVA	Transcriptome variation analysis
UGT	Uridine diphosphate glycosyltransferase
USDA	United States Department of Agriculture
UV	Ultraviolet
ZDS	Zeta-carotene desaturase
ZFNs	Zinc finger nucleases
Z-ISO	Zeta-carotene isomerase
Zn	Zinc
ZSB	Zinc-solubilizing bacteria

On the Road to a Sustainable and Climate-Smart Future: Recent Advancements in Genetics and Genomics of Pulse Crops in the Hills

1

Kanishka R. C., Moatoshi Jamir, Sakuonuo Theunuo, Basavaraja T., Harendra Verma, and Rahul Chandora

1.1 Introduction

Globally, meeting food and nutritional security has become a threat, considering over 828 million people who are enduring acute food insecurity and persistent food deprivation (FAO et al. 2022). The increasing demand for food and the potential impacts of climate change have underscored the urgent need to diversify our food sources. This involves reducing our heavy reliance on cereal crops and exploring alternative food groups to address critical issues like hidden hunger, which can significantly hinder a nation's development. Pulses, as essential food crops, have the potential to play a significant role in addressing global challenges related to food security, nutrition, and environmental sustainability. This is largely due to their extensive genetic diversity, which can be harnessed to develop nutrient-dense and climate-resilient varieties. Currently, more than one billion people are facing malnutrition, particularly protein and micronutrient deficiencies. Therefore, it is crucial to prioritize pulse production, as emphasized by Godfray et al. (2010), to overcome malnutrition, especially in less-developed countries. Pulses stand out because of their high protein, vitamin, and mineral content, as highlighted by Broughton et al.

Kanishka R. C. and Moatoshi Jamir contributed equally to this work.

K. R. C. · R. Chandora (✉)
ICAR-National Bureau of Plant Genetic Resources, Shimla, Himachal Pradesh, India
e-mail: rahul1@icar.gov.in

M. Jamir · S. Theunuo · H. Verma
ICAR Research Complex for NEH Region, Nagaland Centre, Medziphema, Nagaland, India

B. T.
ICAR-Indian Institute of Pulses Research (IIPR), Kanpur, Uttar Pradesh, India

V. Gahlaut, V. Jaiswal (eds.), *Genetics and Genomics of High-Altitude Crops*,
https://doi.org/10.1007/978-981-99-9175-4_1

(2003). Their importance has also been recognized by the United Nations, leading to the celebration of the International Year of Pulses (2016) and (Calles et al. 2019).

Pulses are part of the Leguminosae family, which ranks as the third-largest plant family in the world, second only to cereals in terms of their significance as a source of food for humans and animals. These legumes are often referred to as 'smart food' due to their critical role in various cuisines (such as dal-roti and dal-chawal) and their importance as a source of plant-based protein. Consuming pulses has been linked to a reduced risk of chronic diseases like obesity and diabetes. They hold a vital place in the diets of people worldwide, particularly those with lower incomes, as highlighted by Bressani and Elías (1979). Pulses are not only cost-effective but also nutritionally superior to cereals, providing two to three times more protein. They contribute to a balanced macronutrient intake, especially in countries with rapidly growing populations, like India, where a substantial vegetarian population resides, as noted by Kaur and Singh (2007). Additionally, pulses offer essential minerals (such as iron, zinc, and magnesium), vitamins (including folate, thiamin, and niacin), and beneficial compounds like isoflavonoids in human diets, as mentioned by Kumar et al. (2022) and Cannon et al. (2009).

Pulses are particularly valuable for their ability to fix atmospheric nitrogen through a symbiotic relationship with *Rhizobium* spp. in their root nodules. This nitrogen fixation not only enhances soil fertility and productivity for subsequent cereal crops but also reduces cultivation costs significantly, thereby increasing overall investment efficiency. Pulses demonstrate high water efficiency and can thrive in marginal, rain-fed, and drought-prone areas. Their adaptability to mixed and intercropping systems, as well as multiple cropping, promotes diversification and helps bridge nutritional and production gaps.

In summary, pulses are an essential component of sustainable cropping systems. Introducing them into agricultural practices can enhance resilience to changing environmental conditions, as depicted in Fig. 1.1. More than 75% of the pulses' production is contributed by developing countries, and the rest is fulfilled by developed countries (FAOSTAT 2021). There is a chronic protein deficiency in virtually every developing country. In the post-green revolution era, the areas under pulse cultivation have decreased considerably. In India, the green revolution has not increased the yield of pulses but achieved self-sufficiency in food, and pulse crops still await revolution. So far, improvement in the production of major pulse crops has not transpired over the past decades resulting in low productivity.

This resulted in a large gap between pulse crops' potential and actual yield. In addition, due to the changing climatic scenario, diseases and pests continue to be a great threat to the agricultural system. Evaluation and utilization of genetic resources will be a non-ending process because the development of new and more virulent strains/biotypes of various pathogens is also a non-ending natural phenomenon. It is not a battle you win just once. Pathogens and insects evolve continually seeking out the vulnerability of hosts and developing resistance to pesticides that enable them to penetrate crop defences. Plant genetic resources serve as vital repositories of genes that confer resistance to a wide range of biotic and abiotic stresses, along with numerous traits related to yield and quality. Leveraging these genetic resources

Fig. 1.1 Plants in field conditions: (**a**) French bean, (**b**) chickpea, (**c**) pigeon pea, (**d**) cowpea, (**e**) rice bean, and (**f**) faba bean

through efficient crop breeding programmes holds the potential to significantly enhance both productivity and resilience in the face of increasing biotic and abiotic challenges. This utilization can play a pivotal role in addressing food security issues and countering the adverse impacts of climate change on global agriculture, as discussed by Henry and Nevo (2014) and Tai et al. (2014).

In the realm of pulse crop improvement, there have been limited initiatives focused on leveraging genomics-assisted breeding (GAB). This scarcity of progress in GAB is primarily attributed to the absence of essential genomic tools, as noted by Varshney et al. (2009). However, recent years have witnessed a notable expansion in the availability of genomic tools, technologies, and platforms specifically tailored for pulse crops. This development has created new avenues for the implementation of GAB. While substantial strides have been made in advancing genomics and molecular breeding for chickpea and pigeon pea, comprehensive updates on recent advancements remain somewhat lacking, as discussed by Varshney et al. (2013). This new era of crop improvement, powered by molecular techniques, has greatly improved our understanding of the genetics of traits and genome structure. In the past decade, there has been a remarkable enhancement in the throughput and accuracy of genome sequencing technologies. These advancements have facilitated the creation of chromosome-scale genome assemblies in various crops, thanks to the emergence of third-generation sequencing technologies. The improved availability of genome sequence information has significantly bolstered gene mapping strategies, leading to the discovery of allelic variations at the genetic level. These discoveries regarding genes and traits have empowered us with precision and efficiency in crop breeding programmes. Consequently, a diverse range of enhanced crop varieties has emerged, offering farmers a platform for improved cultivation practices. Furthermore, concurrent advancements in genome editing techniques have further enhanced our ability to make precise and rapid modifications in plant genomes. As a result, breeding methods like genomic selection (GS) have gained more relevance in continually improving populations and enhancing the rate of genetic progress. This is largely attributed to their utilization of genome-wide marker information. The integration of genomics-assisted breeding, bolstered by the growing array of genomic tools and technologies, has ushered in a new era of crop improvement. This transformative approach holds great promise for addressing agricultural challenges and delivering enhanced crop varieties to farmers.

Pulse crops present an opportunity or vehicle through which the livelihood of especially resource-poor small and marginal farmers of the hilly regions can be improved if the potential of these crops is tapped well. In mountain regions particularly, people live in highly dispersed small hamlets and villages and practice subsistence farming by utilizing their rich genetic resources. However, limited information and awareness among people in the hills, with regard to the high genetic diversity found in the mountains and adequate nutrition along with lack of crop diversification, owing to high dependency on cereal crops, have left them vulnerable to nutritional deficiencies which can be attenuated by promoting the increased production of the diverse array of pulse crops found in the region that in turn can also be contributory in increasing the income and livelihood of farmers in the mountains. This chapter's goal is to discuss the genetics and genomics of pulse crops in hilly regions and offer insights into their adaptation, resilience, and potential for improvement. We also explore the research that has the potential to revolutionize pulse crop cultivation in challenging terrains, contributing to global food security and sustainable agriculture.

1.2 Plant Genetic Resources Management

Pulses have a substantial amount of genetic variability existing in the form of released varieties, landraces, CWRs, and other wild forms. Globally, ~1.1 million grain legume germplasm accessions are conserved in different gene banks (Sharma et al. 2013). The genetic resources of pulses are a reservoir of several valuable genes/alleles or are likely to possess many different traits such as agronomic and other qualities, and it is a 'treasure chest' of unknown resources; they contribute in present and future crop improvement programmes of pulses, enhance sustainability, and respond to climate change and meet human needs.

Table 1.1 Global germplasm holdings of French bean, pigeon pea, chickpea, rice bean, faba bean, and cowpea

Name of gene bank	Country name	Accessions
***Phaseolus vulgaris* (French bean); total accessions 56,861**		
Centro Internacional de Agricultura Tropical (COL003)	Colombia	32,359
Western Regional Plant Introduction Station, USDA-ARS, Washington State University (USA022)	United States	13,433
Seed Savers Exchange (USA974)	United States	4619
Estación Experimental Santa Catalina (ECU077)	Ecuador	1578
National Agricultural and Food Centre (NPPC), Research Institute of Plant Production (RIPP) (SVK001)	Slovakia	1175
Genebank, Leibniz Institute of Plant Genetics and Crop Plant Research (DEU146)	Germany	974
Banco Português de Germoplasma Vegetal (PRT001)	Portugal	781
SADC Plant Genetic Resources Centre (ZMB030)	Zambia	747
International Livestock Research Institute (ETH013)	Ethiopia	126
CREA-Centro di RicercaOrticoltura e Florovivaismo - Sede di Monsampolo del Tronto (ITA392)	Italy	171
Institute of Genetic Resources, University of Banja Luka (BIH039)	Bosnia and Herzegovina	166
Genetic Resources Institute (AZE015)	Azerbaijan	114
Albania Gene Bank (ALB017)	Albania	111
CREA-Centro di RicercaGenomica e Bioinformatica - Sede di Montanaso Lombardo (ITA393)	Italy	92
National Plant Genetic Resources Centre (ZMB048)	Zambia	92
Research Institute of Crop Husbandry (AZE003)	Azerbaijan	82
Centro Nacional de RecursosFitogenéticos (ESP004)	Spain	43
National (CYPARI) Genebank, Agricultural Research Institute, Ministry of Agriculture, Rural Development and Environment (CYP004)	Cyprus	31
Millennium Seed Bank - Royal Botanic Gardens Kew (GBR004)	United States	6
Institute of Agriculture (UKR004)	Ukraine	30
Generalidad Valenciana. Universidad Politécnica de Valencia. Escuela Técnica Superior de IngenierosAgrónomos. Banco de Germoplasma (ESP026)	Spain	25
Israel Gene Bank for Agricultural Crops, Agricultural Research Organisation, Volcani Center (ISR002)	Israel	16
AGES Linz - Austrian Agency for Health and Food Safety/ Seed Collection (AUT001)	Austria	6
Gobierno de Aragón. Centro de Investigación y TecnologíaAgroalimentaria. Banco de Germoplasma de Hortícolas (ESP027)	Spain	18
Vegetable Growing Research Institute Public Leqal Entity (AZE005)	Azerbaijan	16

(continued)

Table 1.1 (continued)

Name of gene bank	Country name	Accessions
Comunidad de Madrid. Consejería de Medio Ambiente, Vivienda y Ordenación del Territorio. Instituto Madrileño de Investigación y Desarrollo Rural. Banco de Variedades Locales de Madrid (ESP198)	Spain	13
Institute of Plant Production n.a. V.Y. Yurjev of UAAS (UKR001)	Ukraine	11
Azerbaijan State Agrarian University (AZE014)	Azerbaijan	5
Henry Doubleday Research Association (GBR017)	United Kingdom	4
Ustymivka Experimental Station of Plant Production (UKR008)	Ukraine	4
National Center for Genetic Resources Preservation (USA995)	United States	4
Consejo Superior de InvestigacionesCientíficas. MisiónBiológica de Galicia (ESP009)	Spain	3
Principado de Asturias. Servicio Regional de Investigación y Desarrollo Agroalimentario (ESP032)	Spain	2
Suceava Genebank (ROM007)	Romania	2
Institute of Botany (AZE004)	Azerbaijan	1
Nordic Genetic Resource Center (SWE054)	Sweden	1
***Cajanus cajan* (Pigeon pea); total 14,971**		
International Crop Research Institute for the Semi-Arid Tropics (IND002)	India	13,430
Genetic Resources Research Institute (KEN212)	Kenya	852
Australian Grains Genebank, Agriculture Victoria (AUS165)	Australia	370
International Livestock Research Institute (ETH013)	Ethiopia	159
Agricultural Plant Genetic Resources Conservation and Research Centre (SDNOO2)	Sudan	92
International Institute of Tropical Agriculture (NGA039)	Nigeria	41
Millennium Seed Bank - Royal Botanic Gardens Kew (GBR004)	United Kingdom	2
Banco Português de Germoplasma Vegetal (PRT001)	Portugal	7
National Plant Genetic Resources Centre (ZMB048)	Zambia	6
National Center for Genetic Resources Preservation (USA995)	United States	4
Desert Legume Program (USA971)	United States	2
Comunidad de Madrid. Universidad Politécnica de Madrid. Escuela Técnica Superior de IngenieríaAgronómica, Alimentaria y de Biosistemas. Banco de Germoplasma César Gómez Campo (ESP003)	Spain	2
Genebank, Leibniz Institute of Plant Genetics and Crop Plant Research (DEU146)	Germany	1
Centre for Plant Diversity (HUN003)	Hungary	1
Istituto di Bioscienze e Biorisorse, Consiglio Nazionale delleRicerche (ITA436)	Italy	1

Table 1.1 (continued)

Name of gene bank	Country name	Accessions
Seed Savers Exchange (USA974)	United States	1
***Cicer arietinum* (chickpea); total accessions 87,341**		
International Crop Research Institute for the Semi-Arid Tropics (ICRISAT)	India	20,764
International Centre for Agricultural Research in the Dry Areas (ICARDA)	Lebanon	15,734
Australian Temperate Field Crops Collection (ATFCC)	Australia	8655
Western Regional Plant Introduction Station, USDA-ARS	United States	8038
National Plant Gene Bank of Iran, Seed and Plant Improvement Institute (NPGBI-SPII)	Iran	5700
N.I. Vavilov All-Russian Scientific Research Institute of Plant Industry (VIR)	Russia	2767
Plant Genetic Resources Program (PGRP)	Pakistan	2146
Plant Genetic Resources Department, Aegean Agricultural Research Institute (AARI)	Turkey	2075
Institute of Plant Production nd. a. V. Ya. Yuryev of NAAS	Ukraine	1760
Estación de Iguala, Instituto Nacional de InvestigacionesAgrícolas	Mexico	1600
Institute of Biodiversity Conservation (IBC)	Ethiopia	1173
Institute for Agrobotany (RCA)	Hungary	1170
Uzbek Research Institute of Plant Industry (UzRIPI)	Uzbekistan	1055
***Vigna umbellata* (rice bean); total accessions 5056**		
ICAR—National Bureau of Plant Genetic Resources	India	2235
Institute of Crop Genetic Resources	China	1363
National Agriculture and Food Research Organization	Japan	399
World Vegetable Centre	Taiwan	351
Nepal Agricultural Research Council	Nepal	300
National Plant Genetic Resource Laboratory, UPLB, Laguna	Philippines	161
Centre of Biology, Indonesian Institute of Science Research and Development, Bogor	Indonesia	100
ARS–GRIN	USA	147
***Vicia faba* (faba bean); total accessions 20,824**		
International Centre for Agricultural Research in Dry Areas (LBN002)	Lebanon	13,090
Australian Grains Genebank, Agriculture Victoria (AUS165)	Australia	3022
N.I. Vavilov Research Institute of Plant Industry (RUS001)	Russia	1269
Institute for Plant Genetic Resources 'K.Malkov' (BGR001)	Bulgaria	731
Centre for Genetic Resources, the Netherlands (NLD037)	Netherlands	731
Western Regional Plant Introduction Station, USDA-ARS, Washington State University (USA022)	United States	611
Genebank, Leibniz Institute of Plant Genetics and Crop Plant Research (DEU146)	Germany	281
Estação Nacional Melhoramento Plantas (PRT004)	Portugal	195

(continued)

Table 1.1 (continued)

Name of gene bank	Country name	Accessions
Seed Savers Exchange (USA974)	United States	150
Agricultural Plant Genetic Resources Conservation and Research Centre (SDN002)	Sudan	147
Centro Nacional de Recursos Fitogenéticos (ESP004)	Spain	108
National (CYPARI) Genebank, Agricultural Research Institute, Ministry of Agriculture, Rural Development and Environment (CYP004)	Cyprus	101
Banco Português de Germoplasma Vegetal (PRT001)	Portugal	83
Banco de Germoplasma - Departamento de Recursos Genéticos e Melhoramento, Estação Agronómica Nacional, Instituto Nacional de Investigação Agrária (PRT005)	Portugal	66
Banco de Germoplasma - Universidade da Madeira (PRT102)	Portugal	48
Genetic Resources Institute (AZE015)	Azerbaijan	41
CREA-Centro di Ricerca Zootecnia e Acquacoltura - Sede di Lodi (ITA394)	Italy	29
Agroscope Changins (CHE001)	Switzerland	20
CREA-Centro di Ricerca Orticoltura e Florovivaismo - Sede di Monsampolo del Tronto (ITA392)	Italy	15
Dipartimento di Chimica, Biologia e Biotecnologie, Universitá degli Studi Perugia (ITA363)	Italy	13
Gobierno de Aragón. Centro de Investigación y Tecnología Agroalimentaria. Banco de Germoplasma de Hortícolas (ESP027)	Spain	12
Genetic Resources Unit, Institute of Biological, Environmental & Rural Sciences, Aberystwyth University (GBR016)	United Kingdom	10
AGES Linz - Austrian Agency for Health and Food Safety/ Seed Collection (AUT001)	Austria	7
Sortengarten Erschmatt (CHE100)	Switzerland	7
National Center for Genetic Resources Preservation (USA995)	United States	7
Research Institute of Crop Husbandry (AZE003)	Azerbaijan	6
Gene bank (CZE122)	Czechia	6
Junta de Andalucía. Consejería de Agricultura y Pesca. Instituto Andaluz de Investigación y Formación Agraria, Pesquera, Alimentaria y de la Producción Ecológica. Centro Alameda del Obispo (ESP046)	Spain	6
Nordic Genetic Resource Center (SWE054)	Sweden	6
Agricultural Institute Osijek (HRV021)	Croatia	2
Azerbaijan State Agrarian University (AZE014)	Azerbaijan	1
Comunidad de Madrid. Universidad Politécnica de Madrid. Escuela Técnica Superior de Ingeniería Agronómica, Alimentaria y de Biosistemas. Banco de Germoplasma César Gómez Campo (ESP003)	Spain	1
Henry Doubleday Research Association (GBR017)	United Kingdom	1

Table 1.1 (continued)

Name of gene bank	Country name	Accessions
Research and Development Station for Vegetables - Bacau (ROM055)	Romania	1
***Vigna unguiculata* (cowpea); total accessions 33,580**		
International Institute of Tropical Agriculture (NGA039)	Nigeria	17,090
Plant Genetic Resources Conservation Unit, Southern Regional Plant Introduction Station, University of Georgia, USDA-ARS (USA016)	United States	8248
N.I. Vavilov Research Institute of Plant Industry (RUS001)	Russia	1356
Genetic Resources Research Institute (KEN212)	Kenya	893
Institute of Plant Breeding-National Plant Genetic Resources Laboratory (PHL129)	Philippines	1081
Australian Grains Genebank, Agriculture Victoria (AUS165)	Australia	889
International Livestock Research Institute (ETH013)	Ethiopia	697
SADC Plant Genetic Resources Centre (ZMB030)	Zambia	638
Agricultural Plant Genetic Resources Conservation and Research Centre (SDN002)	Sudan	474
Banco Português de Germoplasma Vegetal (PRT001)	Portugal	342
Botanic Garden Meise (BEL014)	Belgium	332
CREA-Centro di Ricerca Zootecnia e Acquacoltura - Sede di Lodi (ITA394)	Italy	289
Plant Genetic Resources Research Institute (GHA091)	Ghana	178
Suceava Genebank (ROM007)	Romania	183
Centro Agronómico Tropical de Investigación y Enseñanza (CRI001)	Costa Rica	168
Research and Development Station for Plant Culture on Sands Dabuleni (ROM021)	Romania	129
National Plant Genetic Resources Centre (ZMB048)	Zambia	87
Dipartimento di Chimica, Biologia e Biotecnologie, Universitádegli Studi Perugia (ITA363)	Italy	72
Centre for Plant Diversity (HUN003)	Hungary	69
National Center for Genetic Resources Preservation (USA995)	United States	66
Genetic Resources Institute (AZE015)	Azerbaijan	44
Seed Savers Exchange (USA974)	United States	50
Genebank, Leibniz Institute of Plant Genetics and Crop Plant Research (DEU146)	Germany	47
International Center for Biosaline Agriculture (ARE003)	United Arab Emirates	46
Institute of Plant Production n.a. V.Y. Yurjev of UAAS (UKR001)	Ukraine	18
Estação Nacional Melhoramento Plantas (PRT004)	Portugal	19
Desert Legume Program (USA971)	United States	19
National Centre for Genetic Resources and Biotechnology (NGA010)	Nigeria	11

(continued)

Table 1.1 (continued)

Name of gene bank	Country name	Accessions
Institute for Plant Genetic Resources 'K.Malkov' (BGR001)	Bulgaria	12
Millennium Seed Bank - Royal Botanic Gardens Kew (GBR004)	United Kingdom	10
Banco de Germoplasma - Departamento de Recursos Genéticos e Melhoramento, Estação Agronómica Nacional, Instituto Nacional de Investigação Agrária (PRT005)	Portugal	10
Research and Development Station for Vegetables Buzau (ROM068)	Romania	4
Research and Development Station for Vegetables - Bacau (ROM055)	Romania	3
Office of the Styrian Regional Government, Department for Plant Health and Special Crops (AUT025)	Austria	2
National Agricultural and Food Centre (NPPC), Research Institute of Plant Production (RIPP) (SVK001)	Slovakia	2
Agricultural Institute Osijek (HRV021)	Croatia	1
Nordic Genetic Resource Center (SWE054)	Sweden	1

Source: Genesys, https://www.genesys-pgr.org; PGR Portal, pgrportal.nbpgr.ernet.in; Pattanayak et al. 2019

Table 1.2 Status of hilly pulses in the Indian National Gene Bank

Sr. No.	Crop	No. of accessions
1	Chickpea	14,863
2	Pigeon pea	11,940
3	Cowpea	4035
4	French bean	4205
5	Rice bean	2235
6	Faba bean	908
Total		**38,186**

Source: PGR Portal, pgrportal.nbpgr.ernet.in

Conservation of germplasm is one of the foremost activities of the PGR management and further their present and future utilization. In India, systematic plant exploration and collection in India was started in the late 1960s and early 1970s under the IARI PL-480 project. Later, the National Bureau of Plant Genetic Resources came into being in 1976, as the nodal organization for the management of genetic resources in the country and now is the second largest gene bank in the world.

The ex situ seed gene bank at NBPGR holds 68,465 accessions of various pulses accessions that are conserved at −20 °C at the base collections, while the active collections are maintained at 4 °C in medium-term modules across the country at various regional stations of the Bureau and National Active Germplasm Sites. The

Bureau has undertaken multi-crop pulses exploration in collaboration with state agricultural universities (SAUs) and other national crop-based institutes in the different agro-climatic regions especially unexplored areas that exhibit the untapped plant wealth. The major ex situ collections of different pulses are maintained by the Consultative Group on International Agricultural Research (CGIAR) system and other regional or national gene banks of the different countries.

To broaden the genetic base of pulse germplasm, we need to collect the genetic resources from the hilly and remote areas of the country that offer great opportunities for the collection of trait-specific germplasm. India is considered a primary centre of diversity for pigeon pea and a regional diversity centre for chickpea (Arora 1988). Globally, about 97,400 germplasm accessions of chickpea are preserved in more than 30 gene banks worldwide (Chandora et al. 2020). Currently, the largest repositories are contributed by CG centres likely CIAT which holds 37,938 accessions of *Phaseolus* beans; ICRISAT maintains 20,764 germplasm accessions of chickpea and 13,989 accessions of pigeon pea; International Institute of Tropical Agriculture (IITA), Ibadan, Nigeria, maintains 16,558 accessions of cowpea; and International Centre for Agricultural Research in Dry Areas (ICARDA) holds almost 10,000 accessions of faba bean. In addition, many national and regional gene banks also hold substantial collections of genetic resources. ICAR-National Bureau of Plant Genetic Resources, New Delhi, holds 15,084 accessions of chickpea germplasm. The detailed germplasm statuses of pulses are mentioned in Tables 1.1 and 1.2. A wild relative of chickpea, *C. microphyllum*, is endemic to Leh, Spiti, and upper Kinnaur, a good source of genes for large pod size (2.5 cm), seeds/pod (6–8), and tolerance to drought and cold. Crossable with *C. arietinum* but after fertilization embryo abort, hence embryo rescue is required.

The presence of genetic variability, preserved and expanded through diverse plant types, holds immense value for humanity, both in addressing current demands and future requirements. In many regions across the globe, the swift agroecological transformations taking place are leading to the rapid depletion of many species, landraces, and their wild counterparts, which possess valuable and superior genes. It is feared that these forms may become extinct in due course of time. It is therefore essential that concerted efforts be made to collect, consolidate, and conserve the valuable pulse genetic resources, especially the landraces and folk varieties which have evolved in the farmers' field and have been selected and improved over the years by custodian farmers and wild and related species which are growing along with their cultivated crop fields.

To achieve this, extensive characterization and evaluation of the diverse germplasm is a prerequisite step to facilitate their effective utilization in breeding programmes as well as for their efficient management in the gene banks. Collection, evaluation, utilization, and improvement of local types of pulses in the hills and encouragement of their cultivation given their highly unique adaptive traits in nature are more suitable for cultivation on the rolling topography of the hilly regions. There is a need to introduce short-duration verities required for the double cropping in largely rain-fed areas of hills to increase productivity in the region. The Himalayan region possesses a wide range of intra-specific genetic diversity of food legumes

Fig. 1.2 Variability in seed shape, size, and colour in (**a**) French bean, (**b**) chickpea, (**c**) pigeon pea, (**d**) cowpea, (**e**) rice bean, and (**f**) faba bean

(Fig. 1.2). There are about 16 cultivated and 24 wild relatives of food legumes occurring in this region. The important cultivated species are *Phaseolus vulgaris*, *Cicer microphyllum*, *Vicia faba*, *V. unguiculata*, and *V. umbellata*.

Wild relatives of legumes with desirable traits such as resistance to biotic and abiotic stress as well as wider adaptation are also abundant in the region, such as *Cicer microphyllum*, which is a wild type of chickpea. In north-eastern India, *V. umbellata* has performed well in the sub-Himalayan tract and emerged as an alternative for *V. mungo*, especially in high-rainfall areas where disease incidence is very high. Though *Cajanus* might not be a major crop of this region, its distantly related genera like *Arylosia* and *Rhynchosia* are prevalent and have shown tolerance to drought while also possessing genes for wilt resistance. Genus *Vicia* is well represented in the region, and diversity in terms of pod length, seed size, etc. has been observed. Cowpea, for which India is the secondary centre of diversity, also possessed considerable genetic diversity in terms of landraces. Donors for various traits have been identified by evaluating existing germplasm accessions in each crop. Despite retaining this diverse range of genetic resources, their utilization has not

Table 1.3 The consumption pattern and grain micronutrient concentration in hilly pulse crops

Crop	Fe $\mu g\ g^{-1}$	Zn $\mu g\ g^{-1}$	Biofortification approach	Health benefits	References
French bean	40.7–96.7	18.0–42.0	Germplasm screening, conventional breeding, agronomic molecular breeding, genomic-assisted breeding, and genome editing	Exhibits antioxidant, anti-diabetic, anti-obesity, anti-inflammatory, anti-mutagenic, and anti-carcinogenic properties	Ganesan and Xu (2017); Glahn et al. (2020); Langyan et al. (2022)
Pigeon pea	29.2–40.9	24.14–35.68	Germplasm screening and agronomic practices	Seeds and pods are high in phenolic compounds, which have anti-inflammation, antibacterial, antioxidant, anti-carcinogenic, and anti-diabetic properties	Susmitha et al. (2022); Rathod et al. (2022)
Faba bean	37–50	23–32	Germplasm screening, agronomic practices, conventional breeding	Faba bean polyphenols are involved in the protection against the development of human diseases	Khazaei and Vandenberg (2020); Abiodun et al. (2022)
Chickpea	35–70	22–40	Germplasm screening, conventional breeding and agronomic practices	Raw or cooked chickpeas and hummus also contain dietary bio-actives such as phytic acid, sterols, tannins, carotenoids, and other polyphenols	Pushparajah (2012); Misra et al. (2020); Mahto et al. (2022)
Rice bean	25–55	22–30	Germplasm screening and agronomic practices	An important source of dietary antioxidants	Pattanayak et al. (2019)
Cowpea	22–60	15–40	Germplasm screening and agronomic practices	The high fibre content and plant proteins in black-eyed peas reduce hunger hormones and enhance weight loss	Coelho et al. (2021)

been substantial, particularly while considering gene transfer from secondary and tertiary gene pools. The loss of many popular landraces of French bean, chickpea, and cowpea in the recent past is a matter of great concern. However, with recent advancements in biotechnological tools and techniques, the utilization of distant sources of genes from such gene pools should escalate.

Although mountainous regions host a variety of warm and cool-season pulse crops that are integral to local diets, our discussion will centre on six pulse crops: French bean, chickpea, cowpea, pigeon pea, rice bean, and faba bean.

1.3 Biofortification in Hilly Pulses to Enhance Nutrition

Micronutrient deficiencies in iron (Fe) and zinc (Zn) have widespread effects on individuals of all age groups. However, their impact is particularly pronounced in pregnant women and children, especially newborns in the early stages of life. For adults, the recommended daily intake of zinc is typically in the range of 8–11 mg. Pregnant and nursing women have slightly higher zinc requirements, necessitating an intake of 11–13 mg/day. Iron requirements are comparatively higher, ranging from 12 to 28 mg/day for most individuals. Pregnant and lactating women have an even greater need, with recommended daily intakes increasing to a range of 30–38 mg. These nutritional guidelines are based on the Dietary Reference Intakes, 2019, and they highlight the varying dietary needs for iron and zinc across different life stages, with a particular emphasis on the increased demands during pregnancy and lactation. Pulses are high in micronutrients like iron, potassium, magnesium, zinc, and B vitamins like folate, thiamine, and niacin. As a result, strengthening the nutritional makeup of pulses presents an exciting target for combating the 'hidden hunger' of worldwide micronutrient deficiency. Daily mineral requirements can be satisfied by taking 100–200 g pulses (French bean, pigeon pea, rice bean, faba bean, cowpea, and chickpea), and daily iron requirements can be met by consuming 100 g of most dietary legumes. Several essential assumptions were underlying the current strategy for Fe biofortification of French bean. First, the average dry-weight bean Fe concentration is thought to be 50 g/g, with a range of 34–92 g/g (Glahn et al. 2020). It is believed to contain 4–10 times the amount of Fe and 2–3 times the amount of Zn as rice (Pfeiffer and McClafferty 2007). Similarly, the genetic variability for grain nutrients in 600 pigeon pea germplasms preserved at the RS Paroda gene bank, ICRISAT, India, was investigated. Protein (23.35–29.50%), P (0.36–0.50%), K (1.43–1.63%), Ca (1042.36-2099.76 mg/kg), Mg (1311.01-1865.65 mg/kg), Fe (29.23–40.98 mg/kg), Zn (24.14–35.68 mg/kg), Mn (8.56–14.01 mg/kg), and Cu (7.72–14.20 mg/kg) showed substantial variability (Table 1.3) (Srikumar 1993; Susmitha et al. 2022). Besides that, chickpea has been recognized as an effective nutritious crop, with a high concentration of selenium (Se = 15.3–56.3 g/100 g), iron (Fe = 4.6–6.7 mg/100 g), zinc (Zn = 3.7–7.4 mg/100 g), calcium (Ca = 93.4–197.4 mg/100 g), magnesium (Mg = 125.1–158.7 mg/100 g), and potassium (K = 732.2–112 mg/100 g) (Thavarajah 2012; Misra et al. 2020). Furthermore, foliar Zn + Fe fertilization substantially affected nutritional concentration (Zn and Fe) in chickpea grains (Singh et al. 2023). A 100-g serving of biofortified chickpea contains 5.2–6.0 mg of Fe, 2.5–5.3 mg of Zn, and 15.3–56.3 mg of Se, providing a considerable amount of the recommended dietary intake (RDI) of these critical elements (Madurapperumage et al. 2021). Additionally, zinc-solubilizing bacteria (ZSB) are promising and ideal replacements for Zn supplementation in chickpea. *Bacillus altitudinis* strains (BT3 and CT8) have demonstrated a good ability to solubilize insoluble Zn compounds such as oxides, phosphates, and carbonates of Zn, making them a significant source for enhancing Zn uptake and chickpea growth (Kushwaha et al. 2021). Cowpea displays significant genetic diversity when it comes to crucial nutritional components, including protein and micronutrient levels. This diversity opens opportunities for genetic biofortification, a process aimed at enhancing the nutritional content of crops. Biofortified cowpea cultivars have been developed, showing remarkable levels of iron (Fe) and zinc (Zn) content, exceeding 60 mg and 40 mg per kilogram of dry weight, respectively (Coelho et al. 2021; Dhanasekar et al. 2021).

Similarly, faba bean holds a prominent position in the diets of people across regions such as the Middle East, the Mediterranean, China, and Ethiopia. Faba bean is recognized for its high nutritional value, and in many parts of the world, it surpasses the nutritional quality of other legumes like peas (*Pisum sativum* L.) and various grain legumes (Crépon et al. 2010). The seeds had the highest quantities of nitrogen, phosphorus, potassium, zinc, copper, and protein. However, the leaves accumulated the most calcium, magnesium, iron, or manganese (Etemadi et al. 2018). Environmental variation, G × E interaction, and the tannin profile strongly influence the genetic variance for faba bean mineral elements (Khazaei and Vandenberg 2020). Rice bean is an underutilized domesticated legume crop used for nutritional protein in Asia, but little is known about its biofortification (Pattanayak et al. 2019). Seeds of three rice bean accessions had 17.26–21.42% protein, 3.46–4.03% fat, 61.09–64.73% carbohydrates, 3.99–4.58% ash, and 5.22–7.43% fibre (dry weight basis) (Rodriguez and Mendoza 1991).

1.4 Genomic Resources and Their Importance

The fundamental requirement for effectively utilizing germplasm in crop improvement is the characterization of germplasm to identify specific traits. This characterization lays the foundation for the creation of crucial genomic resources, including DNA markers, linkage maps, and genome and transcriptome sequences. These resources are essential for processes like gene tagging, gene mapping, and marker-assisted selection, which expedite and inform breeding efforts for crop improvement. Recent advancements in genome sequencing technology have brought about a revolution in genetics. They have significantly enhanced our understanding of the genetics underlying valuable traits by generating high-throughput genomic resources for several important plants. By applying knowledge gained from genomics, transcriptomics, expression studies, and epigenetics, we can facilitate the development of improved crop varieties. These varieties are characterized by higher yields, improved quality, and enhanced tolerance to various biotic and abiotic stresses. Furthermore, these genomic approaches aid in the identification and annotation of critical and novel genes that control traits of interest. DNA sequencing plays a pivotal role in identifying key genes responsible for various characteristics and assessing variability among cultivars (Thottathil et al. 2016). Additionally, whole-genome resequencing enables the identification of a broad spectrum of intraspecific variability. It also contributes to the recognition and annotation of novel genetic polymorphisms, which is invaluable for the development of numerous molecular markers, including simple sequence repeats (SSRs) and single-nucleotide polymorphisms (SNPs).

1.4.1 Molecular Markers and Genotyping Assays in Pulses

The field of plant breeding underwent a remarkable transformation with the introduction of DNA marker systems in the 1980s, marked by the development of the first molecular marker system known as restriction fragment length polymorphism (RFLP). In legume crops like pigeon pea, lentils, groundnut, and chickpea, which were once categorized as 'orphan crops' due to the lack of sufficient genomic resources, significant progress has been made in the past decade. This progress can be attributed to the evolution of new sequencing technologies that have substantially reduced the cost of DNA sequencing (Varshney et al. 2019).

A wide array of molecular markers and technologies are now available, including thousands of simple sequence repeats (SSRs), single-nucleotide polymorphism (SNP) markers, Diversity Arrays Technology (DArT) markers, various SNP platforms, genotyping by sequencing (GBS), microarray-based markers, next-generation sequencing (NGS)-based markers, and insertion/deletion (InDel) markers, among others. Over the years, thousands of SSRs have been developed, with more than 3000 in chickpea and pigeon pea alone. Collaborative efforts between ICRISAT and DArT Pvt. Ltd., Australia, have resulted in the development of DArT markers, with each set containing more than 15,360 features for chickpea, pigeon pea, and groundnut. Additionally, thousands of SSR and DArT markers, as well as tens of thousands of SNP markers, have been developed through various approaches, leading to the creation of different genotyping platforms and assays.

SNP markers have gained prominence due to their suitability for high-throughput genotyping. Genotyping platforms are available in various densities, including high-density (more than 20,000 SNPs), medium-density (2000–10,000 SNPs), and low-density (1000–10,000 SNPs) platforms (Varshney et al. 2019). High- and medium-density platforms have found extensive use in gene/QTL mapping, background selection, genetic diversity studies, genomic selection (GS), genome-wide association studies (GWAS), and linkage mapping. Notably, high-density genotyping platforms, such as SNP arrays with genome-wide SNPs, have been developed for crops like pigeon pea and chickpea. Moderate-density platforms, including genotyping by sequencing (GBS) and restriction-site-associated sequencing (RAD-Seq), have also been successfully employed in genetic studies, particularly in chickpea.

Low-density genotyping platforms are valuable for applications like marker-assisted selection (MAS), early-generation testing, and hybridity testing. In partnership with Intertek company, ICRISAT has expanded low-density genotyping (with 10 SNPs) for various crop species, including pigeon pea, groundnut, and chickpea, to facilitate the selection of superior individuals in early breeding generations (Varshney et al. 2019).

In compendium, a range of molecular marker systems and genotyping platforms have been employed to study genetic diversity, population structure, and the development of genetic maps and QTL mapping/GWAS in pulse crops like pigeon pea and chickpea. Once identified, these QTLs/genes play a pivotal role in molecular breeding programmes, contributing to the enhancement of targeted traits through approaches such as marker-assisted recurrent selection (MARS), marker-assisted selection (MAS), and genomic selection (GS). These advancements in next-generation technologies hold the promise of developing crop varieties with improved quality traits, higher yields, and enhanced resistance to diseases (Varshney et al. 2021).

1.4.2 Molecular Genetic Maps

Molecular genetic maps are representations of the linear arrangement of molecular markers (loci) on chromosomes, and they are established by estimating recombination fractions between these markers. These maps serve a multitude of purposes, including understanding genome organization, studying species evolution, examining synteny among related species, investigating chromosome/genome rearrangements across different taxa, and, most importantly, identifying genes and quantitative trait loci (QTLs) through QTL interval mapping. The advancement of genomics tools and technologies, particularly marker technologies, has made significant progress in various plants of academic and economic interest. In the case of pulse crops like pigeon pea, genetic mapping faced initial limitations due to a shortage of markers, resulting in relatively sparse genetic maps. However, the introduction of high-density genotyping platforms has substantially increased marker densities in pigeon pea genetic maps. For example, a high-density genetic linkage map of pigeon pea was constructed using SNP markers, featuring 910 marker loci with an average inter-marker distance of 1.11 cM (Saxena et al. 2012). Another high-density genetic map in pigeon pea utilized 6818 SNP loci, covering 974 cM of the genome (Yadav et al. 2019). In the case of chickpea, developing high-density genetic maps was initially challenging due to a narrow genetic base and low intra-specific genetic polymorphism. Nevertheless, next-generation sequencing (NGS) technologies have facilitated the creation of thousands of markers and enabled the construction of high-density marker linkage maps. ICRISAT, for instance, established a comprehensive genetic map for chickpea, comprising 1291 markers across eight linkage groups and spanning a total distance of 845.56 cM (Thudi et al. 2011). Varshney et al. (2014a) identified a limited number of polymorphic markers through extensive screening and mapped 241 and 168 markers on ICCRIL03 and ICCRIL04 mapping populations, respectively. Multiple genetic mapping approaches have led to the development of numerous consensus maps and linkage maps in chickpea (Mallikarjuna et al. 2017). Additionally, integrated physical, genetic, and genome sequence maps have been created in chickpea (Varshney et al. 2014a). Millions of

SNP markers have now been discovered in chickpea, leading to the creation of high-density SNP arrays through NGS-based genome sequencing and resequencing technologies. Similar successes have been achieved in other legume crops like groundnut and lentil, where genetic linkage maps, including high- and medium-density linkage maps, have been established. These expanded genetic linkage maps involved the use of various genotyping platforms and molecular markers, including SNP arrays.

In recent decades, genome-wide association studies (GWAS) have gained prominence and have been employed in nearly all crop plants to discover genes and QTLs associated with important traits. In grain legume crops, several crucial genes/QTLs have been identified for traits such as drought tolerance and yield under drought conditions, utilizing both QTL mapping and GWAS approaches (Varshney et al. 2014b). Additionally, QTLs/genes for drought- and heat-responsive traits have been uncovered using GWAS and gene sequencing methods (Thudi et al. 2014). In chickpea, genes/QTLs associated with significant diseases like botrytis grey mould, *Fusarium* wilt (FW), and ascochyta blight (AB) have been identified (Varshney 2016). Furthermore, genes/QTLs for seed traits, phenology-related traits, and more have been identified (Sivasakthi et al. 2019; Roorkiwal et al. 2020). In pigeon pea, genes/QTLs have been discovered for various traits using different trait mapping approaches. Notably, genes/QTLs have been identified for significant diseases such as sterility mosaic disease (SMD) and *Fusarium* wilt (FW) (Bohra et al. 2020; Saxena et al. 2021) (Table 1.4). QTLs associated with plant height, growth habit, flowering, earliness, and determinacy have been identified through whole-genome

Table 1.4 Identification of trait-specific genes in hilly pulses

Crop	Genes	Roles	References
Faba bean	Putative aquaporin gene *VfPIP1*	Drought tolerance	Cui et al. (2008)
Rice bean	C2H2-type zinc finger transcription factor gene *VuSTOP1*	pH and aluminium tolerance	Fan et al. (2014)
French bean	QTL, FRR3.1^{KM} on Pv03; locus, *rhg1* on chromosome Pv01, QTL, RL2.1^{SA}	Resistance to *fusarium* root rot; soybean cyst nematode; resistance to Al toxicity	Kamfwa et al. (2013); Jain et al. (2019); Njobvu et al. (2020)
Chickpea	QTL, *qDf03* on linkage group CaLG03	Resistance to chickpea pod borer	Barmukh et al. (2021)
Pigeon pea	QTL, qSMD4 on LG07; genes *C. cajan*_03691 and *C. cajan*_18888; genes *C. cajan*_29830 and *C. cajan*_33874	Resistance to sterility mosaic disease; *fusarium* wilt; drought tolerance	Gnanesh et al. (2011); Saxena et al. (2021); Sinha et al. (2016)
Cowpea	QTLs, Fthp28, Fthp87, and Fthp129 on LG2, LG4, and LG6; loci, *Dro-7*, *Dro-10*, *Dro-8*, *Dro-1*, and *Dro-3* on LG1, 3, 4, 7, and 8	Resistance to flower bud thrips; high yield under drought stress	Sobda et al. (2017); Muchero et al. (2013)

scanning approaches and candidate gene sequencing (Mir et al. 2017). Genes have also been pinpointed for traits related to cold tolerance, drought resistance, salinity tolerance, agronomic characteristics, and fertility restoration (Saxena et al. 2020a, 2020b).

Notably, NGS-based high-throughput genotyping approaches are increasingly employed for genetic/trait mapping in legume crops due to their advantages in terms of time and cost-efficiency compared to traditional mapping methods. In chickpea, pigeon pea, and groundnut, sequence-based trait mapping approaches have been used to identify genomic regions associated with rust and late leaf spot resistance (Roorkiwal et al. 2020; Bohra et al. 2020).

1.4.3 Pulse Crop Improvement Through Genomic Breeding

Marker-assisted backcrossing (MABC) has proven to be an effective strategy for introgressing major-effect quantitative trait loci (QTL) that confer resistance to various biotic and abiotic stresses, such as disease resistance and drought tolerance (Varshney et al. 2021). One notable example of the successful application of MABC is in chickpea, where a drought-tolerant line called 'Geletu', derived from an MABC scheme, was released for cultivation in Ethiopia. In the context of field pea, the genomic-assisted breeding (GAB) approach has been demonstrated to be highly efficient for selecting lodging resistance in early segregating generations, surpassing conventional phenotypic selection (Zhang et al. 2006). The GAB approach has also been applied to introduce other valuable traits in field pea, including frost tolerance (Tayeh et al. 2015) and resistance to *Aphanomyces* root rot (Hamon et al. 2013). Similar examples of the successful application of molecular marker-assisted breeding techniques have been observed in common bean, particularly for improving agronomic traits (Blair et al. 2006). Some examples of different approaches in pulse crops are mentioned as follows.

1.4.3.1 French Bean/Snap Bean/Common Bean

Phaseolus vulgaris L. (2n = 22) is one of the major export pulse crops of India. The world production of dried beans was estimated to be 26.6 million tonnes with Myanmar, India, and Brazil as the leading producers (Tongbram et al. 2021). It originated more than 7000 years ago in Latin America and has two primary centres of origin, Meso-America (Mexico and Central America) and Andean regions that are easily distinguished gene pools by molecular means (Blair et al. 2006). French bean is a highly protein-enriched food and is therefore known as 'vegetarian meat'. It contains 1.7% protein in fresh pods and 21.1% in seeds (Alice et al. 2019). It also contains vitamins A, B, C, folates, minerals such as calcium, potassium, selenium, iron, phosphorous, sodium, zinc, omega-3 fatty acids, dietary fibre, and amino acids, and antioxidants such as lutein and beta-carotene, which is anti-inflammatory. Dry seed contains 69.9% carbohydrates, 1.7% fat, 381 mg Ca, 425 mg P, and 12.4 mg Fe per 100 g edible part (Ali and Kushwaha 1987).

Among pulses, a wide range of variability is found in kidney bean for seed shape, colour, and size, plant type, maturity, pod characters, and resistance to anthracnose and BCMV including famous rajmah of Barabanghal, Kalpa and Rogi, Bharmour, Kullu, Kinnaur in Himachal Pradesh Bhaderwah, Kishtwar, Doda in J&K and Sama, Munsyari and Devidhura in Uttarakhand of upper hills of Northwestern Himalayan of India. Related species of French bean, *P. coccineus* and *P. lunatus*, are cross-compatible and found to be resistant to anthracnose, bacterial blight, and CBMV. This crop is not native to the Himalayas but has been introduced at various levels over a few centuries.

Common bean has a genome size of 587 Mb (Schmutz et al. 2014). There are many genetic markers which have been used to identify the location of genes in *Phaseolus*. Some of these are allozymes, first reported by Kami et al. (1995), RFLPs, RAPDs, AFLPs, and ISSRs. There are 10 BAC libraries (Gepts et al. 2008), 83,000 ESTs (Ramírez et al. 2005), and over 25 linkage maps (Kelly et al. 2003) available for *Phaseolus vulgaris* (Table 1.5).

1.4.3.1.1 Biotic Stresses

Understanding the mechanisms of gene action and the inheritance patterns of resistance against pathogens is of paramount importance in devising effective breeding strategies for developing crop varieties that possess robust resistance to diseases and pests. In a study conducted by Agarwal et al. (2021), a total of 645,729 single-nucleotide polymorphisms (SNPs) and 68,713 Insertions and Deletions (Indels) were identified. These genetic markers were used to investigate the genetic basis of resistance to two begomoviruses, namely, cucurbit leaf crumple virus (CuLCrV) and Sida golden mosaic *Florida virus* (SiGMFV), in *Phaseolus vulgaris*, commonly known as common bean. These genetic markers were found to be distributed across the 11 chromosomes of *Phaseolus vulgaris*, with chromosome 2 containing the highest number of variants. Cucurbit leaf crumple virus (CuLCrV) causes the leaves to crumple, curl, and thicken, while Sida golden mosaic Florida virus (SiGMFV) causes leaf mottling, puckering, and curling. Anthracnose, a fungal disease caused by *Colletotrichum lindemuthianum*, is a seed-borne disease which causes lesions on aerial parts of the plant. It causes up to 90–100% yield loss in legumes (Gaudencia et al. 2020). Zuiderveen et al. (2016) reported five QTLs—three major on chromosomes, Pv01, Pv02, and Pv04, and two minor on chromosomes, Pv10 and Pv11, conferring resistance to anthracnose. The resistance was imparted by the *Co-1* gene on Pv01 mapped between 50.16 and 50.30 Mb. The sequencing of single BAC clone has been carried out around the locus Co-4 for resistance to anthracnose (Melotto et al. 2004). Ten QTLs or genes were identified from marker-trait association (MTA) study in common bean. Among these, PVctt1 and BM211 on chromosomes 4 and 8, respectively, are the two major MTAs which explain 20% of the variance for anthracnose resistance (Choudhary et al. 2018). A major QTL, FRR3.1 KM on Pv03, conferring resistance to *Fusarium* root rot (Table 1.4), was mapped between the markers, PVBR109 and PVBR87, in

Table 1.5 Advancements in the generation of genomic resources of hilly pulses

Genomic resources	Faba bean	French bean	Chickpea	Pigeon pea	Cowpea	Rice bean
Mapping resources						
Traditional bi-parental populations	~20 (Arbaoui et al. 2008; Ma et al. 2013; Torres et al. 2006)	~300 (Galeano et al. 2011; Yuste-Lisbona et al. 2012; Zargar et al. 2017; Assefa et al. 2019)	433 (Millan et al. 2010; Verma et al. 2015; Mallikarjuna et al. 2017)	~234 (Varshney et al. 2010; Kinhoégbè et al. 2022)	~30 (including Sesquipedalis group) (Lucas et al. 2011; Muchero et al. 2009a, b; Ouédraogo et al. 2001, 2002, 2012)	~6 (Venkataramana et al. 2016; Guan et al. 2022)
Second-generation populations like MAGIC/NAM	~2 (Sallam and Martsch 2015; Khazaei et al. 2018)	~4 (Diaz et al. 2020, 2021; Keller et al. 2020; Escobar et al. 2022)	~10 (Huang et al. 2015; Attri et al. 2018; Srinivasan et al. 2021; Gaur et al. 2019)	~ 3 (Huang et al. 2015; Saxena et al. 2017)	1 (Huynh et al. (2019)	–
Reverse genetics resources						
TILLING population	–	~7 (Blair et al. 2007; Porch et al. 2009; Tadege et al. 2009)	~ 8 (Gaur et al. 2015; Amri-Tiliouine et al. 2018, 2022)	~3 (Varshney et al. 2010)	–	–
BAC-tools						
BAC libraries	–	10 (Gepts et al. 2008; Pedrosa-Harand et al. 2009; Liu et al. 2010)	~7 (Rajesh et al. 2004; Lichtenzveig et al. 2005; Zhang et al. 2010)	~4 (Varshney et al. 2010; Bohra et al. 2011)	3 (Yu 2012)	
BESs	–	~9 (Fonsêca et al. 2010; Müller et al. 2014; Blair et al. 2014)	~2 (Thudi et al. 2011; Varshney et al. 2014a)	~88,860 (Bohra et al. 2011)	30,527 (Barrera-Figueroa et al. 2011)	–

(continued)

Table 1.5 (continued)

Genomic resources	Faba bean	French bean	Chickpea	Pigeon pea	Cowpea	Rice bean
Physical maps	–	~31 (Schlueter et al. 2008; Córdoba et al. 2010; Blair et al. 2018; Gomes-Messias et al. 2022)	~16 (Varshney et al. 2014a; Ali et al. 2016a, b)	~12 (Yasin et al. 2022; Bohra et al. 2012)	10 × coverage (Close et al. 2011)	–
Genetic markers						
Genomic SSRs		~500 (Yu et al. 2000; Blair et al. 2003, 2009; Gioia et al. 2019; Mir et al. 2021)	300 (Nayak et al. 2010; Varshney 2016)	29,000 (Varshney et al. 2012)		3011 (Chen et al. 2016)
Enriched library based	73 (Zeid et al. 2009)	85 (Métais et al. 2002; Blair et al. 2009)	~78 (Sethy et al. 2006; Nayak et al. 2010)	–	44 (Li et al. 2001)	–
Gene space read (GSR)/BES and NGS based	28,503 (Yang et al. 2012)				1071 (Gupta and Gopalakrishna 2010; 712 (Andargie et al. 2011); 1372 (Xu et al. 2010, 2011a, b)	
EST-SSRs	802 (Kaur et al. 2012); 336 (Kaur et al. (2014)	83,000 (Ramírez et al. 2005; Blair et al. 2011)	265 (Gaur et al. 2011)	~8137 (Pazhamala et al. 2015)	410 (Xu et al. 2010)	~300 (Wang et al. 2016; Chen et al. 2016)
SNPs	75 (Cottage et al. 2012); 14,522 (Kaur et al. 2014)	645,729 (Agarwal et al. 2022)	>3000 (Hiremath et al. 2011; Gaur et al. 2012); 1893 (Gujaria et al. 2011)	35, 000 (Varshney et al. 2012; Saxena et al. 2012; Pazhamala et al. 2015)	1536 (Lucas et al. 2011; Muchero et al. 2009a; Xu et al. 2011a, b)	

Genomic resources	Faba bean	French bean	Chickpea	Pigeon pea	Cowpea	Rice bean
Transcriptome assemblies	1 (Kaur et al. 2012)	~10 (Liao et al. 2013; Wu et al. 2014; Dalla Via et al. 2015)	~29 (Hiremath et al. 2011; Gaur et al. 2015; Iquebal et al. 2017)	~4 (Kudapa et al. 2012)	1 (Muchero et al. 2009a)	1 (Guan et al. 2022)
Genetic linkage maps						
Population specific	~10 (Gutiérrez et al. 2013; Ma et al. 2013; Torres et al. 2010)	~31 (Vallejos and Chase 1991; Vallejos et al. 1992; McConnell et al. 2010; Yuste-Lisbona et al. 2012; Song et al. 2015)	~155 (Paul et al. 2018; Agarwal et al. 2022)	~45 (Varshney et al. 2010; Bohra et al. 2012)	~25 (Lucas et al. 2011; Muchero et al. 2009a, b; Ouédraogo et al. 2001, 2002, 2012)	~4 (Isemura et al. 2010)
Consensus/ composite	4 (Román et al. 2004; Satovic et al. 1996, 2013; Vaz Patto et al. 1999)	~5 (Nodari et al. 1993; Hoyos-Villegas et al. 2016; Gomes-Messias et al. 2022)	~21 (Gowda et al. 2011; Kushwah et al. 2021)	~5 (Saxena et al. 2017)	2 (Muchero et al. 2009a; Lucas et al. 2011)	–
Whole-genome sequence	–	~5 (Schmutz et al. 2014; Song et al. 2015; Vlasova et al. 2016; Lobaton et al. 2018)	4 (Jain et al. 2013; Varshney et al. 2013; Thudi et al. 2016)	~2 (Singh et al. 2012; Varshney et al. 2012)	–	Kaul et al. (2022)

the K132 population and explains 34% of the phenotypic variation (Kamfwa et al. 2013). Angular leaf spot (ALS) caused by the fungus, *Pseudocercospora griseola*, causes up to 80% yield losses in common bean. Oblessuc et al. (2012) identified the major ALS resistance QTL, ALS10.1 DG, located on linkage group B10 from recombinant inbred lines (RILs) derived from a cross between IAC-UNA (susceptible) × CAL 143 (resistant). This QTL was linked to the marker, GATS11b, and explains 16–22% of the phenotypic variation.

Common bacterial blight (CBB) is a seed-borne disease caused by the bacteria *Xanthomonas axonopodis* (syn. campestris). It is a common occurrence in common bean during the summer season and causes yield loss of up to 60% (Zhu et al. 2016). Fourteen SNPs lying on five chromosomes, Pv02, Pv04, Pv08, Pv10, and Pv11, were identified through GWAS by Ambachew et al. (2021). Among these, the SNP S10 1,437,174, lying on chromosome Pv10, was the major marker lying close to a cluster consisting of 18 genes and associated with resistance for CBB. The QTL, LH1.1BE, located on chromosome Pv01 is associated with resistance to common bean leafhoppers, *Empoasca kraemeri* and *Empoasca fabae* (Murray et al. 2004). Leafhoppers are voracious feeders, both adults and nymphs, and cause leaf burn, leaf curling, stunted growth, and chlorosis. Soybean cyst nematode (*Heterodera glycines*) (SCN) can reduce yield by up to 50% in common bean (Poromarto et al. 2010). Jain et al. (2019) conducted GWAS using SNPs and identified the locus, rhg1, which confers resistance to SCN on chromosome Pv01 in Middle American and on chromosome Pv08 in the Andean gene pool of common bean.

1.4.3.1.2 Abiotic Stresses

Aluminium toxicity leads to a reduction in the length and biomass of the root due to impaired cell division in the root apex. Njobvu et al. (2020) carried out a genome-wide association study (GWAS) and identified a QTL, RL2.1 SA, which provides resistance to aluminium toxicity in common bean. This QTL was linked to SNP S2_18220907, which was placed within a genomic region of size 17.5–18.5 Mb. Siamasonta et al. (2021) also identified genomic regions on Pv01, Pv02, Pv05, and Pv06 using GWAS employed SNPs. The SNP on Pv02 was found to be the major QTL, explaining 21.1% of the phenotypic variance for Al toxicity. Yd4.1 on chromosome 4 is a major QTL associated with aluminium and low phosphorous stress in common bean (Diaz et al. 2018). Micronutrients, which include Fe, Zn, Cu, Mn, Mg, Ca, and Mo, determine the growth and development of common bean. Its deficiency can hinder metabolic processes leading to a decrease in seed quality and yield. Genotyping by sequencing (GBS) was used to detect SNPs associated with seed micronutrients. SNPs associated with Fe, Cu, Mn, Mg, and Cu were found on chromosome 3 and SNPs associated with Zn, Mn, Mg, Ca, and Mo on chromosome 11 (Nazir et al. 2022). This can aid in marker-assisted seed enrichment of common bean to overcome nutrient deficiencies.

QTLs for various traits under drought stress in common bean were identified from the F_2 population, derived from a cross between GLP2 (drought susceptible) and KAT B1 (drought tolerant). The major QTLs are NP2.1 KG on chromosome 2 for the of pods per plant, LB2.1 KG on chromosome 2 for leaf biomass, SB1.1 KG on chromosome 1 for stem biomass, and PB1.1 KG on chromosome 1 for pod biomass under drought stress (Langat et al. 2020). Nabateregga et al. (2019) identified pleiotropic QTLs for drought tolerance on Pv10 from RILs which were derived from a cross between the cultivars, BRB 191 × SEQ 1027.

1.4.3.2 Pigeon Pea (*Cajanus cajan* Linn)

Pigeon pea (2n = 22) is an important *kharif* pulse crop in India, accounting for more than 70% of the total production in the world (Singh et al. 2020). Pigeon pea is a good source of carbohydrates such as glucose, fructose, sucrose, stachyose, cellulose and pectin, protein, minerals, fatty acids, amino acids such as lysine, aspartic acid and glutamic acid, etc. The carbohydrate content ranges between 51.4 and 58.8%, protein between 20 and 22%, and lipid between 0.6 and 3.8% (Talari and Shakappa 2018). Pigeon pea (also known as Arhar) is grown in Duggada, Ramganga valley of Uttrakhand, and occasionally grows in the other parts of Shiwalik hills and sporadically in the Una district of Himachal Pradesh. In the north-eastern states of India, mostly the indigenous people grow pigeon pea for green pods as a vegetable (soft pod types) and as a pulse in jhum, backyard/kitchen gardens, and sold in local markets. Local tribal people also use pigeon pea green pods in making local beverages.

Pigeon pea has a genome size of 858 Mb. Some of the genomic developments in pigeon pea include the availability of four transcriptome assemblies (Kudapa et al. 2012), DArT (Yang et al. 2011), and inter-specific genetic maps such as SSR-based genetic map developed from inter-specific F_2 populations from the cross, ICPA 2043 × ICPR 3467, consisting of 140 SSR loci and a map length of 881.57 cM (Bohra et al. 2012), draft genome sequences of the cultivar, Asha (ICPL 87119), covering 511 Mb of the pigeon pea genome (Singh et al. 2012) (Table 1.5).

1.4.3.2.1 Biotic Stresses

Sterility mosaic disease (SMD) is an important disease which can cause yield loss of up to 95% (Reddy and Nene 1980). Gnanesh et al. (2011) identified the QTL qSMD4 on LG07, which provides resistance to SMD. The gene, C.cajan_01839, was identified for resistance to SMD using the next-generation sequencing approach from recombinant inbred lines derived from a cross between CPL 20096 and ICPL 332 (Singh et al. 2016). Other genes identified for resistance to SMD include C.cajan_07858, C.cajan_20995, C.cajan_21801, and C.cajan_17341 (Saxena et al. 2021).

Fusarium wilt is a soil-borne disease caused by *Fusarium udum* which can cause yield losses of 30–100% in pigeon pea (Dhar et al. 2005). Two genes, C.cajan_03691 and C.cajan_18888 (Table 1.4), have been identified by Saxena et al. (2021), for resistance to *Fusarium* wilt (FW). The SSR markers, HASSR8190 and HASSR58180, have been reported by Patil et al. (2017) for resistance to FW through Bulk Segregant Analysis (BSA).

1.4.3.2.2 Abiotic Stresses

Drought in pigeon pea, especially during seedling and reproductive stages, limits plant growth and seed yield (Saxena 2008). The candidate gene, DLP, was mapped for drought stress in pigeon pea through the qRT-PCR approach by Deeplanaik et al. (2013). The gene, CcCDR, is associated with drought and cold stress and was isolated from the subtracted cDNA library of pigeon pea (Tamirisa et al. 2014). Two genes, *C.cajan*_29830 and C.cajan_33874, confer tolerance to drought in pigeon pea and were identified across three pigeon pea genotypes, ICPL 151, ICPL 8755, and ICPL 227 (Sinha et al. 2016). Essential metals such as Fe, Cu, Mn, and Zn and non-essential metals such as Cd, Pb, and Hg, when present in higher concentrations in pigeon pea, inhibit its growth and development. The plants are subjected to oxidation stress and cellular ionic homeostasis is affected (Yadav 2010). The gene, CcMT1, isolated from the cDNA library of *Cajanus cajan* is a probable gene for tolerance to heavy metals (Sekhar et al. 2010).

1.4.3.3 Chickpea (*Cicer arietinum* L.)

Chickpea (2n = 16) is a cool-season leguminous crop that belongs to the Fabaceae family. This nutritious pulse crop is categorized into two main types, the small-seeded desi type and the large-seeded kabuli type, as described by Cubero (1975). Chickpea holds the distinction of being the second most widely cultivated food legume crop globally. It is grown across a vast geographical range, spanning from the Mediterranean basin and West Asia to the Indian subcontinent, Australia, Eastern Africa, and North and South America. The domestication of chickpea took place around 12,000 to 10,000 years ago in the Fertile Crescent, as documented by Zohary and Hopf (1973) and Abbo et al. (2003). It is considered one of the foundational crops in the history of modern agriculture, as highlighted by Zohary and Hopf (1973). Currently, the global production of chickpea exceeds 15.87 million tons, as reported by FAOSTAT (2021). India stands out as the largest producer of chickpea, contributing significantly to global production with an estimated output of 11.91 million tons in 2021, according to FAOSTAT. Chickpea has many nutritional benefits, comprising 50–58% carbohydrates, 18.2% protein in desi type and 18.4% protein in kabuli type, 7–8% moisture, and 3.8–10.2%, lysine, arginine, and minerals such as Fe, Zn, and Se (Madurapperumage et al. 2021). Chickpea is a self-pollinating crop having a genome size of about 738 Mb (Varshney et al. 2013). Genomic advancement in chickpea includes the development of 2000 SSR markers, 15,000 DArT and SNP markers (Varshney 2016). A consensus genetic map, based on 10 populations by using sequence-tagged microsatellite sites (Millan et al. 2010), and a genetic variation map by sequencing 3366 chickpea lines (Varshney et al. 2021) have been developed in chickpea.

1.4.3.3.1 Biotic Stresses

Ascochyta blight in chickpea is a fungal disease caused by *Ascochyta rabiei*. It affects plants growing in cool and wet areas. Symptoms include the appearance of necrotic lesions on the leaves, the breaking of stems, and ultimately the death of the plant. Li et al. (2017) identified 12 candidate genes lying in the AB4.1 region and 20 significant SNPs for resistance against ascochyta blight by employing a whole-genome resequencing approach. Six candidate genes, *CPR02-qAB2.1* on chromosome Ca2; *CPR02-qAB4.1*, *CPR02-qAB4.2*, *CPR02-qAB4.3*, and *CPR02-qAB4.4*, on chromosome Ca4; and *CPR02-qAB7.1* on chromosome Ca7, were mapped by next-generation sequencing-based BSA approach for resistance to ascochyta blight (Deokar et al. 2019). Chickpea pod borer (*Helicoverpa armigera*) is a major pest in chickpea which causes yield loss of up to 90% (Kaur et al. 2022). A QTL, *qDf03*, on linkage group CaLG03, for resistance to *H. armigera* (Table 1.4), was identified from RILs developed from an inter-specific cross between the cultivated species, ICC 4958 (susceptible), and wild species, PI 489777 (resistant), and accounts for 42.49% of the total phenotypic variance for resistance (Barmukh et al. 2021).

1.4.3.3.2 Abiotic Stresses

Li et al. (2018) reported 38 SNPs for yield-related traits, under drought stress, through GWAS-based SNPs. Thirty-two candidate genes encoding for heat shock proteins and pollen-specific leucine-rich repeat extensin-like protein 1 were mined for tolerance to heat in chickpea through RILs derived from a cross between the cultivars, DCP 92–3 (heat sensitive) and ICCV 92944 (heat tolerant), using the genotyping-by-sequencing approach (Jha et al. 2021). The QTLs, qbio-02, qpv-02, qnpp-01, qdff-02, qdff-01, and qdfi-01, located on chromosomes 4 and 1 account for 13.71, 13.59, 12.76, 14.72, and 13.68% of the total variation, respectively, for traits related to heat stress (Kushwah et al. 2021). These QTLs were identified from 187 RILs by an inter-specific cross of the cultivar, GPF 2 (heat tolerant) and ILWC 292 (heat sensitive), developed by using the single seed descent method.

1.4.3.4 Rice Bean (*Vigna umbellata* Thunb.)

Rice bean (*Vigna umbellata* Thunb.) is a diploid plant with a chromosome number of 2n = 22, and it belongs to the subgenus *Ceratotropis* within the Fabaceae family. This warm season, annual legume crop has a multifaceted utility and is extensively cultivated in South and Southeast Asia, as documented by Khadka and Acharya (2009) and Pattanayak et al. (2019). Rice bean is characterized as a non-conventional, underutilized crop, yet it holds significant potential to provide cost-effective nutrition and generate livelihood opportunities, especially for tribal and underprivileged communities residing in hilly regions. The primary cultivation regions for rice bean include the northeastern states, as well as the western and eastern Ghats of India, as noted by Arora et al. (1980). It is also grown sporadically in various regions of the Western Himalayas, such as Pauri, Ramgarh, Lohaghat, Champawat areas in Uttarakhand, and some parts of Jammu and Kashmir and Himachal Pradesh. Additionally, it can be found in the central plains of India, the hills of Chhattisgarh, the southern hills of Odisha, and Kerala.

Rice bean is notable for its nutritional richness and high genetic diversity. It possesses the capacity to thrive in extreme environmental conditions and demonstrates tolerance to various diseases and pests. Additionally, it can withstand high levels of rainfall. This versatile crop offers multiple uses, with its grains used as dal, green pods serving as a vegetable, and biomass utilized as forage. Rice bean is also a source of essential nutrients such as iron (Fe), zinc (Zn), copper (Cu), phosphorus (P), potassium (K), calcium (Ca), and magnesium (Mg), as highlighted by Dhillon and Tanwar (2018). Moreover, it plays a role in livestock feed and can be used as manure, making it a valuable resource with diverse applications.

The crop has been considered an underutilized legume, and so far, only little work has been done to exploit the crop. Rice bean has a genome size of 440 Mb (Kaul et al. 2019). The first genetic linkage map related to the domestication of rice bean was developed by Isemura et al. (2010). Twenty-three genic SSR markers have been developed in rice bean (Chen et al. 2016) which can be used for the construction of genetic linkage maps. A draft genome sequence with 31,276 genes with 96.08% functional coverage has been developed by Kaul et al. (2019) where 97% coding domains of Rho-type GTPase-activating proteins, serine-threonine protein kinases (TAO1), presynaptic morphology proteins, and enhanced disease resistance 1 isoforms in rice bean has been identified. Genetic diversity studies were carried out by using different polymorphic DNA markers, AFLP, RAPD, SSR, and ISSR markers (Bajracharya et al. 2017; Muthusamy et al. 2008; Thakur et al. 2017; Tian et al. 2013; Iangrai et al. 2017). It harbours desirable traits coding for pest resistance as well as tolerance to various abiotic stresses.

1.4.3.4.1 Biotic Stresses

Bruchid or bean weevils (*Callosobruchus maculatus*) attack seeds in storage and can cause destruction within a span of 3–4 months (Banto and Sanchez 1972). Research efforts aimed at building genomic resources to enhance resistance against the primary storage pest, bruchid beetles (*Callosobruchus* spp.), have been initiated, as documented in studies conducted by Tomooka et al. (2000), Kashiwaba et al. (2003), and Somta et al. (2006). Somta et al. (2006) developed the genetic linkage map using the inter-specific mapping population derived from rice bean and *V. nakashimae* and further QTL mapping for traits against bruchid infestation. Venkataramana et al. (2016) identified 11 QTLs controlling bruchid infestation in rice bean, which were distributed over 10 linkage groups (LGs), through QTL mapping analysis. These LGs were mapped from the F_2 populations, derived from crosses between LRB238 and JP100304 (resistant) each to LRB26 (susceptible). Cmpd1.5 and Cmpd1.6 are the major QTLs which were mapped within 11.9 cM and 13.0 cM, respectively, of the flanking markers and correspond to 67.3 and 77.4% of the phenotypic variation for % of seeds damaged by bruchids.

1.4.3.4.2 Abiotic Stresses

The QTL, Sdwa4.4.1 lying on LG4, accounts for 25.1% of the variance for water absorption by seeds which affects seed dormancy (Isemura et al. 2010). PCAS and FDH genes were reported by Fan et al. (2014), corresponding to Al toxicity and tolerance.

1.4.3.5 Faba Bean (*Vicia faba* L.)

Faba bean (2n = 12), also known as broad bean, is a rich source of carbohydrates (58.3%) such as starch; sugars and dietary fibre; proteins (26.1%) such as globulins, albumins, glutelins, and prolamines (Rahate et al. 2021); minerals such as Fe, Ca, P, Mg, Na, and Zn; and vitamins A, C, K, niacin, and folate (Roorkiwal et al. 2021). The genome size of faba bean is quite large, i.e. 13,000 Mb (Ellwood et al. 2008). The first development in faba bean genomics was the identification of 5000 expressed sequence tags (ESTs) from the cultivar 'Windsor', which facilitated data for the embryo transcriptome of faba bean (Ray and Georges 2010). Genetic maps were also developed from different populations and markers. A consensus genetic map was developed by Carrillo-Perdomo et al. (2020), covering a map length of 1548 cM, using three mapping populations. Some of the markers developed for faba bean are SNP (KASP) consisting of 824 loci (Webb et al. 2016), EST- SSR consisting of 647 loci (El-Rodeny et al. 2014), and SNP (Illumina) with 480 loci (Kaur et al. 2014).

1.4.3.5.1 Biotic Stresses

Ascochyta blight (AS) caused by the fungus, *Ascochyta fabae*, is a destructive disease of faba bean which prefers a wet environment. Lesions appear on the leaves, stems, pods, and seeds. Damaged seeds have yellowish-brown stains. This disease causes yield loss of 35–40% in faba bean (Díaz-Ruiz et al. 2009). Kaur et al. (2014) reported four QTLs controlling resistance to ascochyta blight across three linkage groups, Chr-I. A, Chr-II, and Chr-VI, from RILs derived from an intra-specific cross between Icarus (susceptible) and Ascot (resistant) and explained 8–20% of the phenotypic variance. Af2 QTL located on chromosome 2 and Af3 QTL located on chromosome 3 confer resistance to AS and were identified from RILs by crossing the cultivars, 29H and Vf136 (Atienza et al. 2016). Faba bean rust is caused by the fungus *Uromyces viciae-fabae* and can damage up to 80% of the crop (Rashid and Bernier 1991). The two genes, Uvf-2 and Uvf-3, lying on chromosomes 3 and 5, respectively, were mapped from RILs derived by crossing Doza#12034 and Ac1655 (rust-resistant cultivars) with Fiord (rust-susceptible cultivar) and confer resistance to faba bean rust (Ijaz et al. 2021).

1.4.3.5.2 Abiotic Stresses

Frost during the pod-filling stage causes the seeds to have a discoloured, deformed, and wilted appearance. Frost during the flowering stage causes a significant reduction in yield. The SNP marker, VF_Mt3g086600, was reported to be associated with tolerance to frost in faba bean through association mapping (Sallam et al. 2016). Drought is a major concern in faba bean as it reduces photosynthetic activity, chlorophyll content, and leaf water potential. Yield reduction due to drought is up to 40% (Daryanto et al. 2015). The locus, E36M48–279, E41M60–310, and E36M48–304, lying on LG7 at positions 94.4, 27.3 and 112.3 cM, respectively, was found to be associated with tolerance to drought and accounts for 9.31, 6.86, and 6.60% of the phenotypic variation (Ali et al. 2016a, b).

1.4.3.6 Cowpea (*Vigna unguiculata* L. Walp)

Cowpea is a *kharif* crop, categorized as minor pulse in India, which contains protein (23–32% which may vary with each variety), carbohydrates (50–60%), fat (1%), and antioxidants such as polyphenols, flavonoids, anthocyanins, and tannins (Jayathilake et al. 2018). It acts as a cover crop due to its bushy growth and is also used as vegetable and animal feed. It is locally known as *Pereu* and *Ubiyo* in Upper and Perung in Lower Subansiri district of Arunachal Pradesh. Remarkable variability has been found in seed colour, seed length, seed diameter, seed mottling, and seed weight. Cowpea is a diploid (2n = 22), with a genome size of 620 Mb (Boukar et al. 2019). The cowpea accession, IT97K-499-35, was used to develop a BAC physical map and whole-genome shotgun assembly (Muñoz-Amatriaín et al. 2017). The genome assembly of this accession was also developed by Lonardi et al. (2019).

1.4.3.6.1 Biotic Stresses

The Cowpea aphid (*Aphis craccivora*) is an important insect pest of cowpea causing stunted growth, reduced photosynthetic rate, delayed flowering, and ultimately yield. Resistance to cowpea aphid is conferred by a major QTL lying on chromosome 2, which accounts for 39% of antibiosis resistance (Kamphuis et al. 2012). This QTL was identified from the F_2 population by crossing SA30199 (resistant) and Borung (susceptible), through QTL analysis. Cowpea flower bud thrips (*Megalurothrips sjostedti*) feed on reproductive or floral parts of the plant, leading to flower discolouration, distortion, malformation, and abortion. This insect accounts for 80% of yield losses, which can reach 100% in the case of severe infestation (Alabi et al. 2011). Sobda et al. (2017) identified three QTLs, Fthp28, Fthp87, and Fthp129 (Table 1.4), on LG2, LG4, and LG6, respectively, for resistance to flower bud thrips explaining 24.5, 12.2, and 6.5% of the phenotypic variance. These QTLs were identified with the help of SNPs from an F_2 population derived from the cross, SANZI (resistant) x VYA (susceptible). The QTL, QRk-vu11.1, containing the gene Rk, which is the locus for resistance to root-knot nematode in cowpea, was identified through GWAS-based SNPs (Huynh et al. 2018).

1.4.3.6.2 Abiotic Stresses

The SNP markers, C35017374_128, Scaffold87490_630, and Scaffold87490_622, have been reported to account for 12.24, 13.31, and 14.2%, respectively, of the phenotypic variation (Ravelombola et al. 2018) through association mapping and confer tolerance to salinity at germination stages. Ravelombola et al. (2018) also reported the SNP markers, Scaffold68489_600, Scaffold93827_270, Scaffold87490_633, Scaffold87490_640, Scaffold82042_3387, C35069468_1916, and Scaffold93942_1089, accounting for 7.74, 8.30, 9.50, 9.52, 10.31, 12.25, and 11.86%, respectively, conferring salinity tolerance in cowpea at seedling stage. High temperature during floral development is deleterious for cowpea production as this causes male sterility resulting in fruit abortion. QTL analysis was carried out by using RILs derived by crossing CB27 and IT82E-18, and five genomic regions, Cht-1, Cht-2, Cht-3, Cht-4, and Cht-5, on LG2, 7, 6, 10, and 3, respectively, were identified for heat tolerance in cowpea during the reproductive stage and account for 11.5–18.1% of the phenotypic variance (Lucas et al. 2013). High yield under drought stress can be achieved with the help of QTLs having the stay-green trait. The stay-green trait can be described as a phenomenon where the QTLs for delayed senescence, biomass, and grain yield are co-localized, and these traits are enhanced under drought stress. Muchero et al. (2013) reported five loci, Dro-7, Dro-10, Dro-8, Dro-1, and Dro-3, on LG1, 3, 4, 7, and 8, using SNP markers, showing the stay-green trait in cowpea.

References

Abbo S, Berger J, Turner NC (2003) Evolution of cultivated chickpea: four bottlenecks limit diversity and constrain adaptation. Funct Plant Biol 30(10):1081–1087

Abiodun OA, Dauda AO, Fabiyi OA, Akintayo FM (2022) Biofortification: quality improvement of Faba bean. In: Faba bean: chemistry, properties and functionality. Springer, pp 275–299

Agarwal G, Kavalappara SR, Gautam S, Silva A, Simmons A, Srinivasan R, Dutta B (2021) Field screen and genotyping of Phaseolus vulgaris against two begomoviruses in Georgia, USA. Insects 12(1):49

Agarwal C, Chen W, Varshney RK, Vandemark G (2022) Linkage QTL mapping and genome-wide association study on resistance in chickpea to Pythium Ultimum. Front Genet 13:945787

Alabi O, Odebiyi J, Tamò M, Omoloye A (2011) The roles of plant secondary metabolites from cowpea floral structures in resistance to the flower bud thrips. J Agr Sci 1:262–269

Ali M, Kushwaha BL (1987) Cultivation of rabi rajmash in plains (No. RESEARCH)

Ali L, Deokar A, Caballo C, Tar'an B, Gil J, Chen W, Millan T, Rubio J (2016a) Fine mapping for double podding gene in chickpea. Theor Appl Genet 129:77–86

Ali MB, Welna GC, Sallam A, Martsch R, Balko C, Gebser B, Sass O, Link W (2016b) Association analyses to genetically improve drought and freezing tolerance of faba bean (Vicia faba L.). Crop Sci 56(3):1036–1048

Alice AK, Pandey AK, Singh S, Ansari T, Ariina MS (2019) Nutritional profile of French bean. A mini review. Int J Chem Stud 7(1):282–284

Ambachew D, Joshua J, Mmbaga MT, Blair MW (2021) Sources of resistance to common bacterial blight and charcoal rot disease for the production of Mesoamerican common beans in the southern United States. Plan Theory 10(5):998

Amri-Tiliouine W, Laouar M, Abdelguerfi A, Jankowicz-Cieslak J, Jankuloski L, Till BJ (2018) Genetic variability induced by gamma rays and preliminary results of low-cost TILLING on M2 generation of chickpea (Cicer arietinum L.). Frontiers. Plant Sci 9:1568

Amri-Tiliouine W, Jankowicz-Cieslak J, Till BJ, Laouar M (2022) Generation of mutant plants by gamma ray exposure and development of low-cost TILLING population in chickpea (Cicer arietinum L.). Methods Mol Biol 2484:143–159

Andargie M, Pasquet RS, Gowda BS, Muluvi GM, Timko MP (2011) Construction of a SSR-based genetic map and identification of QTL for domestication traits using recombinant inbred lines from a cross between wild and cultivated cowpea (V. Unguiculata (L.) Walp.). Mol Breed 28:413–420

Arbaoui M, Balko C, Link W (2008) Study of faba bean (Vicia faba L.) winter-hardiness and development of screening methods. Field Crop Res 106(1):60–67

Arora R (1988) The Indian gene centre-Priorities and prospects for collection. In: Paroda RS, Arora RK, Chandel KPS (eds) Plant genetic resources–Indian perspectives. NBPGR, New Delhi, pp 66–75

Arora R, Chandel K, Joshi B, Pant K (1980) Rice bean: tribal pulse of eastern India. Econ Botany 34:260

Assefa T, Assibi Mahama A, Brown AV, Cannon EK, Rubyogo JC, Rao IM, Blair MW, Cannon SB (2019) A review of breeding objectives, genomic resources, and marker-assisted methods in common bean (Phaseolus vulgaris L.). Mol Breed 39:1–23

Atienza SG, Palomino C, Gutiérrez N, Alfaro C, Rubiales D, Torres AM, Ávila CM (2016) QTLs for ascochyta blight resistance in faba bean (Vicia faba L.): validation in field and controlled conditions. Crop Pasture Sci 67(2):216–224

Attri H, Jamwal B, Chaudhary N, Sandhu R (2018) Estimation of correlation among the morpho-physiological traits in MAGIC population of chickpea (Cicer arietinum L.). J Pharmacogn Phytochem 7(4):1600–1603

Bajracharya J, Singh S, Dangol B, Hollington P, Witcombe J (2017) Molecular marker diversity in Nepalese and Indian rice bean germplasm. European Commission 6th framework Programme Report 7

Banto S, Sanchez F (1972) The biology and chemical control of Callosobruchus chinensis (Linn.) (Coleoptera: Bruchidae). Philipp Ent 2:167–182

Barmukh R, Roorkiwal M, Jaba J, Chitikineni A, Mishra SP, Sagurthi SR, Munghate R, Sharma H, Varshney RK (2021) Development of a dense genetic map and QTL analysis for pod borer Helicoverpa armigera (Hübner) resistance component traits in chickpea (Cicer arietinum L.). Plant Genome 14(1):e20071

Barrera-Figueroa BE, Gao L, Diop NN, Wu Z, Ehlers JD, Roberts PA, Close TJ, Zhu J-K, Liu R (2011) Identification and comparative analysis of drought-associated microRNAs in two cowpea genotypes. BMC Plant Biol 11:1–11

Blair MW, Pedraza F, Buendia H, Gaitán-Solís E, Beebe SE, Gepts P, Tohme J (2003) Development of a genome-wide anchored microsatellite map for common bean (Phaseolus vulgaris L.). Theor Appl Genet 107:1362–1374

Blair MW, Iriarte G, Beebe S (2006) QTL analysis of yield traits in an advanced backcross population derived from a cultivated Andean× wild common bean (Phaseolus vulgaris L.) cross. Theor Appl Genet 112:1149–1163

Blair MW, Porch T, Cichy K, Galeano CH, Lariguet P, Pankhurst C, Broughton W (2007) Induced mutants in common bean (Phaseolus vulgaris), and their potential use in nutrition quality breeding and gene discovery. Isr J Plant Sci 55(2):191–200

Blair MW, Díaz LM, Buendía HF, Duque MC (2009) Genetic diversity, seed size associations and population structure of a core collection of common beans (Phaseolus vulgaris L.). Theor Appl Genet 119:955–972

Blair MW, Hurtado N, Chavarro CM, Muñoz-Torres MC, Giraldo MC, Pedraza F, Tomkins J, Wing R (2011) Gene-based SSR markers for common bean (Phaseolus vulgaris L.) derived from root and leaf tissue ESTs: an integration of the BMc series. BMC Plant Biol 11:1–10

Blair MW, Cordoba JM, Munoz C, Yuyo DK (2014) BAC-end microsatellites from intra and intergenic regions of the common bean genome and their correlation with cytogenetic features. PLoS One 9(9):e101873

Blair MW, Cortés AJ, Farmer AD, Huang W, Ambachew D, Penmetsa RV, Carrasquilla-Garcia N, Assefa T, Cannon SB (2018) Uneven recombination rate and linkage disequilibrium across a reference SNP map for common bean (Phaseolus vulgaris L.). PLoS One 13(3):e0189597

Bohra A, Dubey A, Saxena RK, Penmetsa RV, Poornima K, Kumar N, Farmer AD, Srivani G, Upadhyaya HD, Gothalwal R (2011) Analysis of BAC-end sequences (BESs) and development of BES-SSR markers for genetic mapping and hybrid purity assessment in pigeonpea (Cajanus spp.). BMC Plant Biol 11(1):1–15

Bohra A, Saxena RK, Gnanesh B, Saxena K, Byregowda M, Rathore A, KaviKishor P, Cook DR, Varshney RK (2012) An intra-specific consensus genetic map of pigeonpea [Cajanus cajan (L.) Millspaugh] derived from six mapping populations. Theor Appl Genet 125:1325–1338

Bohra A, Saxena K, Varshney RK, Saxena RK (2020) Genomics-assisted breeding for pigeonpea improvement. Theor Appl Genet 133:1721–1737

Boukar O, Belko N, Chamarthi S, Togola A, Batieno J, Owusu E, Haruna M, Diallo S, Umar ML, Olufajo O (2019) Cowpea (Vigna unguiculata): genetics, genomics and breeding. Plant Breed 138(4):415–424

Bressani R, Elías LG (1979) Nutritional role of polyphenols in beans. In: Polyphenols in cereals and legumes: proceedings of a symposium. IDRC, Ottawa, ON, Canada

Broughton WJ, Hernández G, Blair M, Beebe S, Gepts P, Vanderleyden J (2003) Beans (Phaseolus spp.)–model food legumes. Plant Soil 252:55–128

Calles T, Xipsiti M and del Castello R (2019) Legacy of the international year of pulses. Environ Earth Sci 78:1–8.

Cannon SB, May GD, Jackson SA (2009) Three sequenced legume genomes and many crop species: rich opportunities for translational genomics. Plant Physiol 151(3):970–977

Carrillo-Perdomo E, Vidal A, Kreplak J, Duborjal H, Leveugle M, Duarte J, Desmetz C, Deulvot C, Raffiot B, Marget P (2020) Development of new genetic resources for faba bean (Vicia faba L.) breeding through the discovery of gene-based SNP markers and the construction of a high-density consensus map. Sci Rep 10(1):6790

Chandora R, Shekhawat N, Malhotra N (2020) Chickpea genetic resources: collection, conservation, characterization, and maintenance. In: Chickpea: crop wild relatives for enhancing genetic gains. Elsevier, pp 37–61

Chen H, Chen X, Tian J, Yang Y, Liu Z, Hao X, Wang L, Wang S, Liang J, Zhang L (2016) Development of gene-based SSR markers in rice bean (Vigna umbellata L.) based on transcriptome data. PloS One 11(3):e0151040

Choudhary N, Bawa V, Paliwal R, Singh B, Bhat MA, Mir JI, Gupta M, Sofi PA, Thudi M, Varshney RK (2018) Gene/QTL discovery for anthracnose in common bean (Phaseolus vulgaris L.) from North-Western Himalayas. PLoS One 13(2):e0191700

Close TJ, Luo M, Lonardi S, Alpert M, Duma D, Bozdag S, Farmer A, May GD, Jackson SA, Ma Y (2011) Genome resources for cowpea genotype IT97K-499-35. In: Plant & animal genome conference

Coelho RC, Barsotti RCF, Maltez HF, Júnior CAL, de Sousa BH (2021) Expanding information on the bioaccessibility and bioavailability of iron and zinc in biofortified cowpea seeds. Food Chem 347:129027

Córdoba JM, Chavarro C, Schlueter JA, Jackson SA, Blair MW (2010) Integration of physical and genetic maps of common bean through BAC-derived microsatellite markers. BMC Genomics 11:1–10

Cottage A, Gostkiewicz K, Thomas JE, Borrows R, Torres AM, O'Sullivan DM (2012) Heterozygosity and diversity analysis using mapped single nucelotide polymorphisms in a faba bean inbreeding programme. Mol Breed 30(4):1799–1809

Crépon K, Marget P, Peyronnet C, Carrouée B, Arese P, Duc G (2010) Nutritional value of faba bean (Vicia faba L.) seeds for feed and food. Field Crop Res 115(3):329–339

Cubero JI (1975) research on the chickpea (Cicer arietinum) in Spain. In International Workshop on Grain Legumes

Cui X-H, Hao F-S, Chen H, Chen J, Wang X-C (2008) Expression of the Vicia faba VfPIP1 gene in Arabidopsis thaliana plants improves their drought resistance. J Plant Res 121:207–214

Dalla Via V, Narduzzi C, Aguilar OM, Zanetti ME, Blanco FA (2015) Changes in the common bean transcriptome in response to secreted and surface signal molecules of rhizobium etli. Plant Physiol 169(2):1356–1370

Daryanto S, Wang L, Jacinthe P-A (2015) Global synthesis of drought effects on food legume production. PLoS One 10(6):e0127401

Deeplanaik N, Kumaran RC, Venkatarangaiah K, Shivashankar SKH, Doddamani D, Telkar S (2013) Expression of drought responsive genes in pigeonpea and in silico comparison with soybean cDNA library. J Crop Sci Biotechnol 16:243–251

Deokar A, Sagi M, Daba K, Tar'an B (2019) QTL sequencing strategy to map genomic regions associated with resistance to ascochyta blight in chickpea. Plant Biotechnol J 17(1):275–288

Dhanasekar P, Souframanien J, Suprasanna P (2021) Breeding cowpea for quality traits: a genetic biofortification perspective. In: Breeding for enhanced nutrition and bio-active compounds in food legumes, pp 157–179

Dhar V, Reddy M, Chaudhary R (2005) Major diseases of pigeonpea and their management. In: Ali M, Kumar S (eds) Advances in Pigeonpea research. Indian Institute of Pulses Research, Kanpur, India, pp 229–261

Dhillon PK, Tanwar B (2018) Rice bean: a healthy and cost-effective alternative for crop and food diversity. Food Security 10(3):525–535

Diaz LM, Ricaurte J, Tovar E, Cajiao C, Teran H, Grajales M, Polania J, Rao I, Beebe S, Raatz B (2018) QTL analyses for tolerance to abiotic stresses in a common bean (Phaseolus vulgaris L.) population. PLoS One 13(8):e0202342

Diaz S, Ariza-Suarez D, Izquierdo P, Lobaton JD, de la Hoz JF, Acevedo F, Duitama J, Guerrero AF, Cajiao C, Mayor V (2020) Genetic mapping for agronomic traits in a MAGIC population of common bean (Phaseolus vulgaris L.) under drought conditions. BMC Genomics 21(1):1–20

Diaz S, Ariza-Suarez D, Ramdeen R, Aparicio J, Arunachalam N, Hernandez C, Diaz H, Ruiz H, Piepho H-P, Raatz B (2021) Genetic architecture and genomic prediction of cooking time in common bean (Phaseolus vulgaris L.). Front Plant Sci 11:622213

Díaz-Ruiz R, Satovic Z, Avila C, Alfaro C, Gutierrez M, Torres A, Román B (2009) Confirmation of QTLs controlling Ascochyta fabae resistance in different generations of faba bean (Vicia faba L.). Crop Pasture Sci 60(4):353–361

Dietary Reference Intakes (2019) Recommended dietary allowances and adequate intakes, elements. National Academies Press. https://www.ncbi.nlm.nih.gov/books/NBK54544

Ellwood SR, Phan HT, Jordan M, Hane J, Torres AM, Avila CM, Cruz-Izquierdo S, Oliver RP (2008) Construction of a comparative genetic map in faba bean (Vicia faba L.); conservation of genome structure with Lens culinaris. BMC Genomics 9(1):1–11

El-Rodeny W, Kimura M, Hirakawa H, Sabah A, Shirasawa K, Sato S, Tabata S, Sasamoto S, Watanabe A, Kawashima K (2014) Development of EST-SSR markers and construction of a linkage map in faba bean (Vicia faba). Breed Sci 64(3):252–263

Escobar E, Oladzad A, Simons K, Miklas P, Lee RK, Schroder S, Bandillo N, Wunsch M, McClean PE, Osorno JM (2022) New genomic regions associated with white mold resistance in dry bean using a MAGIC population. Plant Genome 15(1):e20190

Etemadi F, Barker AV, Hashemi M, Zandvakili OR, Park Y (2018) Nutrient accumulation in faba bean varieties. Commun Soil Sci Plant Anal 49(16):2064–2073

Fan W, Lou H, Gong Y, Liu M, Wang Z, Yang J, Zheng S (2014) Identification of early Al-responsive genes in rice bean (V igna umbellata) roots provides new clues to molecular mechanisms of Al toxicity and tolerance. Plant Cell Environ 37(7):1586–1597

FAO, IFAD, UNICEF, WFP and WHO (2022) The state of food security and nutrition in the world 2022. Repurposing food and agricultural policies to make healthy diets more affordable. FAO, Rome. https://doi.org/10.4060/cc0639en

FAOSTAT (2021) Food and agriculture organization of the United Nations (FAO). http://www.fao.org/faostat/en/-data/QC. Accessed 14 Dec 2022

Fonsêca A, Ferreira J, dos Santos TRB, Mosiolek M, Bellucci E, Kami J, Gepts P, Geffroy V, Schweizer D, dos Santos KG (2010) Cytogenetic map of common bean (Phaseolus vulgaris L.). Chromosom Res 18:487–502

Galeano CH, Fernandez AC, Franco-Herrera N, Cichy KA, McClean PE, Vanderleyden J, Blair MW (2011) Saturation of an intra-gene pool linkage map: towards a unified consensus linkage map for fine mapping and synteny analysis in common bean. PLoS One 6(12):e28135

Ganesan K, Xu B (2017) Polyphenol-rich dry common beans (Phaseolus vulgaris L.) and their health benefits. Int J Mol Sci 18(11):2331

Gaudencia JK, Miriam G, Oliver K, Lexa G (2020) Incidence and severity of anthracnose (Colletotrichum lindemuthianum) on selected common bean (Phaseolus vulgaris L.) genotypes. American. J Agric Res 5:84

Gaur R, Sethy NK, Choudhary S, Shokeen B, Gupta V, Bhatia S (2011) Advancing the STMS genomic resources for defining new locations on the intraspecific genetic linkage map of chickpea (Cicer arietinum L.). BMC Genomics 12:1–18

Gaur R, Azam S, Jeena G, Khan AW, Choudhary S, Jain M, Yadav G, Tyagi AK, Chattopadhyay D, Bhatia S (2012) High-throughput SNP discovery and genotyping for constructing a saturated linkage map of chickpea (Cicer arietinum L.). DNA Res 19(5):357–373

Gaur R, Jeena G, Shah N, Gupta S, Pradhan S, Tyagi AK, Jain M, Chattopadhyay D, Bhatia S (2015) High density linkage mapping of genomic and transcriptomic SNPs for synteny analysis and anchoring the genome sequence of chickpea. Sci Rep 5(1):13387

Gaur PM, Samineni S, Thudi M, Tripathi S, Sajja SB, Jayalakshmi V, Mannur DM, Vijayakumar AG, Ganga Rao NV, Ojiewo C (2019) Integrated breeding approaches for improving drought and heat adaptation in chickpea (Cicer arietinum L.). Plant Breed 138(4):389–400

Gepts P, Aragão FJ, Ed B, Blair MW, Brondani R, Broughton W, Galasso I, Hernández G, Kami J, Lariguet P (2008) Genomics of Phaseolus beans, a major source of dietary protein and micronutrients in the tropics. In: Genomics of tropical crop plants. Springer, pp 113–143

Gioia T, Logozzo G, Marzario S, Spagnoletti Zeuli P, Gepts P (2019) Evolution of SSR diversity from wild types to US advanced cultivars in the Andean and Mesoamerican domestications of common bean (Phaseolus vulgaris). PLoS One 14(1):e0211342

Glahn RP, Wiesinger JA, Lung'aho MG (2020) Iron concentrations in biofortified beans and non-biofortified marketplace varieties in East Africa are similar. J Nutr 150(11):3013–3023

Gnanesh B, Bohra A, Sharma M, Byregowda M, Pande S, Wesley V, Saxena R, Saxena K, Kishor PK, Varshney R (2011) Genetic mapping and quantitative trait locus analysis of resistance to sterility mosaic disease in pigeonpea [*Cajanus cajan* (L.) Millsp.]. Field Crop Res 123(2):53–61

Godfray HCJ, Beddington JR, Crute IR, Haddad L, Lawrence D, Muir JF, Pretty J, Robinson S, Thomas SM, Toulmin C (2010) Food security: the challenge of feeding 9 billion people. Science 327(5967):812–818

Gomes-Messias LM, Vianello RP, Marinho GR, Rodrigues LA, Coelho ASG, Pereira HS, Melo LC, de Souza TLPO (2022) Genetic mapping of the Andean anthracnose resistance gene present in the common bean cultivar BRSMG Realce. Front Plant Sci 13:1033687

Gowda S, Radhika P, Mhase L, Jamadagni B, Gupta V, Kadoo N (2011) Mapping of QTLs governing agronomic and yield traits in chickpea. J Appl Genet 52:9–21

Guan J, Zhang J, Gong D, Zhang Z, Yu Y, Luo G, Somta P, Hu Z, Wang S, Yuan X (2022) Genomic analyses of rice bean landraces reveal adaptation and yield related loci to accelerate breeding. Nat Commun 13(1):5707

Gujaria N, Kumar A, Dauthal P, Dubey A, Hiremath P, Bhanu Prakash A, Farmer A, Bhide M, Shah T, Gaur PM (2011) Development and use of genic molecular markers (GMMs) for construction of a transcript map of chickpea (Cicer arietinum L.). Theor Appl Genet 122:1577–1589

Gupta S, Gopalakrishna T (2010) Development of unigene-derived SSR markers in cowpea (Vigna unguiculata) and their transferability to other Vigna species. Genome 53(7):508–523

Gutiérrez N, Palomino C, Satovic Z, Ruiz-Rodríguez M, Vitale S, Gutiérrez M, Rubiales D, Kharrat M, Amri M, Emeran A (2013) QTLs for Orobanche spp. resistance in faba bean: identification and validation across different environments. Mol Breed 32:909–922

Hamon C, Coyne CJ, McGee RJ, Lesné A, Esnault R, Mangin P, Hervé M, Le Goff I, Deniot G, Roux-Duparque M (2013) QTL meta-analysis provides a comprehensive view of loci controlling partial resistance to Aphanomyces euteichesin four sources of resistance in pea. BMC Plant Biol 13(1):1–19

Henry RJ, Nevo E (2014) Exploring natural selection to guide breeding for agriculture. Plant Biotechnol J 12(6):655–662

Hiremath PJ, Farmer A, Cannon SB, Woodward J, Kudapa H, Tuteja R, Kumar A, BhanuPrakash A, Mulaosmanovic B, Gujaria N (2011) Large-scale transcriptome analysis in chickpea (Cicer arietinum L.), an orphan legume crop of the semi-arid tropics of Asia and Africa. Plant Biotechnol J 9(8):922–931

Hoyos-Villegas V, Song Q, Wright EM, Beebe SE, Kelly JD (2016) Joint linkage QTL mapping for yield and agronomic traits in a composite map of three common bean RIL populations. Crop Sci 56(5):2546–2563

Huang BE, Verbyla KL, Verbyla AP, Raghavan C, Singh VK, Gaur P, Leung H, Varshney RK, Cavanagh CR (2015) MAGIC populations in crops: current status and future prospects. Theor Appl Genet 128:999–1017

Huynh BL, Ehlers JD, Huang BE, Muñoz-Amatriaín M, Lonardi S, Santos JR, Ndeve A, Batieno BJ, Boukar O, Cisse N (2018) A multi-parent advanced generation inter-cross (MAGIC)

population for genetic analysis and improvement of cowpea (Vigna unguiculata L. Walp.). Plant J 93(6):1129–1142
Huynh B-L, Ehlers JD, Close TJ, Roberts PA (2019) Registration of a cowpea [Vigna unguiculata (L.) Walp.] multiparent advanced generation intercross (MAGIC) population. J Plant Regist 13(2):281–286
Iangrai B, Pattanayak A, Khongwir DEA, Pale G, Gatphoh EM, Das A, Chrungoo NK (2017) Development and characterization of a new set of genomic microsatellite markers in rice bean (Vigna umbellata (Thunb.) Ohwi and Ohashi) and their utilization in genetic diversity analysis of collections from north East India. PLoS One 12(7):e0179801
Ijaz U, Sudheesh S, Kaur S, Sadeque A, Bariana H, Bansal U, Adhikari K (2021) Mapping of two new rust resistance genes Uvf-2 and Uvf-3 in faba bean. Agronomy 11(7):1370
Iquebal MA, Soren KR, Gangwar P, Shanmugavadivel P, Aravind K, Singla D, Jaiswal S, Jasrotia RS, Chaturvedi SK, Singh NP (2017) Discovery of putative herbicide resistance genes and its regulatory network in chickpea using transcriptome sequencing. Front Plant Sci 8:958
Isemura T, Kaga A, Tomooka N, Shimizu T, Vaughan DA (2010) The genetics of domestication of rice bean. Vigna umbellata Annals of botany 106(6):927–944
Jain M, Misra G, Patel RK, Priya P, Jhanwar S, Khan AW, Shah N, Singh VK, Garg R, Jeena G (2013) A draft genome sequence of the pulse crop chickpea (C icer arietinum L.). Plant J 74(5):715–729
Jain S, Poromarto S, Osorno JM, McClean PE, Nelson BD (2019) Genome wide association study discovers genomic regions involved in resistance to soybean cyst nematode (Heterodera glycines) in common bean. PLoS One 14(2):0212140
Jayathilake C, Visvanathan R, Deen A, Bangamuwage R, Jayawardana BC, Nammi S, Liyanage R (2018) Cowpea: an overview on its nutritional facts and health benefits. J Sci Food Agric 98(13):4793–4806
Jha UC, Nayyar H, Palakurthi R, Jha R, Valluri V, Bajaj P, Chitikineni A, Singh NP, Varshney RK, Thudi M (2021) Major QTLs and potential candidate genes for heat stress tolerance identified in chickpea (Cicer arietinum L.). Front Plant Sci 12:655103
Kamfwa K, Mwala M, Okori P, Gibson P, Mukankusi C (2013) Identification of QTL for fusarium root rot resistance in common bean. J Crop Improv 27(4):406–418
Kami J, Velásquez VB, Debouck DG, Gepts P (1995) Identification of presumed ancestral DNA sequences of phaseolin in Phaseolus vulgaris. Proc Natl Acad Sci 92(4):1101–1104
Kamphuis LG, Gao L, Singh KB (2012) Identification and characterization of resistance to cowpea aphid (Aphis craccivora Koch) in Medicago truncatula. BMC Plant Biol 12:1–12
Kashiwaba K, Tomooka N, Kaga A, Han O-K, Vaughan D (2003) Characterization of resistance to three bruchid species (Callosobruchus spp., Coleoptera, Bruchidae) in cultivated rice bean (Vigna umbellata). J Econ Entomol 96(1):207–213
Kaul T, Eswaran M, Thangaraj A, Meyyazhagan A, Nehra M, Raman NM, Bharti J, Gayacharan, Badapanda C, Balamurali B (2019) Rice Bean (Vigna umbellata) draft genome sequence: unravelling the late flowering and unpalatability related genomic resources for efficient domestication of this underutilized crop. bioRxiv:816595
Kaul T, Easwaran M, Thangaraj A, Meyyazhagan A, Nehra M, Raman NM, Verma R, Sony SK, Abdel KF, Bharti J (2022) De novo genome assembly of rice bean (Vigna umbellata)—a nominated nutritionally rich future crop reveals novel insights into flowering potential, habit, and palatability centric–traits for efficient domestication. Front Plant Sci 13:739654
Kaur M, Singh N (2007) Characterization of protein isolates from different Indian chickpea (Cicer arietinum L.) cultivars. Food Chem 102(1):366–374
Kaur S, Pembleton LW, Cogan NO, Savin KW, Leonforte T, Paull J, Materne M, Forster JW (2012) Transcriptome sequencing of field pea and faba bean for discovery and validation of SSR genetic markers. BMC Genomics 13(1):1–12
Kaur S, Kimber RB, Cogan NO, Materne M, Forster JW, Paull JG (2014) SNP discovery and high-density genetic mapping in faba bean (Vicia faba L.) permits identification of QTLs for ascochyta blight resistance. Plant Sci 217:47–55

Kaur A, Sharma U, Singh S, Singh R, Vikal Y, Singh S, Malik P, Kaur K, Singh I, Bindra S (2022) Introgressing cry1Ac for pod borer resistance in chickpea through marker-assisted backcross breeding. Front Genet 13:847647

Keller B, Ariza-Suarez D, De la Hoz J, Aparicio JS, Portilla-Benavides AE, Buendia HF, Mayor VM, Studer B, Raatz B (2020) Genomic prediction of agronomic traits in common bean (Phaseolus vulgaris L.) under environmental stress. Frontiers. Plant Sci 11:1001

Kelly J, Gepts P, Miklas P, Coyne D (2003) Tagging and mapping of genes and QTL and molecular marker-assisted selection for traits of economic importance in bean and cowpea. Field Crop Res 82(2–3):135–154

Khadka K, Acharya B (2009) Cultivation practices of ricebean. Research and Development (LI-BIRD), Local Initiatives for Biodiversity, Pokhara

Khazaei H, Vandenberg A (2020) Seed mineral composition and protein content of faba beans (Vicia faba L.) with contrasting tannin contents. Agronomy 10(4):511

Khazaei H, Stoddard FL, Purves RW, Vandenberg A (2018) A multi-parent faba bean (Vicia faba L.) population for future genomic studies. Plant genetic. Resources 16(5):419–423

Kinhoégbè G, Djèdatin G, Saxena RK, Chitikineni A, Bajaj P, Molla J, Agbangla C, Dansi A, Varshney RK (2022) Genetic diversity and population structure of pigeonpea (Cajanus cajan [L.] Millspaugh) landraces grown in Benin revealed by genotyping-by-sequencing. Plos One 17(7):e0271565

Kudapa H, Bharti AK, Cannon SB, Farmer AD, Mulaosmanovic B, Kramer R, Bohra A, Weeks NT, Crow JA, Tuteja R (2012) A comprehensive transcriptome assembly of pigeonpea (Cajanus cajan L.) using sanger and second-generation sequencing platforms. Mol Plant 5(5):1020–1028

Kumar Y, Basu S, Goswami D, Devi M, Shivhare US, Vishwakarma RK (2022) Anti-nutritional compounds in pulses: implications and alleviation methods. Legume Sci 4(2):e111

Kushwah A, Bhatia D, Rani U, Yadav IS, Singh I, Bharadwaj C, Singh S (2021) Molecular mapping of quantitative trait loci for ascochyta blight and botrytis grey mould resistance in an inter-specific cross in chickpea (Cicer arietinum L.) using genotyping by sequencing. Breed Sci 71(2):229–239

Kushwaha P, Srivastava R, Pandiyan K, Singh A, Chakdar H, Kashyap PL, Bhardwaj AK, Murugan K, Karthikeyan N, Bagul SY (2021) Enhancement in plant growth and zinc biofortification of chickpea (Cicer arietinum L.) by Bacillus altitudinis. J Soil Sci Plant Nutr 21:922–935

Langat CK, Ombori O, Cheruiyot R, Gathaara M, Karanja D, Leley P (2020) Mapping of quantitative trait loci (QTL) related to drought tolerance in common bean (Phaseolus vulgaris L.) using F2 population from (KATB1 XGLP2). International. Ann Sci 8(1):75–86

Langyan S, Yadava P, Khan FN, Bhardwaj R, Tripathi K, Bhardwaj V, Bhardwaj R, Gautam RK, Kumar A (2022) Nutritional and food composition survey of major pulses toward healthy, sustainable, and biofortified diets. Front Sustain Food Syst 6:878269

Li CD, Fatokun CA, Ubi B, Singh BB, Scoles GJ (2001) Determining genetic similarities and relationships among cowpea breeding lines and cultivars by microsatellite markers. Crop Sci 41(1):189–197

Li Y, Ruperao P, Batley J, Edwards D, Davidson J, Hobson K, Sutton T (2017) Genome analysis identified novel candidate genes for ascochyta blight resistance in chickpea using whole genome re-sequencing data. Front Plant Sci 8:359

Li Y, Ruperao P, Batley J, Edwards D, Khan T, Colmer TD, Pang J, Siddique KH, Sutton T (2018) Investigating drought tolerance in chickpea using genome-wide association mapping and genomic selection based on whole-genome resequencing data. Front Plant Sci 9:190

Liao D, Cram D, Sharpe AG, Marsolais F (2013) Transcriptome profiling identifies candidate genes associated with the accumulation of distinct sulfur γ-glutamyl dipeptides in Phaseolus vulgaris and Vigna mungo seeds. Front Plant Sci 4:60

Lichtenzveig J, Scheuring C, Dodge J, Abbo S, Zhang H-B (2005) Construction of BAC and BIBAC libraries and their applications for generation of SSR markers for genome analysis of chickpea, Cicer arietinum L. Theor Appl Genet 110:492–510

Liu S, Yu K, Huffner M, Park S, Banik M, Pauls K, Crosby W (2010) Construction of a BAC library and a physical map of a major QTL for CBB resistance of common bean (Phaseolus vulgaris L.). Genetica 138:709–716

Lobaton JD, Miller T, Gil J, Ariza D, de la Hoz JF, Soler A, Beebe S, Duitama J, Gepts P, Raatz B (2018) Resequencing of common bean identifies regions of inter–gene pool introgression and provides comprehensive resources for molecular breeding. Plant Genome 11(2):170068

Lonardi S, Muñoz-Amatriaín M, Liang Q, Shu S, Wanamaker SI, Lo S, Tanskanen J, Schulman AH, Zhu T, Luo MC (2019) The genome of cowpea (Vigna unguiculata [L.] Walp.). Plant J 98(5):767–782

Lucas MR, Diop NN, Wanamaker S, Ehlers JD, Roberts PA, Close TJ (2011) Cowpea–soybean synteny clarified through an improved genetic map. Plant Genome 4(3):218

Lucas MR, Ehlers JD, Huynh B-L, Diop N-N, Roberts PA, Close TJ (2013) Markers for breeding heat-tolerant cowpea. Mol Breed 31:529–536

Ma Y, Bao SY, Yang T, Hu JG, Guan JP, He YH, Wang XJ, Wan Y, Sun X, Jiang JY (2013) Genetic linkage map of Chinese native variety faba bean (Vicia faba L.) based on simple sequence repeat markers. Plant Breed 132(4):397–400

Madurapperumage A, Tang L, Thavarajah P, Bridges W, Shipe E, Vandemark G, Thavarajah D (2021) Chickpea (Cicer arietinum L.) as a source of essential fatty acids–a biofortification approach. Front Plant Sci 12:734980

Mahto RK, Singh C, Chandana BS, Singh RK, Verma S, Gahlaut V, Manohar M, Yadav N, Kumar R (2022) Chickpea biofortification for cytokinin dehydrogenase via genome editing to enhance abiotic-biotic stress tolerance and food security. Front Genet 13:900324

Mallikarjuna BP, Samineni S, Thudi M, Sajja SB, Khan AW, Patil A, Viswanatha KP, Varshney RK, Gaur PM (2017) Molecular mapping of flowering time major genes and QTLs in chickpea (Cicer arietinum L.). Frontiers. Plant Sci 8:1140

McConnell M, Mamidi S, Lee R, Chikara S, Rossi M, Papa R, McClean P (2010) Syntenic relationships among legumes revealed using a gene-based genetic linkage map of common bean (Phaseolus vulgaris L.). Theor Appl Genet 121:1103–1116

Melotto M, Coelho M, Pedrosa-Harand A, Kelly J, Camargo L (2004) The anthracnose resistance locus co-4 of common bean is located on chromosome 3 and contains putative disease resistance-related genes. Theor Appl Genet 109:690–699

Métais I, Hamon B, Jalouzot R, Peltier D (2002) Structure and level of genetic diversity in various bean types evidenced with microsatellite markers isolated from a genomic enriched library. Theor Appl Genet 104:1346–1352

Millan T, Winter P, Jüngling R, Gil J, Rubio J, Cho S, Cobos M, Iruela M, Rajesh P, Tekeoglu M (2010) A consensus genetic map of chickpea (Cicer arietinum L.) based on 10 mapping populations. Euphytica 175:175–189

Mir RR, Rather IA, Bhat MA, Parray G, Varshney RK (2017) Molecular mapping of genes and QTLs in pigeonpea. In: The pigeonpea genome, pp 55–64

Mir RR, Choudhary N, Bawa V, Jan S, Singh B, Bhat MA, Paliwal R, Kumar A, Chitikineni A, Thudi M (2021) Allelic diversity, structural analysis, and genome-wide association study (GWAS) for yield and related traits using unexplored common bean (Phaseolus vulgaris L.) germplasm from Western Himalayas. Front Genet 11:609603

Misra G, Joshi-Saha A, Salaskar D, Reddy K, Dixit G, Srivastava A, Jayalakshmi V, Pithia M, Gaur P (2020) Baseline status and effect of genotype, environment and genotype× environment interactions on iron and zinc content in Indian chickpeas (Cicer arietinum L.). Euphytica 216:1–16

Muchero W, Diop NN, Bhat PR, Fenton RD, Wanamaker S, Pottorff M, Hearne S, Cisse N, Fatokun C, Ehlers JD (2009a) A consensus genetic map of cowpea [Vigna unguiculata (L) Walp.] and synteny based on EST-derived SNPs. Proc Natl Acad Sci 106(43):18159–18164

Muchero W, Ehlers JD, Close TJ, Roberts PA (2009b) Mapping QTL for drought stress-induced premature senescence and maturity in cowpea [Vigna unguiculata (L.) Walp.]. Theor Appl Genet 118:849–863

Muchero W, Roberts PA, Diop NN, Drabo I, Cisse N, Close TJ, Muranaka S, Boukar O, Ehlers JD (2013) Genetic architecture of delayed senescence, biomass, and grain yield under drought stress in cowpea. PLoS One 8(7):e70041

Müller BSF, Sakamoto T, Menezes IPP, Prado GS, Martins WS, Brondani C, Barros EG, Vianello RP (2014) Analysis of BAC-end sequences in common bean (Phaseolus vulgaris L.) towards the development and characterization of long motifs SSRs. Plant Mol Biol 86:455–470

Muñoz-Amatriaín M, Mirebrahim H, Xu P, Wanamaker SI, Luo M, Alhakami H, Alpert M, Atokple I, Batieno BJ, Boukar O (2017) Genome resources for climate-resilient cowpea, an essential crop for food security. Plant J 89(5):1042–1054

Murray J, Michaels T, Cardona C, Schaafsma A, Pauls K (2004) Quantitative trait loci for leafhopper (Empoasca fabae and Empoasca kraemeri) resistance and seed weight in the common bean. Plant Breed 123(5):474–479

Muthusamy S, Kanagarajan S, Ponnusamy S (2008) Efficiency of RAPD and ISSR markers system in accessing genetic variation of rice bean (Vigna umbellata) landraces. Electron J Biotechnol 11(3):32–41

Nabateregga M, Mukankusi C, Raatz B, Edema R, Nkalubo S, Alladassi B (2019) Quantitative trait loci (QTL) mapping for intermittent drought tolerance in BRB 191× SEQ 1027 Andean Intragene cross recombinant inbred line population of common bean (Phaseolus vulgaris L.). Afr J Biotechnol 18(21):452

Nayak SN, Zhu H, Varghese N, Datta S, Choi H-K, Horres R, Jüngling R, Singh J, Kavi Kishor P, Sivaramakrishnan S (2010) Integration of novel SSR and gene-based SNP marker loci in the chickpea genetic map and establishment of new anchor points with Medicago truncatula genome. Theor Appl Genet 120:1415–1441

Nazir M, Mahajan R, Mansoor S, Rasool S, Mir RA, Singh R, Thakral V, Kumar V, Sofi PA, El-Serehy HA (2022) Identification of QTLs/candidate genes for seed mineral contents in common bean (Phaseolus vulgaris l.) through genotyping-by-sequencing. Front Genet 13:750814

Njobvu J, Hamabwe SM, Munyinda K, Kelly JD, Kamfwa K (2020) Quantitative trait loci mapping of resistance to aluminum toxicity in common bean. Crop Sci 60(3):1294–1302

Nodari R, Tsail S, Gilbertson R, Gepts P (1993) Towards an integrated linkage map of common bean 2. Development of an RFLP-based linkage map. Theor Appl Genet 85:513–520

Oblessuc PR, Baroni RM, Garcia AAF, Chioratto AF, Carbonell SAM, Camargo LEA, Benchimol LL (2012) Mapping of angular leaf spot resistance QTL in common bean (Phaseolus vulgaris L.) under different environments. BMC Genet 13:1–9

Ouédraogo J, Maheshwari V, Berner D, St-Pierre C-A, Belzile F, Timko M (2001) Identification of AFLP markers linked to resistance of cowpea (Vigna unguiculata L.) to parasitism by Striga gesnerioides. Theor Appl Genet 102:1029–1036

Ouédraogo JT, Gowda BS, Jean M, Close TJ, Ehlers JD, Hall AE, Gillaspie AG, Roberts PA, Ismail AM, Bruening G, Gepts P (2002) An improved genetic linkage map for cowpea (Vigna unguiculata L.) combining AFLP, RFLP, RAPD, biochemical markers, and biological resistance traits. Genome 45(1):175–188

Ouedraogo JT, Ouedraogo M, Gowda BS, Timko MP (2012) Development of sequence characterized amplified region (SCAR) markers linked to race-specific resistance to Striga gesnerioides in cowpea (Vigna unguiculata L.). Afr J Biotechnol 11(62):12555–12562

Patil PG, Bohra A, Dubey J, Saabale PR, Mishra RK, Pandey G, Das A, Rathore M, Singh F, Singh NP (2017) Genetic analysis and molecular resistance to race 2 of fusarium wilt in pigeonpea [Cajanus cajan (L.) Millsp.]. Crop Prot 100:117–123

Pattanayak A, Roy S, Sood S, Iangrai B, Banerjee A, Gupta S, Joshi DC (2019) Rice bean: a lesser known pulse with well-recognized potential. Planta 250:873–890

Patto MV, Torres A, Koblizkova A, Macas J, Cubero J (1999) Development of a genetic composite map of Vicia faba using F2 populations derived from trisomic plants. Theor Appl Genet 98:736–743

Paul PJ, Samineni S, Thudi M, Sajja SB, Rathore A, Das RR, Khan AW, Chaturvedi SK, Lavanya GR, Varshney RK (2018) Molecular mapping of QTLs for heat tolerance in chickpea. Int J Mol Sci 19(8):2166

Pazhamala L, Saxena RK, Singh VK, Sameerkumar C, Kumar V, Sinha P, Patel K, Obala J, Kaoneka SR, Tongoona P (2015) Genomics-assisted breeding for boosting crop improvement in pigeonpea (Cajanus cajan). Front Plant Sci 6:50

Pedrosa-Harand A, Kami J, Gepts P, Geffroy V, Schweizer D (2009) Cytogenetic mapping of common bean chromosomes reveals a less compartmentalized small-genome plant species. Chromosom Res 17:405–417

Pfeiffer WH, McClafferty B (2007) HarvestPlus: breeding crops for better nutrition. Crop Sci 47:S-88–S-105

Porch TG, Blair MW, Lariguet P, Galeano C, Pankhurst CE, Broughton WJ (2009) Generation of a mutant population for TILLING common bean genotype BAT 93. J Am Soc Hortic Sci 134(3):348–355

Poromarto SH, Nelson BD, Goswami RS (2010) Effect of soybean cyst nematode on growth of dry bean in the field. Plant Dis 94(11):1299–1304

Pushparajah D, Thavarajah T (2012) Evaluation of chickpea (Cicer arietinum L.) micronutrient composition: biofortification opportunities to combat global micronutrient malnutrition. Food Res Int 49(1):99–104. https://doi.org/10.1016/j.foodres.2012.08.007

Rahate KA, Madhumita M, Prabhakar PK (2021) Nutritional composition, anti-nutritional factors, pretreatments-cum-processing impact and food formulation potential of faba bean (Vicia faba L.): a comprehensive review. Lwt 138:110796

Rajesh P, Coyne C, Meksem K, Sharma KD, Gupta V, Muehlbauer F (2004) Construction of a Hin dIII bacterial artificial chromosome library and its use in identification of clones associated with disease resistance in chickpea. Theor Appl Genet 108:663–669

Ramírez M, Graham MA, Blanco-López L, Silvente S, Medrano-Soto A, Blair MW, Hernández G, Vance CP, Lara M (2005) Sequencing and analysis of common bean ESTs. Building a foundation for functional genomics. Plant Physiol 137(4):1211–1227

Rashid K, Bernier C (1991) The effect of rust on yield of faba bean cultivars and slow-rusting populations. Can J Plant Sci 71(4):967–972

Rathod PS, Patil D, Bellad S, Haveri RV (2022) Biofortification of Zn and Fe in chickpea through agronomic intervention in medium black soils of Karnataka. Legum Res 45(8):981–987

Ravelombola W, Shi A, Weng Y, Mou B, Motes D, Clark J, Chen P, Srivastava V, Qin J, Dong L (2018) Association analysis of salt tolerance in cowpea (Vigna unguiculata (L.) Walp) at germination and seedling stages. Theor Appl Genet 131:79–91

Ray H, Georges F (2010) A genomic approach to nutritional, pharmacological and genetic issues of faba bean (Vicia faba): prospects for genetic modifications. GM Crops 1(2):99–106

Reddy M, Nene Y (1980) Estimation of yield loss in pigeonpea due to sterility mosaic. In: Proceedings of international workshop on pigeonpeas, pp 15–19

Rodriguez MS, Mendoza EMT (1991) Nutritional assessment of seed proteins in rice bean [Vigna Umbellata (thumb.) Ohwi and Ohashi]. Plant Foods Hum Nutr 41:1–9

Román B, Satovic Z, Pozarkova D, Macas J, Dolezel J, Cubero J, Torres A (2004) Development of a composite map in Vicia faba, breeding applications and future prospects. Theor Appl Genet 108:1079–1088

Roorkiwal M, Bharadwaj C, Barmukh R, Dixit GP, Thudi M, Gaur PM, Chaturvedi SK, Fikre A, Hamwieh A, Kumar S (2020) Integrating genomics for chickpea improvement: achievements and opportunities. Theor Appl Genet 133:1703–1720

Roorkiwal M, Pandey S, Thavarajah D, Hemalatha R, Varshney RK (2021) Molecular mechanisms and biochemical pathways for micronutrient acquisition and storage in legumes to support biofortification for nutritional security. Front Plant Sci 12:682842

Sallam A, Martsch R (2015) Association mapping for frost tolerance using multi-parent advanced generation inter-cross (MAGIC) population in faba bean (Vicia faba L.). Genetica 143(4):501–514

Sallam A, Arbaoui M, El-Esawi M, Abshire N, Martsch R (2016) Identification and verification of QTL associated with frost tolerance using linkage mapping and GWAS in winter faba bean. Front Plant Sci 7:1098

Satovic Z, Torres A, Cubero J (1996) Genetic mapping of new morphological, isozyme and RAPD markers in Vicia faba L. using trisomics. Theor Appl Genet 93:1130–1138

Satovic Z, Avila CM, Cruz-Izquierdo S, Díaz-Ruíz R, García-Ruíz GM, Palomino C, Gutiérrez N, Vitale S, Ocaña-Moral S, Gutiérrez MV (2013) A reference consensus genetic map for molecular markers and economically important traits in faba bean (Vicia faba L.). BMC Genomics 14:1–15

Saxena K (2008) Genetic improvement of pigeon pea—a review. Trop Plant Biol 1:159–178

Saxena RK, Singh VK, Kale SM, Tathineni R, Parupalli S, Kumar V, Garg V, Das RR, Sharma M, Yamini KN, Muniswamy S, Ghanta A, Rathore A, Kumar CVS, Saxena KB, Kishor PBK, Varshney RK (2017) Construction of genotyping-by-sequencing based high-density genetic maps and QTL mapping for fusarium wilt resistance in pigeonpea. Sci Rep. 7(1):1911

Saxena RK, Varma Penmetsa R, Upadhyaya HD, Kumar A, Carrasquilla-Garcia N, Schlueter JA, Farmer A, Whaley AM, Sarma BK, May GD (2012) Large-scale development of cost-effective single-nucleotide polymorphism marker assays for genetic mapping in pigeonpea and comparative mapping in legumes. DNA Res 19(6):449–461

Saxena RK, Rathore A, Bohra A, Yadav P, Das RR, Khan AW, Singh VK, Chitikineni A, Singh IP, Kumar CS (2018) Development and application of high-density axiom Cajanus SNP array with 56K SNPs to understand the genome architecture of released cultivars and founder genotypes. Plant Genome 11(3):180005

Saxena RK, Kale S, Mir RR, Mallikarjuna N, Yadav P, Das RR, Molla J, Sonnappa M, Ghanta A, Narasimhan Y (2020a) Genotyping-by-sequencing and multilocation evaluation of two interspecific backcross populations identify QTLs for yield-related traits in pigeonpea. Theor Appl Genet 133:737–749

Saxena RK, Molla J, Yadav P, Varshney RK (2020b) High resolution mapping of restoration of fertility (Rf) by combining large population and high density genetic map in pigeonpea [Cajanus cajan (L.) Millsp]. BMC Genomics 21:1–8

Saxena RK, Hake A, Bohra A, Khan AW, Hingane A, Sultana R, Singh IP, Naik SS, Varshney RK (2021) A diagnostic marker kit for fusarium wilt and sterility mosaic diseases resistance in pigeonpea. Theor Appl Genet 134:367–379

Schlueter JA, Goicoechea JL, Collura K, Gill N, Lin J-Y, Yu Y, Kudrna D, Zuccolo A, Vallejos CE, Muñoz-Torres M (2008) BAC-end sequence analysis and a draft physical map of the common bean (Phaseolus vulgaris L.) genome. Trop Plant Biol 1:40–48

Schmutz J, McClean PE, Mamidi S, Wu GA, Cannon SB, Grimwood J, Jenkins J, Shu S, Song Q, Chavarro C (2014) A reference genome for common bean and genome-wide analysis of dual domestications. Nat Genet 46(7):707–713

Sekhar K, Priyanka B, Reddy VD, Rao KV (2010) Isolation and characterization of a pigeonpea cyclophilin (CcCYP) gene, and its over-expression in Arabidopsis confers multiple abiotic stress tolerance. Plant Cell Environ 33(8):1324–1338

Sethy NK, Shokeen B, Edwards KJ, Bhatia S (2006) Development of microsatellite markers and analysis of intraspecific genetic variability in chickpea (Cicer arietinum L.). Theor Appl Genet 112:1416–1428

Sharma S, Upadhyaya HD, Varshney RK, Gowda CL (2013) Pre-breeding for diversification of primary gene pool and genetic enhancement of grain legumes. Front Plant Sci 4:309

Siamasonta F, Njobvu J, Hamabwe SM, Munyinda K, Kamfwa K (2021) Genome-wide association analysis of resistance to aluminum toxicity in the roots of common bean seedlings. Crop Sci 61(4):2271–2279

Singh NK, Gupta DK, Jayaswal PK, Mahato AK, Dutta S, Singh S, Bhutani S, Dogra V, Singh BP, Kumawat G (2012) The first draft of the pigeonpea genome sequence. J Plant Biochem Biotechnol 21:98–112

Singh VK, Khan AW, Saxena RK, Kumar V, Kale SM, Sinha P, Chitikineni A, Pazhamala LT, Garg V, Sharma M (2016) Next-generation sequencing for identification of candidate genes for fusarium wilt and sterility mosaic disease in pigeonpea (Cajanus cajan). Plant Biotechnol J 14(5):1183–1194

Singh D, Mathimaran N, Boller T, Kahmen A (2020) Deep-rooted pigeon pea promotes the water relations and survival of shallow-rooted finger millet during drought—despite strong competitive interactions at ambient water availability. PLoS One 15(2):e0228993

Singh G, Brar HS, Virk HK, Khokhar A, Kaur C, Singh K, Gupta RK (2023) Effect of foliar-applied urea on symbiotic parameters, yield and monetary returns of irrigated chickpea. J Plant Nutr 46(8):1668–1682

Sinha P, Pazhamala LT, Singh VK, Saxena RK, Krishnamurthy L, Azam S, Khan AW, Varshney RK (2016) Identification and validation of selected universal stress protein domain containing drought-responsive genes in Pigeonpea (Cajanus cajan L.). Front Plant Sci 6:1065

Sivasakthi K, Marques E, Kalungwana NA, Carrasquilla-Garcia N, Chang PL, Bergmann EM, Bueno E, Cordeiro M, Sani SGA, Udupa SM (2019) Functional dissection of the chickpea (Cicer arietinum L.) stay-green phenotype associated with molecular variation at an ortholog of Mendel's I gene for cotyledon color: implications for crop production and carotenoid biofortification. Int J Mol Sci 20(22):5562

Sobda G, Boukar O, Tongoona PB, Ayertey J, Offei KS (2017) Quantitative trait loci (QTL) for cowpea resistance to flower bud thrips (Megalurothrips sjostedti Trybom). Int J Plant Breed Genet

Somta P, Kaga A, Tomooka N, Kashiwaba K, Isemura T, Chaitieng B, Srinives P, Vaughan D (2006) Development of an interspecific Vigna linkage map between Vigna umbellata (Thunb.) Ohwi & Ohashi and V. Nakashimae (Ohwi) Ohwi & Ohashi and its use in analysis of bruchid resistance and comparative genomics. Plant Breed 125(1):77–84

Song Q, Jia G, Hyten DL, Jenkins J, Hwang E-Y, Schroeder SG, Osorno JM, Schmutz J, Jackson SA, McClean PE (2015) SNP assay development for linkage map construction, anchoring whole-genome sequence, and other genetic and genomic applications in common bean. G3: genes, genomes. Genetics 5(11):2285–2290

Srikumar T (1993) The mineral and trace element composition of vegetables, pulses and cereals of southern India. Food Chem 46(2):163–167

Srinivasan S, Sajja SB, Biswajit M, Uttam C, Mahendar T, Varshney RK, Gaur PM (2021) MAGIC lines in chickpea: development and exploitation of genetic diversity. Euphytica 217(7)

Susmitha D, Kalaimagal T, Senthil R, Vetriventhan M, Manonmani S, Jeyakumar P, Anita B, Reddymalla S, Choudhari PL, Nimje CA (2022) Grain nutrients variability in Pigeonpea Genebank collection and its potential for promoting nutritional security in dryland ecologies. Front Plant Sci 13:934296

Tadege M, Wang TL, Wen J, Ratet P, Mysore KS (2009) Mutagenesis and beyond! Tools for understanding legume biology. Plant Physiol 151(3):978–984

Tai AP, Martin MV, Heald CL (2014) Threat to future global food security from climate change and ozone air pollution. Nat Clim Chang 4(9):817–821

Talari A, Shakappa D (2018) Role of pigeon pea (Cajanus cajan L.) in human nutrition and health: a review. Asian J Dairy Food Res 37(3):212–220

Tamirisa S, Vudem DR, Khareedu VR (2014) Overexpression of pigeonpea stress-induced cold and drought regulatory gene (CcCDR) confers drought, salt, and cold tolerance in Arabidopsis. J Exp Bot 65(17):4769–4781

Tayeh N, Aubert G, Pilet-Nayel M-L, Lejeune-Hénaut I, Warkentin TD, Burstin J (2015) Genomic tools in pea breeding programs: status and perspectives. Front Plant Sci 6:1037

Thakur S, Bhardwaj N, Chahota R (2017) Evaluation of genetic diversity in ricebean [Vigna umbellata (Thunb.) Ohwi and Ohashi] germplasm using SSR markers. Electronic journal of. Plant Breed 8(2):674–679

Thavarajah P (2012) Evaluation of chickpea (Cicer arietinum L.) micronutrient composition: biofortification opportunities to combat global micronutrient malnutrition. Food Res Int 49(1):99–104

Thottathil GP, Jayasekaran K, Othman AS (2016) Sequencing crop genomes: a gateway to improve tropical agriculture. Trop Life Sci Res 27(1):93

Thudi M, Bohra A, Nayak SN, Varghese N, Shah TM, Penmetsa RV, Thirunavukkarasu N, Gudipati S, Gaur PM, Kulwal PL (2011) Novel SSR markers from BAC-end sequences, DArT arrays and a comprehensive genetic map with 1,291 marker loci for chickpea (Cicer arietinum L.). PLoS One 6(11):e27275

Thudi M, Gaur PM, Krishnamurthy L, Mir RR, Kudapa H, Fikre A, Kimurto P, Tripathi S, Soren KR, Mulwa R (2014) Genomics-assisted breeding for drought tolerance in chickpea. Funct Plant Biol 41(11):1178–1190

Thudi M, Khan AW, Kumar V, Gaur PM, Katta K, Garg V, Roorkiwal M, Samineni S, Varshney RK (2016) Whole genome re-sequencing reveals genome-wide variations among parental lines of 16 mapping populations in chickpea (Cicer arietinum L.). BMC Plant Biol 16(1):53–64

Tian J, Isemura T, Kaga A, Vaughan D, Tomooka N (2013) Genetic diversity of the rice bean (Vigna umbellata) genepool as assessed by SSR markers. Genome 56(12):717–727

Tomooka N, Kashiwaba K, Vaughan DA, Ishimoto M, Egawa Y (2000) The effectiveness of evaluating wild species: searching for sources of resistance to bruchid beetles in the genus Vigna subgenus Ceratotropis. Euphytica 115:27–41

Tongbram K, Singh YC, Ram D, Singh NG, Singh RKH, Singh OK (2021) An economic analysis of French bean (Phaseolus vulgaris L.) production in Bishnupur district of Manipur. Asian J Agric Ext Econ Sociol 39(8):33–39

Torres AM, Roman B, Avila CM, Satovic Z, Rubiales D, Sillero JC, Cubero JI, Moreno MT (2006) Faba bean breeding for resistance against biotic stresses: towards application of marker technology. Euphytica 147:67–80

Torres AM, Avila CM, Gutierrez N, Palomino C, Moreno MT, Cubero JI (2010) Marker-assisted selection in faba bean (Vicia faba L.). Field Crop Res 115(3):243–252

Vallejos C, Chase C (1991) Extended map for the phaseolin linkage group of Phaseolus vulgaris L. Theor Appl Genet 82:353–357

Vallejos CE, Sakiyama NS, Chase CD (1992) A molecular marker-based linkage map of Phaseolus vulgaris L. Genetics 131(3):733–740

Varshney RK (2016) Exciting journey of 10 years from genomes to fields and markets: some success stories of genomics-assisted breeding in chickpea, pigeonpea and groundnut. Plant Sci 242:98–107

Varshney RK, Nayak SN, May GD, Jackson SA (2009) Next-generation sequencing technologies and their implications for crop genetics and breeding. Trends Biotechnol 27(9):522–530

Varshney R, Penmetsa R, Dutta S, Kulwal P, Saxena R, Datta S, Sharma T, Rosen B, Carrasquilla-Garcia N, Farmer A (2010) Pigeonpea genomics initiative (PGI): an international effort to improve crop productivity of pigeonpea (Cajanus cajan L.). Mol Breed 26:393–408

Varshney RK, Chen W, Li Y, Bharti AK, Saxena RK, Schlueter JA, Donoghue MT, Azam S, Fan G, Whaley AM (2012) Draft genome sequence of pigeonpea (Cajanus cajan), an orphan legume crop of resource-poor farmers. Nat Biotechnol 30(1):83

Varshney RK, Song C, Saxena RK, Azam S, Yu S, Sharpe AG, Cannon S, Baek J, Rosen BD, Tar'an B (2013) Draft genome sequence of chickpea (Cicer arietinum) provides a resource for trait improvement. Nat Biotechnol 31(3):240–246

Varshney RK, Mir RR, Bhatia S, Thudi M, Hu Y, Azam S, Zhang Y, Jaganathan D, You FM, Gao J (2014a) Integrated physical, genetic and genome map of chickpea (Cicer arietinum L.). Funct Integr Genomics 14:59–73

Varshney RK, Thudi M, Nayak SN, Gaur PM, Kashiwagi J, Krishnamurthy L, Jaganathan D, Koppolu J, Bohra A, Tripathi S (2014b) Genetic dissection of drought tolerance in chickpea (Cicer arietinum L.). Theor Appl Genet 127:445–462

Varshney RK, Pandey MK, Bohra A, Singh VK, Thudi M, Saxena RK (2019) Toward the sequence-based breeding in legumes in the post-genome sequencing era. Theor Appl Genet 132(3):797–816

Varshney RK, Bohra A, Yu J, Graner A, Zhang Q, Sorrells ME (2021) Designing future crops: genomics-assisted breeding comes of age. Trends Plant Sci 26(6):631–649

Venkataramana PB, Gowda R, Somta P, Ramesh S, Mohan Rao A, Bhanuprakash K, Srinives P, Gireesh C, Pramila C (2016) Mapping QTL for bruchid resistance in rice bean (Vigna umbellata). Euphytica 207:135–147

Verma S, Gupta S, Bandhiwal N, Kumar T, Bharadwaj C, Bhatia S (2015) High-density linkage map construction and mapping of seed trait QTLs in chickpea (Cicer arietinum L.) using genotyping-by-sequencing (GBS). Sci Rep 5(1):17512

Vlasova A, Capella-Gutiérrez S, Rendón-Anaya M, Hernández-Oñate M, Minoche AE, Erb I, Câmara F, Prieto-Barja P, Corvelo A, Sanseverino W (2016) Genome and transcriptome analysis of the Mesoamerican common bean and the role of gene duplications in establishing tissue and temporal specialization of genes. Genome Biol 17(1):1–18

Wang L, Do Kim K, Gao D, Chen H, Wang S, Lee S, Jackson SA, Cheng X (2016) Analysis of simple sequence repeats in rice bean (Vigna umbellata) using an SSR-enriched library. Crop J 4(1):40–47

Webb A, Cottage A, Wood T, Khamassi K, Hobbs D, Gostkiewicz K, White M, Khazaei H, Ali M, Street D (2016) A SNP-based consensus genetic map for synteny-based trait targeting in faba bean (Vicia faba L.). Plant Biotechnol J 14(1):177–185

Wu J, Wang L, Li L, Wang S (2014) De novo assembly of the common bean transcriptome using short reads for the discovery of drought-responsive genes. PLoS One 9(10):e109262

Xu P, Wu X, Wang B, Liu Y, Qin D, Ehlers JD, Close TJ, Hu T, Lu Z, Li G (2010) Development and polymorphism of Vigna unguiculata ssp. unguiculata microsatellite markers used for phylogenetic analysis in asparagus bean (Vigna unguiculata ssp. sesquipedalis (L.) Verdc.). Mol Breed 25:675–684

Xu P, Hu T, Yang Y, Wu X, Wang B, Liu Y, Qin D, Ehlers J, Close T, Lu Z (2011a) Mapping genes governing flower and seedcoat color in asparagus bean (Vigna unguiculata ssp. sesquipedalis) based on single nucleotide polymorphism and simple sequence repeat markers. HortScience 46(8):1102–1104

Xu P, Wu X, Wang B, Liu Y, Ehlers JD, Close TJ, Roberts PA, Diop N-N, Qin D, Hu T (2011b) A SNP and SSR based genetic map of asparagus bean (Vigna. Unguiculata ssp. sesquipedalis) and comparison with the broader species. PLoS One 6(1):e15952

Yadav S (2010) Heavy metals toxicity in plants: an overview on the role of glutathione and phytochelatins in heavy metal stress tolerance of plants. S Afr J Bot 76(2):167–179

Yadav P, Saxena K, Hingane A, Kumar C, Kandalkar V, Varshney RK, Saxena RK (2019) An "axiom Cajanus SNP Array" based high density genetic map and QTL mapping for high-selfing flower and seed quality traits in pigeonpea. BMC Genomics 20(1):1–10

Yang SY, Saxena RK, Kulwal PL, Ash GJ, Dubey A, Harper JD, Upadhyaya HD, Gothalwal R, Kilian A, Varshney RK (2011) The first genetic map of pigeon pea based on diversity arrays technology (DArT) markers. J Genet 90:103–109

Yang T, Bao S-y, Ford R, Jia T-j, Guan J-p, He Y-h, Sun X-l, Jiang J-y, Hao J-j, Zhang X-y (2012) High-throughput novel microsatellite marker of faba bean via next generation sequencing. BMC Genomics 13(1):1–11

Yasin JK, Mishra BK, Arumugam Pillai M, Chinnusamy V (2022) Physical map of lncRNAs and lincRNAs linked with stress responsive miRs and genes network of pigeonpea (Cajanus cajan L.). J Plant Biochem Biotechnol 31(2):271–292

Yu K (2012) Bacterial artificial chromosome libraries of pulse crops: characteristics and applications. BioMed Res Int

Yu K, Park S, Poysa V, Gepts P (2000) Integration of simple sequence repeat (SSR) markers into a molecular linkage map of common bean (Phaseolus vulgaris L.). J Hered 91(6):429–434

Yuste-Lisbona FJ, Santalla M, Capel C, García-Alcázar M, De La Fuente M, Capel J, De Ron AM, Lozano R (2012) Marker-based linkage map of Andean common bean (Phaseolus vulgaris L.) and mapping of QTLs underlying popping ability traits. BMC Plant Biol 12:1–16

Zargar SM, Mahajan R, Nazir M, Nagar P, Kim ST, Rai V, Masi A, Ahmad SM, Shah RA, Ganai NA (2017) Common bean proteomics: present status and future strategies. J Proteome 169:239–248

Zeid M, Mitchell S, Link W, Carter M, Nawar A, Fulton T, Kresovich S (2009) Simple sequence repeats (SSRs) in faba bean: new loci from Orobanche-resistant cultivar 'Giza 402'. Plant Breed 128(2):149–155

Zhang C, Tar'an B, Warkentin T, Tullu A, Bett KE, Vandenberg B, Somers DJ (2006) Selection for lodging resistance in early generations of field pea by molecular markers. Crop Sci 46(1):321–329

Zhang X, Scheuring CF, Zhang M, Dong JJ, Zhang Y, Huang JJ, Lee M-K, Abbo S, Sherman A, Shtienberg D (2010) A BAC/BIBAC-based physical map of chickpea, Cicer arietinum L. BMC Genom 11(1):1–10

Zhu J, Wu J, Wang L, Blair MW, Zhu Z, Wang S (2016) QTL and candidate genes associated with common bacterial blight resistance in the common bean cultivar Longyundou 5 from China. Crop J 4(5):344–352

Zohary D, Hopf M (1973) Domestication of pulses in the Old World: legumes were companions of wheat and barley when agriculture began in the near east. Science 182(4115):887–894

Zuiderveen GH, Padder BA, Kamfwa K, Song Q, Kelly JD (2016) Genome-wide association study of anthracnose resistance in Andean beans (Phaseolus vulgaris). PLoS One 11(6):e015

Genetics and Genomics of Kodo Millet (*Paspalum scrobiculatum* L.) 2

Pooja Shukla, Shivani Shukla, Kajal Pandey,
Pooja Choudhary, Ravikesavan Rajasekaran,
and Mehanathan Muthamilarasan

2.1 Introduction

Cereals are known to supply major calories to the human population for ages and are considered vital part of daily diet globally. However, continuously changing climatic conditions have adversely influenced the production of major cereals such as rice, wheat, and maize. Further, continuous application of fertilisers has degraded the soil quality, further lowering cereal productivity. These factors put an onus on the agricultural sector and increase global food insecurity challenges. Furthermore, the continuously rising human population threatens global food security. Climate change impacts are worse in arid and semi-arid regions (Deshpande et al. 2015). Like major cereals, millets have been domesticated for thousands of years, and few evidence shows the 4000 years of millets domestication (Shahidi and Chandrasekara 2013). In the last decade, significant awareness about millets' nutritional benefits and climate resilience traits has occurred, which has paved a new roadmap to strengthen millet production, which declined substantially due to the cultivation of major cereal crops like wheat and rice (Deshpande et al. 2015). The millets are known as "nutricereals" due to their nutraceutical properties, whereas the

Pooja Shukla, Shivani Shukla, Kajal Pandey, and Pooja Choudhary contributed equally to this work.

P. Shukla · S. Shukla · M. Muthamilarasan (✉)
Department of Plant Sciences, School of Life Sciences, University of Hyderabad,
Hyderabad, Telangana, India
e-mail: muthu@uohyd.ac.in

K. Pandey · P. Choudhary
Department of Biotechnology, Jaypee Institute of Information Technology,
Noida, Uttar Pradesh, India

R. Rajasekaran
Department of Millets, Centre for Plant Breeding and Genetics, Tamil Nadu Agricultural
University, Coimbatore, Tamil Nadu, India

V. Gahlaut, V. Jaiswal (eds.), *Genetics and Genomics of High-Altitude Crops*,
https://doi.org/10.1007/978-981-99-9175-4_2

climate-resilient traits have proved their potential to attain global food security. Given this, the year 2023 has been declared the "International Year of Millets" (IYM 2023) by FAO, which will extend the awareness and expedite the research work in millets.

Kodo millet (*Paspalum scrobiculatum* L.) is one of the neglected and underutilised small millets, known to have its origin in Africa and widely cultivated in Indonesia, Thailand, New Zealand, and Vietnam. Reported to be domesticated over 3000 years ago in India, the crop is primarily and indigenously cultivated in Rajasthan, Madhya Pradesh, Uttar Pradesh, Tamil Nadu (highest productivity), Chhattisgarh, Gujarat, Karnataka, and Maharashtra (leading producer), making India the largest producer of the millet in minor food crop category. Kodo millet is climate resilient, adaptive to various stresses, and gives good yield on marginal soil in arid and semi-arid regions.

Kodo millet is rich in many antioxidants and essential micronutrients, thus considered an amenable crop for nutritional security. An ESR spectroscopic analysis of various small millets demonstrated higher DPPH (2,2-diphenyl-1-picrylhydrazyl) quenching activity of kodo millet (Hegde and Chandra 2005). Being a C_4 crop, it has a higher photosynthetic ability and produces more biomass. Hence, kodo millet is also considered an important fodder crop in India. However, substantial research must be performed to exploit the variability in this crop by employing novel strategies. Further, the changing climatic conditions and continuously increasing population demand for enhancing the yield and productivity of kodo millet (Vetriventhan and Upadhyaya 2019). Considering this, a better knowledge of genetic variability in agronomically important traits is crucial to identify kodo millet germplasm for successful crop improvement programmes (Vetriventhan and Upadhyaya 2019). This chapter will provide an overview of the current research status and highlight the significant gaps in exploiting the desirable traits of this crop for future reliability and attaining food security.

2.2 History, Origin, and Taxonomy of Kodo Millet

Known to have undergone domestication about 3000 years ago and continued to be cultivated since then, kodo millet is reported mostly to be found in damp areas of Old World tropics and sub-tropics (de Wet 1986). The crop's domestication studies reveal its introduction to Australia from Zimbabwe. However, owing to its ability to grow well in most areas, it can be found in places like marginally wetted soils, waterlogged, flooded, and marshy soils (Galinato 1999; Hariprasanna 2017). The crop is harvested as wild cereal in most of West Africa and India. In India, it is widely spread in the southern states like Tamil Nadu and Kerala and northern states like Rajasthan and West Bengal (de Wet 1986).

All millets belong to the order Poales. Kodo millet, locally known as kodon, dhan, goicha, and vagaru in different parts of the Indian sub-continent, belongs to the family Poaceae (formerly known as Gramineae) and subfamily Panicoideae (Paniceae tribe). Cronquist, in 1988, classified kodo under the division

Magnoliophyta, class Liliopsida with subclass Commelinidae. APG IV (2016) places kodo millet under clade Commelinids (Simpson 2019). Kodo millet originated from *Paspalum sanguinale* Lamk and gradually domesticated into subspecies, namely, *P. scrobiculatum* var. scrobiculatum (annually grown) and *P. scrobiculatum* var. *commensonii* (Perennial), widely preferred for its grain in India and Africa (de Wet et al. 1983; Heuze et al. 2015). Paspalum genus consists of 350 species (both annual and perennial) of high herbage value, of which some are as follows: *P. notatum* Fluegge, *Paspalum scrobiculatum* L., *P. compactum* Roth, *P. conjugatum* Bergius, and *P. dilatatum* poir (Baki and Ipor 1992; Galinato 1999).

In the case of inflorescence, kodo millet consists of cleistogamous flower, which is normally self-pollinating (Gupta et al. 2012). However, maturation of female organs before male organs (protogyny) has been reported to be inherited by the wild species of kodo millet to increase the chances of cross-pollination, which eventually leads to the emergence and spread to different races, namely, regularis, irregularis, and variabilis, named after considering the spikelet serial arrangement variation in *P. scrobiculatum* (Upadhyaya et al. 2008).

2.3 Habit and Habitat

Kodo millet is an annual or perennial, self-pollinating, mostly tufted grass that grows up to 90 cm high. Rooting mostly occurs from the lower nodes. Adventitious and branched roots are shallow, being functional throughout their life. The stem can be observed as erect, drooping, or prostrate, in addition to swollen nodes. Nodal bands are pigmented purple at later stages. Culms are erect or geniculate ascending, mostly succulent. Leaf-sheaths may be shorter than adjacent culm internode. Ligule is an eciliated membrane, brown and about 2 mm long. Leaf blades are 5–40 cm long and 3–15 mm wide, glabrous, and pale green. Leaf blades are the apex, which can be attenuate or filiform in nature (https://powo.science.kew.org).

Inflorescence comprises spike or spike-like racemes bearing 1–20 racemes: digitate or borne along the central axis, up to 15 cm long. Rachis is broadly winged, consisting of a sharp midrib, 1–2.5 mm wide. The spike/panicle exertion style can be free, semi-compact, or compact; the spikelets can be seen arranged regularly, irregularly, or variably (Fig. 2.1). Spikelets are glabrous, orbicular, packing abaxial, regular, and 2-rowed, lacking awn. Each spikelet has one or two flowers along with a bract or glume at the base, where one is placed slightly above and opposite the other. Fertile spikelets are pedicelled that are oblong. The glumes are membranous, lacking the lower glume, whereas the upper glume is long as the spikelet. The lower lemma is flat, membranous, and without a palea or floret (de Wet et al. 1983; Galinato 1999; Yadava and Jain 2006; Hariprasanna 2017). The upper lemma is observed to be crustaceous, consisting of a floret in its axil, which turns brown at maturity. Stamens are three in number and filamentous in shape. In addition, two locules are present, which open by longitudinal suture. Gynoecium is a monocarpellary consisting of two bifid, purple, feathery stigma and a distinct style, along with

Fig. 2.1 Florets of kodo millet. Kodo millet panicle compactness, spikelet arrangement, seed size, seed shape, and hull colour are shown

a superior single-celled ovary and one ovule (de Wet et al. 1983; Galinato 1999; Hariprasanna 2017; Ravikesavan et al. 2023).

The flowers of the plant show a cleistogamous mechanism, and thus, are only self-pollinated. However, a few cultivars like Ips 147, Ips 197, and Ips 427 favour outcrossing by showing protogynous flowering. Complete flowering of the spikelet takes about a week and remains open for about 3 days post-emergence. Anthesis (flower blooming) (Fig. 2.1) for the crop is generally observed between 6.00 and 11.30 a.m. in the morning; however, the best time reported for the same is from 5:45 to 7:30 a.m. Anther colour changes from white to light yellow, marking anther dehiscence (Harinarayana 1989). Tillers that mature early show seed sets, whereas sterile seeds are produced by tillers that show late maturation (Hariprasanna 2017).

Grain maturity can be observed after 30 days post-flowering. The grain consists of palea and a scutellum and is elliptical, orbicular in shape being 1.5–2.5 mm long (de Wet et al. 1983) enclosed in the hard, corneous, persistent husk, making it the most challenging millet to de-husk. The thousand-grain weight ranges between 3 and 6.7 g, wherein about 37% of the grain weight is contributed by bran and husk. The grains may vary from light red to dark grey in colour (Fig. 2.1), which are sweet and bitter in taste, respectively (de Wet et al. 1983; Galinato 1999; Yadava and Jain 2006).

2.4 Genetics and Genetic Diversity

Kodo millet is an allotetraploid with chromosome numbers ranging from 20 to 60. While many kodo millet species like *P. commersonii*, *P. conjugatum*, *P. dilatatum*, *P. lanceolatum*, *P. longifolium*, and *P. setaceum* contain a diploid set of chromosomes (2n = 40), there are some relatives with chromosome number as low as 12 (*P. hexastachyum*, 2n = 12) to as high as 120 (*P. giganteum*, knot grass, 2n = 120) (Chandrasekharan and Parthasarathy 1960; Yadava and Jain 2006; Hariprasanna 2017).

Reports about the karyotype and meiotic behaviour of kodo millet confirm the chromosomal number to be 2n = 40. However, due to significantly less difference in chromosome size, categorising the somatic chromosomes as short, medium, or long is difficult. Therefore, categories are based on the three types of observed chromosomes, showing sub-median centromere with a satellite on the long arm, median centromere, and sub-median centromere. In kodo millet, the meiotic behaviour revealed 20 bivalents both at diakinesis and metaphase I. Mostly ring-shaped bivalents were observed, and about 10% of pollen mother cells show one to two quadrivalents. Further, systematic tetrad formation is reported. Twenty bivalent formations indicate the allotetraploid nature of kodo millet. However, the segmental allotetraploid nature of the crop is indicated by a small percentage of quadrivalents formed (Hiremath and Dandin 1975; Magoon and Manchanda 1961).

The genetic diversity of kodo millet is immense and, therefore, is preserved in national and international seed banks and repositories like International Crop Research Institute for the Semi-Arid Tropics (ICRISAT), All India Coordinated Research Project on Small Millets (AICSMIP), National Bureau of Plant Genetic

Resources (NBPGR), and Tamil Nadu Agricultural University (TNAU). Among the worldwide accessions, important varieties and accessions have been identified as core collections by Upadhyaya et al. (2016) for better utilisation of the kodo millet genetic diversity. Genetic diversity for various critical agronomic traits such as 50% flowering, days to maturity, plant height, straw yield, and number of tillers per plant, flag leaf length, peduncle length, and panicle length as the important parameters for grain yield have been studied in kodo millet through stringent screening (Hariprasanna 2017). Further, clustering in kodo millet successfully identified non-lodging genotypes in cluster I, which had higher culm parameters (Ragini et al. 2015). Additionally, the genetic diversity among *P. scrobiculatum* and *P. polystachyum* studied through RAPD analysis revealed a higher polymorphism among the individual accessions suggesting a greater genetic diversity between the genotypes (M'Ribu and Hilu 1996). The number of productive tillers, the number of spikes present per plant, panicle weight, and the number of grains obtained per panicle, fodder yield, and grain yield were suggested as the highly heritable traits that can be considered for selection (Ravikesavan et al. 2023). In all the variability studies that were conducted in various research programmes, these traits recorded a higher phenotypic coefficient of variation (PCV) and genotypic coefficient of variation (GCV), along with higher heritability coupled with the genetic advance (Dhagat et al. 1978; Nirubana et al. 2017).

2.5 Kodo Millet as Functional Food and Nutraceuticals

Given the nutritional qualities of this crop, kodo millet can be considered as one of the important crops for nutrition research; thus, breeding programmes focusing on crop improvement accelerated (Deshpande et al. 2015; Hariprasanna 2017). Kodo millet is cooked and eaten like rice in Africa and India. Additionally, since the grain's integrity is not lost upon cooking, all the rice recipes, including pudding and bakery products, can be prepared from kodo millet grains. Besides higher amounts of protein, healthy fats, fibre, minerals, calcium, and iron, kodo millet is also rich in various bioactive compounds such as folic acid, niacin (vitamin B), thiamine, and biotin and amino acids like methionine, lysine, tryptophan, flavonoids, and phenolics (Deshpande et al. 2015). Also, significantly higher content of micronutrients like iron, zinc, magnesium, and potassium is present in kodo millet [Nutritive value of Indian foods (2007), National Institute of Nutrition, India]. Suggestively, due to its superior nutritional qualities in comparison to rice, kodo millet holds the potential to replace rice (https://www.dhan.org/smallmilletfoundation/document/processing/Receipe-kodo-millet.pdf).

Interestingly, with optimum conditions like soaking time (13.81 h), germination temperature (38.75 °C), and germination time (35.82 h), significantly enhanced protein, crude fibre, minerals, and GABA contents were observed. These conditions also aided in the synthesis of new bioactive compounds, namely, hexa-decanoicacid (8.19%), 9-Octadecenoicacid (5.00%), butyl-6,9,12,15-octadecatetraenoate (4.03%), Pregan-20-one, 2-hydroxy-5,6-epoxy-15-methyl (3.45%), hexadecanoicacid-methylester (1.43%), and [n-propyl-9,12,15-octadecatrienoate (0.86%), along with a significant decrease in phytates and tannins during germination (Sharma et al. 2017a, b).

In the case of the antioxidant activity of millets, beta-glucans extracted from germinated kodo millet show higher (2,2-diphenyl-1-picrylhydrazyl) DPPH-radical scavenging and FRAP (ferric reducing antioxidant power) activity as compared to foxtail millet (Sharma et al. 2018). The flour from germinated kodo millet showed a decrease in viscoelasticity, pasting properties, and gelatinisation: however, significant improvements in in vitro antioxidant activity, metal chelating activity, and hydrogen peroxide scavenging activity were observed (Sharma et al. 2021). Kodo millet grown in Himachal Pradesh contains high amounts of protein, carbohydrates, crude fibres, and minerals. Additionally, TLC and HPLC analysis showed the presence of ferulic acid and cinnamic acid, along with efficient anti-microbial potential against *Staphylococcus aureus*, *Leuconostoc mesenteroides*, *Bacillus cereus*, and *Enterococcus faecalis* (Sharma et al. 2017a, b). Polyphenol-rich extracts (PREs) from kodo millet showed higher taxifolin and catechin content, which are known to mitigate weight gain, hepatic steatosis, adipose tissue hypertrophy, and systemic inflammation, significantly (Khare et al. 2020). Additionally, the hydroxyl and peroxyl radical inhibition and anti-proliferative activities of kodo millet phenolic extracts were also confirmed with the highest activity in the extract from the hull, followed by whole grain and dehulled grain extracts (Chandrasekara and Shahidi 2011).

Consumption of pigmented millets was suggested to help prevent and manage diabetes and its complications due to its anti-amylase activity along with middle-stage antiglycation and glycation-reversing activity (Senevirathne et al. 2021). Given this, it has been demonstrated that the methanolic extract of kodo millet inhibits glycation and improves cross-linking of collagen (Hegde et al. 2002). Further, the aqueous paste of kodo millet flour (300 mg) was reported to accelerate the process of wound healing by improving protein and collagen contents near the wound and increasing the rate of contraction and epithelialisation (Hegde et al. 2002).

2.6 Antinutritional Factors

The main antinutritional factors in kodo millets are phytic acid, oxalic, tannins, free phenolics, haemagglutinin, etc. They are termed so because of their properties, such as reducing the activity of enzymes and the chelation of metals. Millet seed coats are rich sources of polyphenols, mainly tannins and phenolic acid, which play a major role in maintaining homeostasis and improving human health. They do not add any value to the nutritional content of millets but possess many other pharmacological effects on human health, conferring anticarcinogenic, anti-inflammatory, and anti-oestrogenic effects (Ferguson 2001).

Antinutrients are organic compounds that inhibit grains' bioavailability and nutritional value. The excessive presence of phytates, tannins, and enzyme inhibitors in millet seeds can deteriorate their activity by binding themselves to the nutritional components of seeds, forming complexes poorly absorbed in the gastrointestinal tract and hence not digested. Therefore, the concentration of these antinutrients must be at appropriate levels in millet grains and can be controlled by strategies such as heating, soaking, decortication, and fermentation. They offer many therapeutic effects to human health if taken in the right amount.

When present in high concentrations, these biologically active components lower seeds' nutritional content (Shyam and Singh 2018). Therefore, it is essential to ascertain the toxic effects of the sample of seeds used in the dietary system. Phytates present in kodo millets are considered antinutritional components as they chelate some metallic compounds, like Ca and Fe, produces poorly soluble salts that are not digested, thus, lowering the bioavailability (Wasagu et al. 2015). On the other hand, tannins and phenolic compounds interfere with digestion by inhibiting the activity of enzymes, like α-amylase and trypsin, diminishing the digestion of proteins and carbohydrates, respectively (Rehman and Shah 2001; Ertop and Bektaş 2018). Some studies reveal that intake of excessive tannins can result in consequences such as depression, oral oesophageal cancer, and abnormal digestive tract functioning.

2.7 Kodo Poisoning

"Kodo poisoning" is a condition occurring in kodo millets, resulting in cyclopiazonic acid formation. CPA is a crystalline, odourless secondary metabolite with metal-chelating activity produced by certain fungi strains such as *Penicillium* and *Aspergillus* (Deepika et al. 2022). The condition has led to a significant financial barrier in the production, consumption, and popularisation of kodo millet. Farmers in kodo millet cultivating regions believe the poisonous effect is due to the fungal infection resulting from rainfall in the area. A report in 1922 described four cases of acute poisoning due to the intake of bread manufactured using kodo millet flour (Swarup 1922).

The primary reason for poisoning is the consumption of kodo grains infected by mycotoxin. CPA is the principal factor for toxicity leading to anorexia, depression, mucosal hyperaemia, and several other symptoms (Deepika et al. 2022). A study revealed increased GOT and GPT activities triggering hepatotoxicity in rats after exposure to CPA (Antony et al. 2003). Many findings indicate that exposure of kodo millets to CPA can inhibit its nutritional quality, leading to several harmful conditions like nausea and tumours. After significantly affecting the millet grains, CPA leads to disease Paspalum ergot, which can even cause animal death. Analytical methods such as TLC, HPLC, and immunoassays can be used to detect CPA in millet grains.

2.8 Advanced Omics Approaches for Trait Improvement

Advanced tools such as molecular markers, gene editing, and multi-omics approaches have substantially replaced the conventional breeding programmes for crop improvement. Additionally, significant studies demonstrate the application of these techniques in small millets improvement and provide a roadmap for less studied millets such as kodo millet (Fig. 2.2). In kodo millet, limited genetic resources are available, and genome is not yet sequenced (Muthamilarasan and Prasad 2021).

2.8.1 Genomics

The availability of genome sequence has become an essential resource in deciphering the mechanisms for plant growth, development, and response against stimuli like biotic and abiotic stress by aiding direct access to the coding and non-coding regions. Not only it supports the development of markers for better understanding and identification of diversity, variability, and origin, but it also aids the mapping of sequence polymorphism linked to a particular trait of interest which further helps in the construction of markers for genomics-assisted genetic improvement.

Polymorphism in genome-wide RAPD markers developed was utilised to understand the genetic diversity and geographic variation pattern. The study led to the identification of four major groups, one for African accessions and three for Indian accessions. Additionally, genome-wide SNP marker data sets for 190 accessions of kodo millet have been developed, identifying 3461 SNPs (Johnson et al. 2019). Seven putative sub-populations of kodo millet identified under population genetic analysis were confirmed by the heritability values for each phenotype data, thereby confirming genome-wide population markers (Johnson et al. 2019). Another study performed shotgun metagenome sequencing of kodo millet rhizosphere, leading to the characterisation of alpha-diversity of 107 species, which confirmed the dominance of actinobacteria (Prabha et al. 2018). The study demonstrated the multifunctionality of various rhizosphere microorganisms and their role in assisting plant growth and development under stressed and nutrient-deprived conditions (Prabha et al. 2019).

Third-generation sequencing methods and analysis through high-throughput tools have accelerated genome research in crops. Small millets like foxtail millet, finger millet, proso millet, Japanese barnyard millet, and tef millet are in the limelight since their genome has been decoded (Vetriventhan et al. 2020). Millets, like kodo millet, can be studied by comparative genomics, which could pave the way for marker-assisted selection, where agronomically important gene homologues from other minor millets and cereals could be utilised.

2.8.2 Transcriptomics

Integrated omics approaches are reported in crop research for trait-based breeding programmes. Transcriptome-based gene profiling is one of the omics techniques that helps unravel the patterns and pattern variations in candidate genes playing a role in various cellular processes' regulation. These identifications are essential for developing strategies for crop variety improvement through breeding programmes. De novo transcriptome of kodo millet cultivar CO3 under dehydration stress has

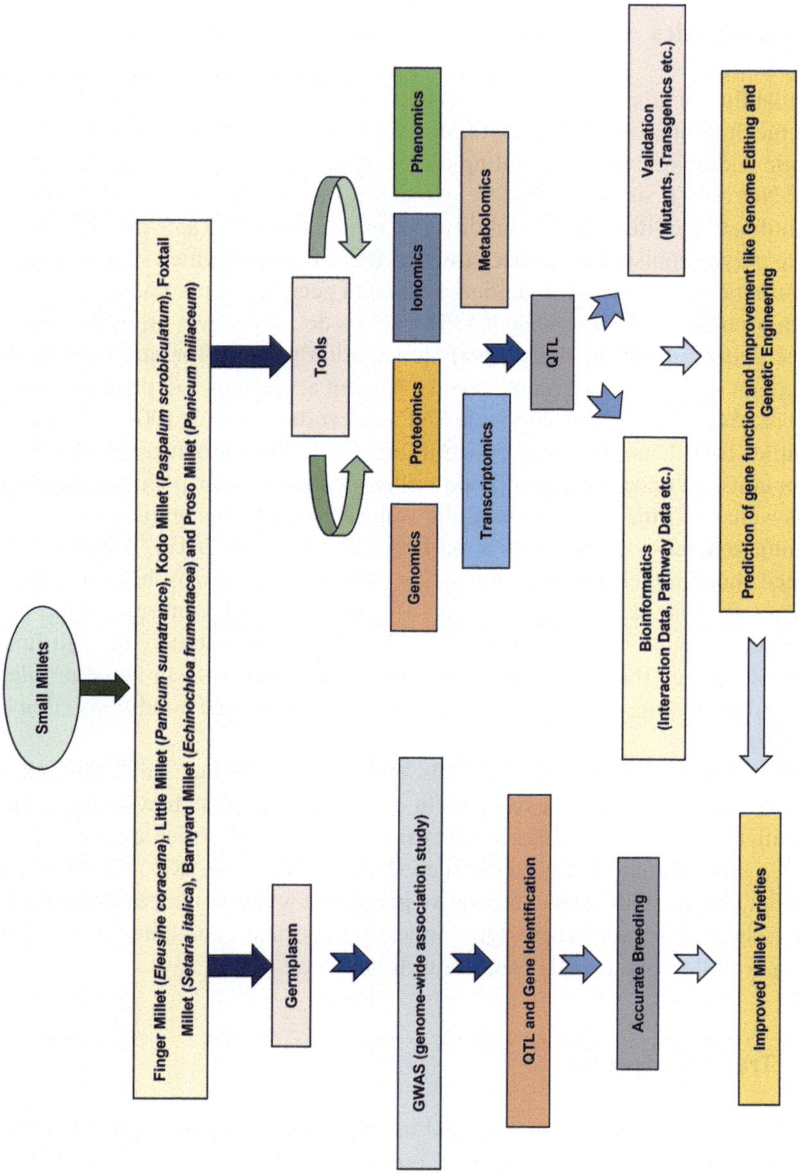

Fig. 2.2 Proposed roadmap for kodo millet improvement by employing multi-omics strategies

been recently analysed to decipher the candidate genes playing a role in the susceptibility response (Suresh et al. 2022). Genes identified were involved in photosynthesis, cellular metabolism, hormone metabolism and signalling, and antioxidant scavenging. Additionally, expression profiling of two candidate genes associated with the lodging effect and seven photosynthetic genes demonstrated their function in lodging resistance and photosynthetic efficiency in kodo millet cv. CO3 mutant lines, respectively (Jency et al. 2020).

2.8.3 Proteomics

Proteomic analysis has emerged as a new technology to understand the role of proteins and their post-translational modifications in regulating various biological processes, including defence response. High-throughput mass spectrometry-based proteomics approaches are now available, which provide a comprehensive overview of signalling pathways (Mustafa and Komatsu 2021). Further, proteomics approaches have also accelerated the marker-assisted genetic augmentation studies in different crops for developing improved cultivars. A substantial number of studies have been performed on various small millets (Karthikeyan et al. 2022). However, no such comprehensive study has been reported on kodo millet. A 20 kDa prolamin from kodo millet has been identified, isolated, and characterised to have homology with barnyard, little and foxtail millet. The protein contains 152–155 amino acid residues per polypeptide chain. The amino acid composition was reported to be high in glutamic acid, alanine, leucine, and serine, and a lower quantity of lysine and methionine was reported (Parameswaran and Thayumanavan 1997).

2.8.4 Metabolomics

Metabolomics is the study of metabolites and their response to specific conditions. With the advancement of novel approaches, significant efforts have been made to understand the metabolic variations in small millets (Li et al. 2021; Padhiyar et al. 2022). Metabolomics is used in interpreting the structure and function of phenolics in millet grains. It can be combined with various omics technologies like proteomics and transcriptomics to investigate their pharmacological effect on human health further. To study the genetic component of phenolic in kodo millets, GWAS and QTL analysis can be used along with metabolomics to explore metabolites and identify biomarkers. A recent study performed metabolic profiling of modified cereal bran from finger millet, kodo millet, and rice (Devi et al. 2023). Further, significant metabolomic studies must be performed in kodo millet to understand the nutritional qualities under various environmental conditions. However, substantial efforts have been made in other small millets. For example, finger millets are found to have more nutrition because of the presence of procyanidins B1 and B2 and higher concentrations of dietary fibres, proteins, calcium, and kaempferol glycoside (Hassan et al. 2020). A study analysed and compared the content of ferulic and isoferulic in white and coloured proso millets using mass spectrometry (Yadav et al. 2021). Detailed metabolomic investigation using analytical techniques is required to better understand the biological functions and processes in kodo millet and analyse their pharmacological and nutritive effects on human health.

2.8.5 Phenomics

Crop phenomics is new integrated science which amalgamates information from life sciences, information science, engineering, and agricultural science with artificial intelligence to understand the multifactorial phenotypes of crop performance in a complex environment. This enables the comprehensive integration of omics-based big data and illustrates the interaction between genotype, environment, and phenotype (Zhao et al. 2019). Advancement in techniques of phenomics and genetic tools has helped characterise diverse germplasm of millets leading to crop improvement. Many integrated approaches such as GWAS, bi-parental QTL mapping, association mapping, TILLING, and genome-wide prediction are widely used to produce improved cultivars of crops (Rhowell Jr et al. 2021). These techniques have been successfully exploited in major crops such as rice and wheat; however, limited success has been achieved in millets. Various phenomics techniques available are magnetic resonance imaging, fluorescence imaging, spectral reflectance, and others which help in appropriate phenotyping and in identifying the function of genes in crops (Baldaniya et al. 2017). The development of plant phenotyping methods has made it possible to collect data swiftly and accurately on agricultural characteristics. However, the genetics of agronomically important traits in kodo millet is poorly understood. It needs extensive efforts and the application of multi-omics approaches to identify genes for climate resilience and nutritional traits. Given this, applying these phenomics approaches in kodo millet research could assist in mining genes associated with agronomically important traits, thus, accelerating the breeding programmes for kodo millet improvement.

2.9 Improved Kodo Millet Varieties

Breeding programmes for improving kodo millet were started long ago, which resulted in the release of several cultivars. The first improved cultivar of kodo millet, PLR 1, was released in 1942 after all efforts of crop improvement programmes (Yadava and Jain 2006). Following this, various improved kodo millet cultivars have been released in different states of India, such as TNAU-86 in Madhya Pradesh, GPUK-3 in Karnataka, RBJ-155 in Chhattisgarh, JK-65 in Gujarat, and KMV 20 in Tamil Nadu (https://www.millets.res.in/farmer/Small_Millet_Pub/2018/4KODOMILLET.pdf). However, genetic modifications are required to increase the yield and tolerance of kodo millets to biotic and abiotic stresses. Somatic embryogenesis and plant regeneration can be applied to shoot tips of kodo millet using 2,4-D, NAA, and IBA as growth regulators to perform regeneration studies with an aim to produce better millet grains (Arockiasamy et al. 2001; Ceasar and Ignacimuthu 2010). Kodo millet DK-127 is non-lodging, non-shattering, ellipsoidal, erect with green foliage, with milling recovery up to 53.4% (Pali et al. 2019). This variety of kodo millet was found to be better than GK-2 and GPUK-3. High-efficiency transformation methods using *Agrobacterium* and microprojectile bombardment are some approaches for genetic engineering in millets and other crops. However, no successful report of such genetic transformation has been reported in kodo millet. The most crucial requirement for crop improvement is better knowledge and conservation of available germplasm resources. Efforts are now being made to collect, conserve, and characterise the genetic diversity of germplasm. However, the application of advanced genetic engineering tools is still limiting in kodo millet research due to minimal genomic resources, emphasising the utilisation of related millet species to explore kodo millet.

Recognising millets as the potential solution for the major current global problems, FAO declared 2023 as the International Year of Millets (IYM). IYM 2023 aims to raise awareness about the multi-level benefits of incorporating millets into our diet, including better production, nutrition, environment, and life. Through IYM, stronger science-policy interaction can be established and serve as a platform to empower stakeholders to act and make new ones. The goals of IYM overlap with the UN 2030 agenda for sustainable development. It directly targets 6 out of 17 Sustainable Development Goals (SDGs), i.e. SDG 2 (end hunger), SDG 3 (good health and well-being), SDG8 (decent work and economic growth), SDG 12 (sustainable consumption and production), SDG 13 (climate action), and SDG 15 (life on earth) (FAO 2022) (Fig. 2.3).

2.10 Conclusions and Future Perspectives

Diversity in millets is significant, thus making these crops suitable for promoting them as mainstream crops. Due to superior nutritional qualities and climate resilience traits, kodo millet could provide a potential solution for various agricultural and health challenges ranging from drought, world hunger, nutritional insecurities, and poverty. Given this, recognising the potential solution for almost all the current global problems, FAO declared 2023 as the International Year of Millets. IYM 2023 aims to raise awareness about the multi-level benefits of consuming millets, leading to better production, nutrition, and a better life. IYM promotes stronger science-policy interaction to establish a platform to empower stakeholders and

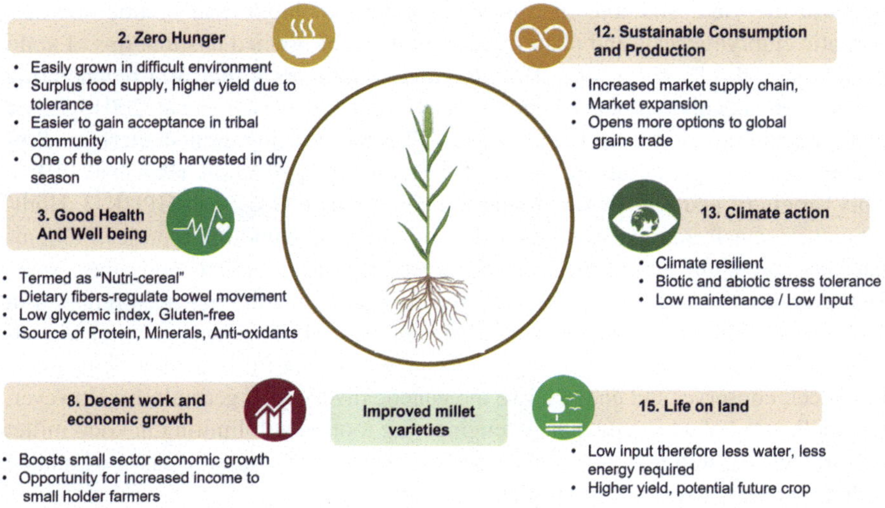

Fig. 2.3 Sustainable Development Goals (SDGs) of UN 2030 overlapping with the goals of International Year of Millets 2023. Cultivation and consumption of millets can assist in attaining 6 out of 17 SDGs

farmers to adopt millet farming. These efforts will also promote extensive studies in less studied millets, including kodo millet. Though millets are superior in nutritional aspects, they also contain several antinutrients, limiting the biological availability of essential nutrients. Therefore, extensive research and application of novel multi-omics approaches associated with new gene editing techniques are required to overcome these traits and promote agronomically important features. Integrating large-scale data from different multi-omics approaches would assist in gene mining that could be edited using genome-editing tools to develop improved kodo millet cultivars. Recently, new-generation genome sequencing platforms could enable the genome sequencing of kodo millet and identify candidate genes for different important traits. Further, genomic selection and genome editing tools could provide a plethora of information and help understand this crop's complex but essential traits.

Acknowledgements Authors' work in the area of millet genetics and genomics is funded by the Institute of Eminence grant (Project No.: UoH-IoE-RC2-21-014) awarded to the University of Hyderabad by Ministry of Education, Govt. of India (Ref. No.: F11/9/2019-U3(A)).

References

Antony M, Shukla Y, Janardhanan KK (2003) Potential risk of acute hepatotoxicity of kodo poisoning due to exposure to cyclopiazonic acid. J Ethnopharmacol 87:211–214

Arockiasamy S, Prakash S, Ignacimuthu S (2001) High regenerative nature of Paspalum scrobiculatum L., an important millet crop. Curr Sci 80:496–498

Baldaniya JD, Kale BH, Mungra KS, Modha KG (2017) Phenomics for crop improvement. Agribios 16:85–89

Baki BB, Ipor IB (1992) Paspalum scrobiculatum L. In: plant Resources of South-East Asia (PROSEA), pp 185–187

Ceasar SA, Ignacimuthu S (2010) Effects of cytokinins, carbohydrates and amino acids on induction and maturation of somatic embryos in kodo millet (Paspalum scrobiculatum Linn.). Plant Cell Tissue Organ Cult 102:53–162

Chandrasekharan SN, Parthasarathy SV (1960) Cytogenetics and plant breeding. P. Varadachary & Co.

Chandrasekara A, Shahidi F (2011) Bioactivities and antiradical properties of millet grains and hulls. J Agric Food Chem 59:9563–9571

de Wet JMJ (1986) Origin, Evolution and Systematics Of Minor Cereals. In: Proceedings of the first international small millets workshop, 29th Oct.–2nd Nov. 1986, Bangalore, India

de Wet JMJ, Brink DE, Rao KP, Mengesha MH (1983) Diversity in kodo millet, *Paspalum scrobiculatum*. Eco Bot 37:159–163

Deepika C, Hariprasanna K, Das IK, Jacob J, Ronanki S, Ratnavathi CV, Bellundagi A, Sooganna D, Tonapi VA (2022) 'Kodo poisoning': cause, science and management. J Food Sci Technol 59:2517–2526

Deshpande SS, Mohapatra D, Tripathi MK, Sadvatha RH (2015) Kodo millet-nutritional value and utilization in Indian foods. J Grain Process Storage 2:16–23

Devi K, Kumar V, Kumar V, Mahajan N, Kaur J, Sharma S, Kumar A, Khan R, Bishnoi M, Kondepudi KK (2023) Modified cereal bran (MCB) from finger millet, kodo millet, and rice bran prevents high-fat diet-induced metabolic derangements. Food Funct 14:1459–1475

Dhagat NK, Goswami U, Narsinghani VG (1978) Genetic variability, character correlations and path analysis in barnyard-millet [Echinochloa crus-galli (Linn.) Beauv, India]. Indian J Agric Sci 48:211–214

Ertop MH, Bektaş M (2018) Enhancement of bioavailable micronutrients and reduction of Antinutrients in foods with some processes. Food Health 4:159–165

FAO (2022) https://www.fao.org/millets-2023/en. Accessed on March 15, 2023

Ferguson LR (2001) Role of plant polyphenols in genomic stability. Mutat Res 475:89–111

Galinato MI (1999) Upland rice weeds of south and Southeast Asia. Int Rice Res Inst

Gupta A, Sood S, Agrawal PK, Bhatt JC (2012) Floral biology and pollination system in small millets. Eur J Plant Sci Biotechnol 6:80–86

Harinarayana G (1989) Breeding and varietal improvement of small millets in India. In: Seetharam A, Riley KW, Harinarayana G (eds) Small millets in global agriculture. Oxford & IBH Publishing Co. Pvt. Ltd., pp 59–70

Hariprasanna K (2017) Kodo millet, Paspalum scrobiculatum L. In: Patil V (ed) Millets and sorghum: biology and genetic improvement. Wiley, pp 199–225

Hassan ZM, Sebola NA, Mabelebele M (2020) Assessment of the phenolic compounds of pearl and finger millets obtained from South Africa and Zimbabwe. Food Sci Nutr 8:4888–4896

Hegde P, Chandrakasan G, Chandra T (2002) Inhibition of collagen glycation and crosslinking in vitro by methanolic extracts of finger millet (Eleusine coracana) and Kodo millet (Paspalum scrobiculatum). J Nutr Biochem 13:517

Hegde PS, Chandra TS (2005) ESR spectroscopic study reveals higher free radical quenching potential in kodo millet (Paspalum scrobiculatum) compared to other millets. Food Chem 92:177–182

Heuzé V, Tran G, Giger-Reverdin S (2015) Scrobic (Paspalum scrobiculatum) forage and grain. Feedipedia, a programme by INRAE, CIRAD, AFZ and FAO. Retrieved from https://www.feedipedia.org/node/401. Accessed on March 15, 2023

Hiremath SC, Dandin SB (1975) Cytology of Paspalum scrobiculatum Linn. Curr Sci 44:20–21

Jency JP, Rajasekaran R, Singh RK, Muthurajan R, Prabhakaran J, Mehanathan M, Prasad M, Ganesan J (2020) Induced mutagenesis enhances lodging resistance and photosynthetic efficiency of kodomillet (Paspalum scrobiculatum). Agronomy 10:227

Johnson M, Deshpande S, Vetriventhan M, Upadhyaya HD, Wallace JG (2019) Genome-wide population structure analyses of three minor millets: Kodo millet, little millet, and proso millet. Plant Genome 12:190021

Karthikeyan A, Renganathan VG, Senthil N (2022) Role of proteomics in understanding the abiotic stress tolerance in minor millets. In: Omics of climate resilient small millets, pp 125–139

Khare P, Maurya R, Bhatia R, Mangal P, Singh J, Podili K, Bishnoi M, Kondepudi KK (2020) Polyphenol rich extracts of finger millet and kodo millet ameliorate high fat diet-induced metabolic alterations. Food Funct 11:9833–9847

Li W, Wen L, Chen Z, Zhang Z, Pang X, Deng Z, Liu T, Guo Y (2021) Study on metabolic variation in whole grains of four proso millet varieties reveals metabolites important for antioxidant properties and quality traits. Food Chem 357:129791

M'Ribu HK, Hilu KW (1996) Application of random amplified polymorphic DNA to study genetic diversity in Paspalum scrobiculatum L. (Kodo millet, Poaceae). Genet Resour Crop Evol 43:203–210

Magoon ML, Manchanda PL (1961) A cytological study of some species in the genus Paspalum. Indian J Genet Pl Br 21:212–221

Mustafa G, Komatsu S (2021) Plant proteomic research for improvement of food crop under stresses. A Review Mol Omics 17:860

Muthamilarasan M, Prasad M (2021) Small millets for enduring food security amidst pandemics. Trends Plant Sci 26:33–40

Nirubana V, Ganesamurthy K, Ravikesavan R, Chitdeshwari T (2017) Genetic variability studies for yield and yield components in kodo millet (Paspalum scrobiculatum L.). Electron J Plant Breed 8:704–707

Padhiyar SM, Kheni J, Singh HV (2022) Metabolomic characterization of barnyard millet (Echinochloa frumentacea L.) involved in different stages of spike development. Pharma Innov 11:604–611

Pali V, Ramani M, Patil HE, Patil S, Patel BK (2019) High yielding Kodo millet variety 'GAK-3' for cultivation in hilly region of Gujarat. J Pharmacogn Phytochem 8:533–537

Parameswaran KP, Thayumanavan B (1997) Isolation and characterization of a 20 kD prolamin from kodo millet (Paspalum scrobiculatum) (L.): homology with other millets and cereals. Plant Foods Hum Nutr 50:359–373

Prabha R, Singh DP, Verma MK, Sahu P, Kumar P (2018) Bacterial diversity in rhizosphere of Paspalum scrobiculatum L.(kodo millet) is revealed with shotgun metagenome sequencing and data analysis. Data Brief 20:1653–1657

Prabha R, Singh DP, Gupta S, Gupta VK, El-Enshasy HA, Verma MK (2019) Rhizosphere metagenomics of Paspalum scrobiculatum l. (kodo millet) reveals rhizobiome multifunctionalities. Microorganisms 7:608

Ragini S, Subramanian A, Nirmalakumari A, Bapu JK (2015) Genetic diversity in kodo millet (Paspalum scrobiculatum) relative to culm strength and yield. J Crop Improv 29:420–431

Ravikesavan R, Jeeva G, Jency JP, Muthamilarasan M, Francis N (2023) Kodo millet (Paspalum scrobiculatum L.). In neglected and underutilized crops, academic press 279-304

Rehman ZU, Shah WH (2001) Tannin contents and protein digestibility of black grams (Vigna mungo) after soaking and cooking. Plant Foods Hum Nutr 56:265–273

Rhowell NT Jr, Fernie AR, Sreenivasulu N (2021) Meeting human dietary vitamin requirements in the staple rice via strategies of biofortification and post-harvest fortification. Trends Food Sci Tech 109:65–82

Senevirathne IG, Abeysekera WK, Abeysekera WP, Jayanath NY, Galbada Arachchige SP, Wijewardana DC (2021) Antiamylase, antiglucosidase, and antiglycation properties of millets and sorghum from Sri Lanka. Evid Based Complementary Altern Med 18:5834915

Shahidi F, Chandrasekara A (2013) Millet grain phenolics and their role in disease risk reduction and health promotion: a review. J Funct Foods 5:570–581

Sharma S, Saxena DC, Riar CS (2017a) Using combined optimization, GC–MS and analytical technique to analyze the germination effect on phenolics, dietary fibers, minerals and GABA contents of Kodo millet (Paspalum scrobiculatum). Food Chem 233:20–28

Sharma S, Sharma N, Handa S, Pathania S (2017b) Evaluation of health potential of nutritionally enriched Kodo millet (Eleusine coracana) grown in Himachal Pradesh, India. Food Chem 214:162–168

Sharma S, Saxena D, Riar CS (2018) Characteristics of β-glucan extracted from raw and germinated foxtail (Setaria italica) and kodo (Paspalum scrobiculatum) millets. Int J Biol Macromol 118:141–148

Sharma S, Jan R, Riar C (2021) Analyzing the effect of germination on the pasting, rheological, morphological and in-vitro antioxidant characteristics of kodo millet flour and extracts. Food Chem 361:130073

Shyam R, Singh RP (2018) Evaluation of nutritional value and anti-nutritional factors of kodo millet (*Paspalum scrobiculatum* l.) germplasm grown in eastern (U.P.). Plant Arch 18:247–250

Simpson MG (2019) Plant systematics. Elsevier

Suresh BV, Choudhary P, Aggarwal PR, Rana S, Singh RK, Ravikesavan R, Prasad M, Muthamilarasan M (2022) De novo transcriptome analysis identifies key genes involved in dehydration stress response in kodo millet (Paspalum scrobiculatum L.). Genomics 114:110347

Swarup A (1922) Acute "Kodon" poisoning. Indian Med Gazette 57:257

Upadhyaya HD, Gowda CLL, Reddy VG, Sube S (2008) Diversity of small millets germplasm in genebank at ICRISAT. In: 5th International Symposium on New Crops and Uses: their role in a rapidly changing world, 3-4 September, 2007, University of Southampton, Southampton, UK

Upadhyaya HD, Vetriventhan M, Dwivedi SL, Pattanashetti SK, Singh SK (2016) Proso, barnyard, little, and kodo millets. In: Genetic and genomic resources for grain cereals improvement. Academic Press, pp 321–343

Vetriventhan M, Upadhyaya HD (2019) Variability for productivity and nutritional traits in germplasm of Kodo millet, an underutilized nutrient-rich climate smart crop. Crop Sci 59:1095–1106

Vetriventhan M, Azevedo VC, Upadhyaya HD, Nirmalakumari A, Kane-Potaka J, Anitha S, Ceasar SA, Muthamilarasan M, Bhat BV, Hariprasanna K, Bellundagi A (2020) Genetic and genomic resources, and breeding for accelerating improvement of small millets: current status and future interventions. Nucleus 63:217–239

Wasagu R, Lawal M, Galadima L, Aliero A (2015) Nutritional composition, antinutritional factors and elemental analysis of Nymphaea lotus (water lily). Bayero J Pure Appl Sci 8:1

Yadav CB, Srivastava RK, Gangashetty PI, Yadav R, Mur LA, Yadav RS (2021) Metabolite diversity and metabolic genome-wide marker association studies (MGWAS) for health benefiting nutritional traits in pearl millet grains. Cell 10:3076

Yadava HS, Jain AK (2006) Advances in Kodo millet research. Directorate of Information and Publications of Agriculture, ICAR, New Delhi India

Zhao C, Zhang Y, Du J, Guo X, Wen W, Gu S, Wang J, Fan J (2019) Crop phenomics: current status and perspectives. Front Plant Sci 10:714

Genetic and Molecular Advancements in Saffron (*Crocus sativus* L.)

3

Vishek Choudhary, Anita Choudhary, Vijay Gahlaut, and Vandana Jaiswal

3.1 Introduction

Saffron is high-value, industrial herb that can reach up to 20,000 euro per kg. Saffron is valuable since time for its odorous, colouring, and remedial assets (Plessner et al. 1989). It possesses multiple primary and secondary metabolites that enhance its market price. It is native to the Middle East and East cultivated about 3600 years ago (Fernández 2004). Usage of saffron recorded from the Southern Aegean Islands of Crete and Santorini in 3600-year-old Minoan frescos (Nemati et al. 2019). Main cultivated variety, i.e. saffron *Crocus* (*C. sativus*), is a male-sterile triploid, whereas its nearby relatives are obligate outbreeding diploids (Caiola 1999). The word "saffron" is taken from the Arabic word zafaran which means yellow (Winterhalter and Straubinger 2000). Primeval Greeks named it "Koricos", while Romans used the term "Crocum". "Kum Kum" "Kesar" "avarakta" "saurab" "mangalya" "agnishikha" "kumkuma" "mangal" "kusrunam" in Sanskrit and "Koung" in Kashmiri language, "safran" in France and German. Genus *Crocus*, according to IUCN (The International Union for Conservation of Nature) (Anonymous 2019), includes 320 species, and of these, 94 are the principal interest of researchers. Its genome size was found to be very large by flow cytometry up to 10.5 Gb (Busconi et al. 2018). *Crocus sativus* is a monocotyledonous perennial, stemless, aromatic, cormous,

V. Choudhary · A. Choudhary · V. Jaiswal (✉)
Biotechnology Division, CSIR-Institute of Himalayan Bioresource Technology, Palampur, Himachal Pradesh, India

Academy of Scientific and Innovative Research (AcSIR), Ghaziabad, Uttar Pradesh, India

V. Gahlaut
Biotechnology Division, CSIR-Institute of Himalayan Bioresource Technology, Palampur, Himachal Pradesh, India

Department of Biotechnology and University Center for Research and Development, Chandigarh University, Mohali, Punjab, India

V. Gahlaut, V. Jaiswal (eds.), *Genetics and Genomics of High-Altitude Crops*, https://doi.org/10.1007/978-981-99-9175-4_3

geophytic, genetically sterile herb with corms that remain dormant during summer and shows flowering only during autumn season (Güner et al. 2000). The plant grows to 20–30 cm in height, sprouting 5–11 white and non-photosynthetic leaves known as cataphylls (Mehraj et al. 2005). For blooming it requires cold environmental cues. Saffron is recognized as a subhysteranthous herb, so before or after the appearance of the leaves, the flowers will appear. This phenomenon can be defined as a technique of adaptation to resolve the drought period (Molina et al. 2005). A flower is monoecious having both male and female sex organs; true sepals and petals are absent and tepals (perianth) present; anther is short and produces unviable pollens that limit its ability to reproduce sexually. Vegetative reproduction occurs by corm multiplication through the development of new corms that replace the old ones (Fernàndez et al. 2011). New corm production occurs seasonally once a year, generating 2–3 new corms per parent corm. In saffron, infertility is mainly related to male gametophyte, and it is confirmed by molecular and cytological studies (Caiola 2003). *Crocus* sterility also limits the use of old breeding methods for further improvement. The highlight of the plant is the presence of three-lobed stigmas with a short yellow style (Mathew 1999; Harpke et al. 2013). Average size of the stigmas are around 20-40 mm and are trumpet shaped notched or concave at the distal end, free or joined in pairs or threes at the end of the style. Stigma gains commercial importance as it contains active components/active ingredients/compounds such as crocin, picrocrocin, and safranal. Corm and perianth also constitute a variety of chemicals such as (Frusciante et al. 2014) anthocyanins; carotenoids, i.e. zeaxanthin; lycopene; α- and β-carotenes; volatile components, e.g. flavanol glycosides; and phenolics (vanillic, syringic, gallic, caffeic and salicylic acids). Apart from the carotenoids, corm contains various other useful constituents like glucose and amino acids, e.g. glutamic acids, cysteine, serine, glycine, threonine, tyrosine, alanine, arginine, histidine, lysine, proline, phenylalanine, leucine, valine, and methionine (Rubio-Moraga et al. 2010).

3.2 Origin of *Crocus sativus*

The Middle East is the centre of origin for *Crocus sativus* (Vavilov 1951). It was suggested that Asia Minor or the Southwest Greek Islands should be their potential place of origin (Tammaro 1990). *Crocus sativus* was chosen and grown in Crete during the late Bronze Era (Negbi 1999). It expanded from here to India, China, and the countries in the Middle East. The spreading of saffron to all of Mediterranean Europe later is contributed by the Arabs (Ingram 1969). Caiola and Canini (2010) has updated comprehensive archaeological and ancient research on the nature and diversity of saffron and its related species. The genetic background of *Crocus* is a widely discussed and contentious subject. It was found that *C. cartwrightianus* is the most probable progenitor, whereas utmost morphologically dissimilar include *C. thomasii* (Mathew 1999). By analysing cytological and morphological basis, the closest relative of *C. sativus* was discovered to be *C. cartwrightianus* (Beiki et al. 2010). No major differences in the DNA of related species, namely, *C. thomasii*,

C. cartwrightianus, as well as *C. hadriaticus*, was found (Frello et al. 2004) in investigation when analysing their tandemly repeated DNA sequence from five plastid regions. DNA barcode study reveals that evolution of *C. sativus* species happened autonomously under the impact of numerous agroecogeographical pressures and also showed up that *C. thomasii* and *C. hadriaticus* are similar to each other (Gismondi et al. 2013). Comparative investigation with AFLP on genomes of *C. thomasii* and *C. cartwrightianus*, together with *C. sativus*, didn't identify any genomic changes (Zubor et al. 2004). Some studies were conducted to find the similarities and dissimilarities among different *Crocus* species which were based on ISSR (inter-simple sequence repeat) data (Sik et al. 2008; Caiola and Canini 2010; Moraga et al. 2010), five plastid of barcode regions (Petersen et al. 2008), and IRAP (inter-retroelement amplified polymorphism) data (Alsayied et al. 2015). Frizzi et al. (2007) checked allozyme-based variations; interprimer binding site (iPBS) markers in 17 different species of crocus analysis showed *C. cartwrightianus*, which is similar to *C. sativus*, which was found to be the closest wild descendant or originator of saffron, accompanied by *C. thomasii* and *C. pallasii* subsp. *pallasii.* By analysis data from different studies (Nemati et al. 2019) such as GBS (genotyping by sequencing), chloroplast genomes, and genome size, nuclear single copy loci stated that *C. cartwrightianus* is the single originator of *C. sativus.*

3.3 Distribution

Saffron has been grown in the East and the Middle East for around 3600 years (Fernández 2004). In Iran, India, Greece, Spain, Italy, Turkey, France, Switzerland, Israel, Pakistan, Azerbaijan, China, Egypt, Japan, Afghanistan, Iraq, and, as of late, Australia, saffron is currently being grown very extensively. The world's total saffron production is calculated to be about 205 tons yearly. About 80% (160 t) of the total world's yearly saffron production is contributed by Iran out of a total 205 tons cultivated yearly. Khorasan province located in Iran solely documented 137 tons of saffron on 46,000 hectare area (Ehsanzadeh et al. 2003). Earlier, Spain has been considered the world's traditional leader and most alleged producer of saffron for centuries, but production nowadays is only around 0.3–0.5 tons. Greek production per year accounts for 4–6 tons. Sardinia, Aquila, and Cascia provinces in Italy produce around 100 kg of saffron; Safranbolu in turkey produces 10 kg; Gâtinais, Quercy, in France harvests around 4–5 kg and almost negligible amount in mund region of Switzerland, i.e. around 1 kg (Ahmad et al. 2011). European developed countries such as Italy and Greece are facing a severe decrease in saffron production, despite the fact that Iran has seen a significant rise in the last three decades (Fernández 2004).

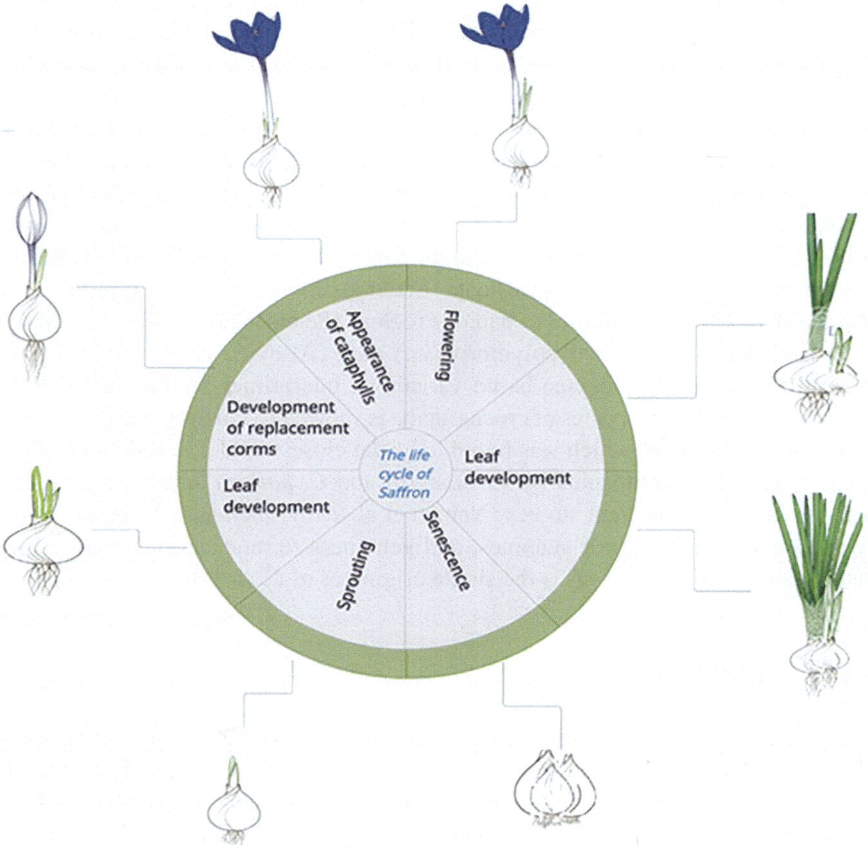

Fig. 3.1 Different stages involved in life cycle of saffron

3.4 Life Cycle of Saffron

The pattern of growth in saffron is broken down into three different stages: flowering, vegetative growth, and corm formation. Various developmental stages in life cycle of saffron are shown in Fig. 3.1. Blossoming takes place in the autumn and vegetative growth in the winter. Kalia et al. (2023) found ten PEBP gene members in saffron and described each gene's function during saffron flowering using the online and transcriptome data. At the base of the corm, there occur formations of the substitute corms. Underground corm/crocus is a compact, ovoid to sub-globose mass of the starch, covered in sheaths known as tunics (Srivastava 1963). This parallel fibre tunic cover is well suited for cool winters and warm summers (Agayev and Zarifi 2009). Each parent corm produces 2–3 daughter corms from apical buds and several from axillary buds, depending on their size.

3.5 Allo or Autotriploid Nature of *Crocus sativus*

Origin and mode of evolution of saffron is a matter of concern from ages. There is a conflict over the decades over the allo or autotriploid nature of saffron and its ancestors. Induction of polyploidy is considered as important tool for evolution of many plants (Ramsey and Schemske 1998; Madlung 2013). Among the species, polyploid origin is detrimental for knowing the phylogenetic relationship (Estep et al. 2013; Kalendar et al. 2011). Saffron a combination of two or more hybrid genomes (allopolyploidy) has previously been considered and causes variability in the use of new genomes or phenotypes (Agayev 2002; Beiki et al. 2010; Buggs et al. 2010; Caiola and Canini 2010; Tsaftaris et al. 2011; Harpke et al. 2013; Alsayied et al. 2015). Some studies also established autotriploidy (the chromosomes are doubled, and this may involve a single genome) nature of saffron (Karasawa 1932; Caiola et al. 2004). Nemati et al. (2019) based on chloroplast genome and genome-wide DNA polymorphism confirmed its autotripiloid origin from *C. cartwrightianus*. Recently (Schmidt et al. 2019) utilized FISH (fluorescent in situ hybridization) technique while comparing chromosome analysis of different *C. cartwrightianus* gave conclusive evidence about autotriploid nature of saffron.

3.6 Karyotype Analysis

Cytological studies show that it has three sets of eight chromosomes, i.e. 2n = 24, x = 8 (Karasawa 1933; Mathew 1999). Being triploid, during meiosis, there is irregular segregation of chromosomes thus resulting in anomalies in gametophytic development and limiting its potential to gain variations with generations (Chichiriccò 1984). It produces no fertilizable gametes (Ghaffari 1986). The karyotype of *C. sativus* was examined (Agayev 2002) and concluded that the first triad entails the largest chromosomes, sub-acrocentric, and one of the last (leftward) includes an allocyclic area adjacent to the telomere on the broad arm (Stebbins 1971). The subsequent triplet, sub-acrocentric, is distinguished by the existence of satellites on all three chromosomes on the broad arm. Various sizes of satellites are seen. third, fourth, and eighth triplets are metacentric; sixth and seventh are sub-metacentric. About the fifth triplet, it is heteromorphic, which means all the chromosomes are of different morphology, i.e. one is metacentric (5.1), and other two are sub-acrocentric (chromosomes 5.2 and 5.3) (Agayev 2002). Comparison of karyogram of *C. sativus* L., *C. sativus* L. '*Kashmirianus*', *C. cartwrightianus* Herb., *C. oreocreticus* B.L. Burtt, and *C. hadriaticus* herb showed mutual relations among themselves. Recently carried out investigation through fluorescent in situ hybridization (FISH) also showed relationship among *Crocus* species (Schmidt et al. 2019). With the help of FISH reference, they gave the model for the evolutionary origin of saffron.

3.7 Biosynthesis of Apocarotenoids

Saffron is perennial flowering herb being utilized in different industries, i.e. as a food colourant, dyeing agent, medicine, and a flavouring agent. Saffron possess various components like fat, moisture, minerals, proteins, crude fibres, and sugars. More than 150 advantageous volatile compounds have been reported in saffron stigmas out of which three major bioactive compounds, namely, crocin, picrocrocin, and safaranal provide it unique organoleptic properties (Mollazadeh et al. 2015). Crocin is a glycoside of crocetin, a mixture of dicrocin and tricrocin and is accountable for the reddish-orange colour of stigma (Poma et al. 2012; Bolhasani et al. 2005). Picrocrocin, a monoterpenoid colourless glycoside, consists of a ten-carbon backbone (two isoprene units) that imparts bitterness. On dehydration picrocrocin formulates safranal (Pfander and Schurteberger 1982; Sampathu et al. 1984). 80 °C is considered to be ideal temperature for safranal generation (Gregory et al. 2005). It constitutes about 1–13% of saffron's dry weight. Safranal, a monoterpenol aldehyde (degradation product of zeaxanthin (carotenoid) via picrocrocin intermediate), provides distinctive fragrance (Melnyk et al. 2010; Rezaee and Hosseinzadeh 2013; Cossignani et al. 2014; Nescatelli et al. 2017). Out of 150 aromatic and volatile components, few of them are completely authenticated (Bathaie and Mousavi 2010). Apart from the major components, some minor components were also isolated from tepals. These are crocusatins a terpenoid having antityrosinase activity (Li et al. 2004), flavonoids, and all glycosidic by-products of kaempferol that act as a lateral constituent for imparting the bitterness to saffron (Carmona et al. 2007), anthraquinones, and anthocyanin (Saito et al. 1960; Gao et al. 1999). Zeaxanthin acts as a precursor to the production of a major stigma compound that makes it a promising candidate for functional food. Figure 3.2 represents an overview of the biosynthetic pathway of major metabolites in saffron.

Firstly, zeaxanthinin undergo oxidative cleavage to give rise cyclocitral and a dialdehyde. NADP-dependent oxidoreductase gives rise to diacid crocetin from crocetin dialdehyde. Cyclocitral undergo a series of biochemical reactions and converted to safranal. Crocin biosynthesis has been catalysed with the aid of various enzymes in many subcellular compartments, i.e. chromoplasts, ER (Endoplasmic reticulum) surface, and the cytosol (Demurtas et al. 2018). Biosynthesis begin with the plastid in which zeaxanthin is present (Ahrazem et al. 2016) and ends inside the vacuole in which crocin accumulates (Bouvier et al. 2003; Gómez-Gómez et al. 2017). Carotenoid cleavage dioxygenase 2 (CCD2) executed the first dedicated step of zeaxanthin cleavage in crocin biosynthesis (Pfander and Schurteberger 1982; Bouvier et al. 2003; Frusciante et al. 2014). Aldehyde dehydrogenase (ALDH) carries out the next step in crocin biosynthesis, i.e. dehydrogenation of crocetin and dialdehyde uridine diphosphate glycosyltransferases (UGT) carry out final step (Demurtas et al. 2018). Two UGTs are involved, i.e. one is GjUGT75L6 which formulates crocin 1 and 2, and other one is GjUGT94E5 which carries out production of crocins 2, 3, and 4 (Nagatoshi et al. 2012).

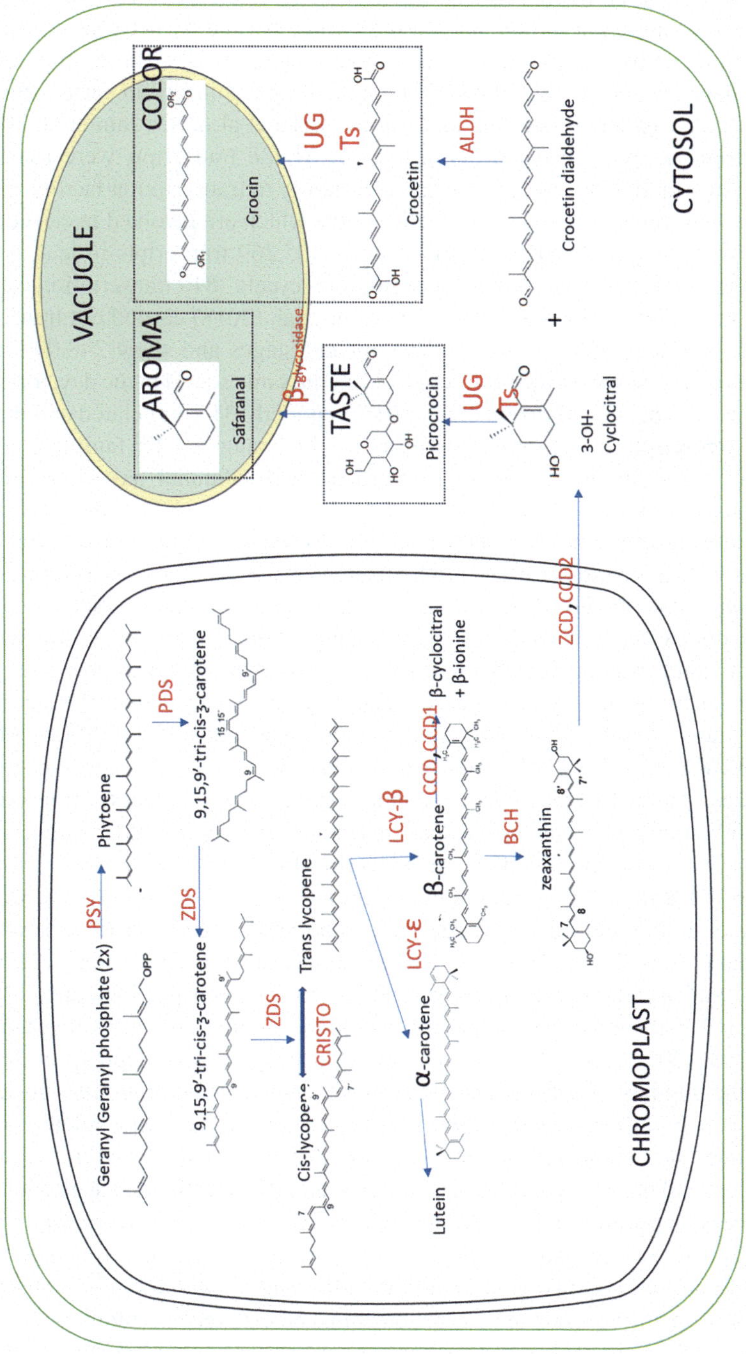

Fig. 3.2 An overview of biosynthetic pathway of major metabolites in saffron

3.8 Gene Expression Analysis

In order to unravel the molecular beginning of carotenoids/apocarotenoids biosynthesis, the genomic organization, and corresponding regulatory network, transcriptome study of whole *Crocus sativus* is a must to give insight about crop. Many scientists have created de novo transcriptome plant assemblies from various tissues leaves, corm, roots, tepals, stamen, and stigma (Baba et al. 2015; Jain et al. 2016). Transcriptome analysis revealed that a total of 64,438 transcripts were found in flower out of which 5789 transcripts shows similarity to transcription factors, along with many new genes discovered for the first time which are involved in carotenoid biosynthesis (Baba et al. 2015). In other study 105,269 transcripts in leaf, corm, tepal, stigma, and stamen as well as transcriptome reveals 3819 transcription factors encoding transcripts (Jain et al. 2016). Ahrazem et al. (2018) carried out transcriptome of *Crocus sieberi* tepal at two development stages and show 248,099 transcripts. In comparison with transcriptomics from the same stages in the development of other *Crocus* species, Ahrazem et al. (2019) reported 131 upregulated and downregulated transcription factors, representing a wider range of TF families, in the analysis of transcriptomes. Crocin, picrocrocin, and safaranal have tremendous pharmacological properties that drew attention of researchers to gain insight regarding their biosynthetic pathways. Jain et al. (2016) sequenced the transcriptome of five different *Crocus* sativus tissues to understand the molecular base of apocarotenoid. Many genes involved in various biological processes and molecular functions were discovered as a result of functional annotation. *Crocus sativus* transcripts were functionally annotated as being involved in various biological processes and molecular functions in 54% of the cases. The differential expression of transcripts encoding transcription factors involved in secondary metabolism (MYB, MYB related, WRKY, C2C2-YABBY, and bHLH) was discovered. When compared to other tissues, stamen had the highest number of differentially expressed transcripts. At least 4828 transcripts were upregulated in stamen when compared to stigma, followed by 4781 transcripts when compared to leaf. In 13,407 (12.7%) transcripts, a total of 16,721 SSRs (2–6 nt) with a minimum length of 12 bp were found, with a frequency of 1 SSR per 6.6 kb. Large-scale genotyping studies in *C. sativus* can benefit greatly from the availability of a large number of SSRs for a variety of applications. Tissue-specific expression was found in 1075 transcripts, including 124 in the corm, 161 in the tepal, 304 in the leaf, 144 in the stigma, and 342 in the stamen. Nemati et al. (2018) used RNA sequencing to investigate gene expression differences in the stigmas of saffron and its close relative *C. cartwrightianus* to learn about the genetic differences that distinguish saffron from its closest relatives. For the first time, they compared the transcriptomes of cultivated and wild saffron to better understand the molecular basis of apocarotenoid biosynthesis. He also looked at the expression profiles of all transcripts in *C. sativus* and *C. cartwrightianus* stigmas and found several genes with differential gene expression. Apocarotenoid biosynthesis-related putative genes were also discovered, which could be used to improve saffron further. For apocarotenoid biosynthesis and accumulation, transcriptomics and crocin data from *C. sativus*, *C. ancyrensis*, and *C. cartwrightianus* are sequenced and assembled from two stigma developmental stages. Previous transcriptomes of saffron tissues were perfectly complemented by the six transcriptomes obtained (Ahrazem et al. 2018). DXS-CLA1, PDS, ZDS, Z-ISO, CrtISO, LYC-B, BCH-2, CCD2, and UGT74AD2 expression levels and apocarotenoid levels were found to be positively correlated in all three species. The researchers compared data from six transcriptomes in order to find potential regulators of apocarotenoid accumulation, with a focus on transcription factors. Eleven TFs from the ARF, bHLH, C2H2, HB, CBF/DREB1, ALFIN, and NF-YC families are expressed in the stigma of the three species and are related to apocarotenoid levels. TFs involved in flower and fruit development, retrograde signalling, epigenetic modifications, and light and cold responses were also discovered. In *C. sativus*, Jain et al. (2016) discovered 78 genes involved in the apocarotenoid biosynthetic pathway. Out of 78 (10%) genes involved in apocarotenoid biosynthesis, Nemati et al. (2018) discovered seven that differed in expression between samples.

Frusciante et al. (2014) perform transcriptome sequencing of 6 different developmental stages of stigma and found out 120,000,454 reads in each development stage. Furthermore, profound transcriptomics analysis also identified carotenoid cleavage dioxygenase (CCD2), a novel dioxygenase that catalyses the first step in crocin biosynthesis from carotenoid zeaxanthin.

3.9 Therapeutic Significance

Plants have been used for medicinal purposes for as long as humanity has existed. The Sumerians (4000 B.C.) and the Chinese (3750 B.C.) are widely thought to be among the first to report medicinal plant uses. Saffron was known in Anatolia during the Hittites period as “A-Zupiru”, which implies “queen of herbal drugs” (Koç 2012). Traditional medicine has influenced the development of modern medicine, which makes extensive use of technological advancements. Saffron is known for its culinary applications and being utilized in food industries effectively, apart from having tremendous medicinal importance. Since ancient times, saffron has been used as a natural herb to improve health care, particularly in Asia and Middle Eastern nations (Javadi et al. 2013). Saffron was used as a tranquillizer in Ancient Egypt. In Ancient Greece, saffron was used to treat insomnia and hangovers brought on by excessive wine consumption. Saffron’s high content of antioxidants including flavonoids and carotenes, as well as vitamins and minerals, has been linked to a variety of health benefits. It’s used in Persian, Indian, European, Arabian, and Turkish dishes as a food colourant, as well as a traditional medicine to combat 90 diseases (Mousavi and Bathaie 2011). Figure 3.3 shows the effect of saffron and its constituents on different parts of the body. Secondary metabolites derived from saffron and their possible effects on overcoming various diseases and health problems are shown in Table 3.1.

Saffron extracts have antitumor, anticarcinogenic, and antimutagenic properties (Abdullaev and Frenkel 1992, Abdullaev 2003). Anti-cancer, antimutagenic, antitumor, antioxidant, antigenotoxic, memory and learning enhancer, anti-inflammatory,

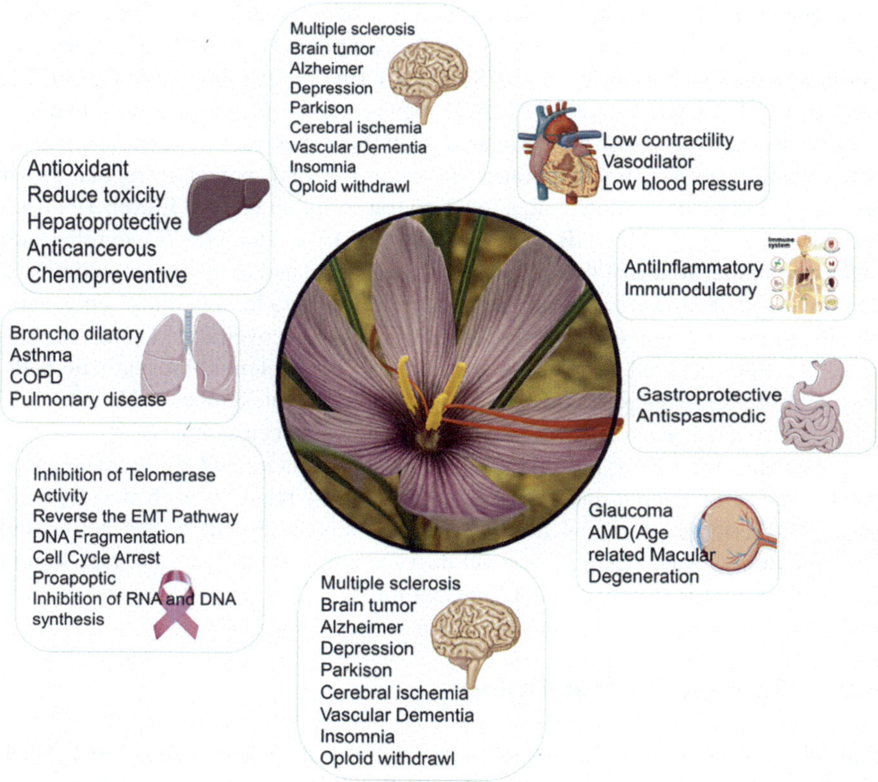

Fig. 3.3 The effect of saffron and its constituents on different parts of body

anticonvulsants, antidepressants, blood pressure regulators, oxygen boosting tissues, bronchodilator, and other properties of the active ingredients in saffron were also observed (Hashtjini et al. 2018). Furthermore, research has shown that saffron or its active ingredients can significantly reduce oxidative damage in ischemic kidneys (Hosseinzadeh et al. 2005), skeletal muscle (Hosseinzadeh et al. 2009), heart (Joukar et al. 2010), and brain (Ochiai et al. 2007). Saffron, like other antidepressants, may exert its antidepressant effect by modulating the levels of certain chemicals in the brain, including serotonin (Hausenblas et al. 2013). It was also demonstrated that decoction made of *Crocus sativus* is used to cure headache (Mousavi and Bathaie 2011). Studies on model animals have reported neuroprotective effect of saffron (Zheng et al. 2007; Deslauriers et al. 2011; Sarshoori et al. 2014). A number of bioactive ingredients are considered to be causing their health promotion properties such as asthma treatment, atherosclerosis, menstrual pain, and depression; their role as an antioxidant with anti-cancer and memory-related properties; their effectiveness in treating moderate to mild depression; and their high antioxidant level. *Crocus sativus* quality studies are growing in popularity, owing to the spice's antioxidant properties and their positive impact on human health (Shahi et al. 2016).

Table 3.1 Different biological activities performed by different secondary metabolites produced from *Crocus sativus*

Biological activity	Secondary metabolite	References
Anti-inflammatory	Crocin and crocetin	Ashktorab et al. (2019); Hosseinzadeh and Younesi (2002)
Antileukaemic	Crocin and crocetin	Moradzadeh et al. (2018)
Hepatoprotective	Crocin and crocetin	Khorasany and Hosseinzadeh (2016); Mohammad et al. (2011)
Antihypertensive and hypolipidemic	Crocins, crocetin, and safranal	Ghaffari and Roshanravan (2019)
Antidepressant and anxiolytic	Safranal and crocin	Lopresti and Drummond (2017)
Neuroprotective	Crocin, crocetin, and safranal	Hatziagapiou et al. (2019)
Alzheimer's	Crocin	Akhondzadeh et al. (2010)
Parkinson's	Crocetin	Ahmad et al. (2005)
Ocular disease (macular degeneration	Crocin and crocetin	Broadhead et al. (2019)
Cerebral ischemia	Crocin, crocetin, and safranal	Abdel-Rahman et al. (2020)
Improvement of memory and learning skills	Crocin	Ghaffari et al. (2015)
Anti-aging and skin diseases	Crocin and crocetin	Mzabri et al. (2019)
Antisolar and moisturizing	Crocin, kaempherol, quercetin, phenolic components such as tannic, gallic, caffeic, cinnamic, chlorogenic, ferulic, and vanillic acids	Golmohammadzadeh et al. (2010)
Anti-bacterial (*Salmonella enterica*)	Safranal and crocin	Pintado et al. (2011)
Antifungal (*Phytophthora infestans*)	Saponins	Khoulati et al. (2019)
Allelopathic: -on lettuce plant	Safranal	Mardani et al. (2019)
Reproductive/estrogenic activity	Saffron aqueous extract (SAE)	Tavana et al. (2012)
Aphrodisiac activity		Hosseinzadeh et al. (2008)
Spasmodic activity	Hydroalcoholic extracts	Sadraei et al. (2003)
Relaxant activity	Safranal	Boskabady et al. (2011)
Antitussive activity	Safranal and crocin	Hosseinzadeh and Ghenaati (2006)
Anxiolytic activity	Crocin and safranal	Sarris et al. (2013)
Anticonvulsant activity	Aqueous and ethanolic extracts of stigma	Sadeghnia et al. (2008)

(continued)

Table 3.1 (continued)

Biological activity	Secondary metabolite	References
Antinociceptive activity	Stigma and petal extracts	Hosseinzadeh and Younesi (2002)
Adaptogenic activity	Safranal, saffron aqueous extract	Hooshmandi et al. (2011)
Effect on menstrual distress	Saffron odour	Fukui et al. (2011)
Antivenin activity	Crocin	Santhosh et al. (2013)
Antiplatelet activity	Crocetin	Ayatollahi et al. (2014)
Wound healing activity	Saffron extract	Khorasani et al. (2008)
Sleep enhancement activity	Crocin	Masaki et al. (2012)
Immunomodulatory activity	Saponins	Kianbakht and Ghazavi (2011)
Antigenotoxic/ genoprotective activity	Stigma extract and crocin	Hosseinzadeh et al. (2008)
Amyloid-beta aggregation inhibitory activity and Alzheimer's disease	Crocetin, dimethylcrocetin, and safranal	Geromichalos et al. (2012)
Analgesic activity	Safranal and crocin	Amin and Hosseinzadeh (2012)
Protective activity against extremity ischemia-reperfusion injury	Crocin and safranal	Hosseinzadeh et al. (2009)
Allergy problems	Saffron extract	Gómez-Gómez et al. (2010)

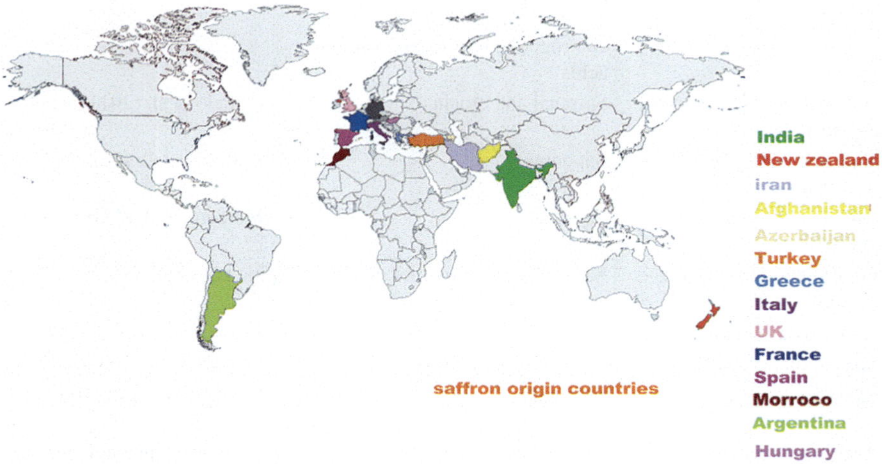

Fig. 3.4 Different countries which cultivate saffron throughout the world

3.10 Genetic Resources: Ex Situ Germplasm Collection

Saffron is a highly nutritional food product that is also one of the oldest crops and a plant with a long history of medicinal use. Figure 3.4 shows different countries/places that cultivate saffron. The exact genetic diversity of *Crocus sativus* is currently unknown on a global scale. One of the decided essential procedures for the protection and use of genetic sources in industrial vegetation has been the conservation and evaluation of germplasm (Bhatia 2015). Saffron crops have lost a significant amount of land in European countries, which may have led to significant genetic deterioration of the crop. Fourteen assemblies from European and non-European recently came up and taken the responsibility for the preservation of saffron genetic diversity. The European Commission has approved a project called "Genetic Resources of Saffron and Other Crocus spp: CROCUS-BANK". CROCUSBANK structured Action 018 AGRI GEN RES 870/2004 in the direction of generating, characterizing, and exploiting a world germplasm series of *Crocus* species. Currently, the germplasm assembly of saffron comprises 443 accessions representing 54 specific crocus taxa (which includes saffron crocus) and is predicted to grow in more than 500 accessions with the aid of the cease of CROCUS-BANK action. The preserved genetic fabric of saffron presently includes 197 accessions from 15 international locations. The *Crocus* spp. collection currently contains 246 genotypes, including 142 from the natural environment (representing 12 international locations of the genus' natural distribution), 24 from public and private donations, and 80 from commercial nurseries (www.crocusbank.org).

3.11 Closest Relative of *Crocus sativus* Based on Molecular Markers

Genetic markers are identified DNA sequences existing on chromosomes that are beneficial for individual or species identification. Diverse sorts of markers are available, i.e. morphological markers are markers that are associated with variations in the form, size, colour, and surface of various plant parts. Biochemical markers linked to protein variation and amino acid banding pattern. Non-PCR-based markers (RFLP) and PCR-based markers (RAPD, AFLP, SSR, SNP, etc.) are the two types of DNA-based markers. As PCR methods improved in the last 15 years, plenty of new marker technologies emerged, widening the scope of high-density molecular maps for all major crop species. In addition, molecular markers have been extensively used in crop genetic diversity and phylogeny studies. Molecular marker tools have aided the creation of genetic maps for several species, including essential crop species. Analysis of genetic markers will calculate the effect on the allelic level of the individual quantitative trait loci. Genetic markers may be useful in plant breeding programmes for parental selection, donor trait, and recurrent parent selection in backcross breeding, trait selection in segregating populations, and finding new positive alleles in exotic germplasm. A comparison of *C. sativus* to other *Crocus* species based on multiple DNA markers is shown in Table 3.2.

Table 3.2 Molecular markers and their attributes

Type of marker studied	No. of markers	Result of study	References
RAPD		*C. sativus* and *C. cartwrightianus* are very closely related, and *C. sativus* also shows similarity with *C. thomasii*. *C. thomasii* and *C. cartwrightianus*, the nearest one and *C. pallasii* and *C. asumaniae* are distant relative	Caiola et al. (2004)
AFLP		Found less genetic difference in comparison of C*rocus sativus* with *C. thomasii* and *C. cartwrightianus*	Zubor et al. (2004)
Microsatellite markers	27	Found that *C. sativus* turns similar to *C. speciosus* and *C. hausknechtii*	Nemati et al. (2014)
ISSR and ITS		Similar genetic association between *C. pallasii* and *C. sativus*	Sheidai et al. (2018)
ISSR	12	*C. cartwrightianus* cv. *albus* is more closely related to *C. sativus* instead of *C. cartwrightianus*	Moraga et al. (2010)
iPBS-retrotransposon	83	*C. pallasii* subsp. *pallasii* and *C. cartwrightianus* were a close relative to *C. sativus*	Gedik et al. (2017)
IRAP	11	*C. pallasii* subsp. *pallasii* and *C. cartwrightianus* were a close relative to *C. sativus*	Alsayied et al. (2015)
AFLP and MS-AFLP		Analyse saffron to look for genetic (DNA sequence) and epigenetic (cytosine methylation) variation	Busconi et al. (2015)
RAPD	38	Distinction and variability of *C. sativus* from several geographic areas, variability, and the presence of markers in the molecular pattern of different *Crocus* sp.	Pardo et al. (2003)
ESTs	6,603	Studied Expressed Sequence Tags (ESTs) from a saffron stigma cDNA library	D'Agostino et al. (2007)
SSRs	27	To assess the molecular variability and discriminating capacity of these markers regarding their effectiveness in establishing genetic relationships in these *Crocus* ecotypes	Nemati et al. (2014)
SRAP and RAPD	75 and 98	Markers with regard to detection of polymorphism, number of loci scored and PIC (polymorphic information content) values	Keify and Beiki (2012)
Microsatellite markers	12	To study population and conservation genetics of this economically and medically important species	Nemati et al. (2012)
ISSR	12	*C. cartwrightianus* cv. *albus* is more related to *C. sativus* than to *C. cartwrightianus* and may be an albinic saffron	Rubio-Moraga et al. (2010)

(continued)

Table 3.2 (continued)

Type of marker studied	No. of markers	Result of study	References
IRAP		*C. sativus*, Iranian *Crocus* genus showed high within and between species diversity. In some cases, genetic variation was high among ecotypes of the same species from different geographical regions	Alavi-Kia et al. (2008)
SSRs	16,721	Expression profiling in *Crocus sativus* to gain insights into apocarotenoid biosynthesis	Jain et al. (2016)

Based on RAPD (random amplified polymorphic DNA) pattern and phylogenetic studies, *C. sativus* and *C. cartwrightianus* are very closely related, and *C. asumaniae* and *C. pallasii* are the most dissimilar to *C. sativus* (Caiola et al. 2004). When *C. sativus*, with *C. thomasii*, and *C. cartwrightianus* were compared using AFLP (amplified fragment length polymorphism) markers, it was discovered that there was less genetic diversity between them (Zubor et al. 2004). The use of 27 microsatellite markers in phylogenetic studies for 67 accessions from four species of the genus *Crocus* collected from Iran was reported that *C. sativus* was becoming more similar to *C. speciosus* and *C. hausknechtii* (Nemati et al. 2014). For investigations on species affinities in the genus *Crocus*, ISSR (inter-simple sequence repeats) and ITS (internal transcribed spacer) markers were employed to examine five separate species and found a similar genetic association between *C. pallasii* and *C. sativus* (Sheidai et al. 2018). ISSR-based data showed *C. cartwrightianus* cv. *albus* as more closely related to *C. sativus* instead of *C. cartwrightianus* (Moraga et al. 2010). Eighty-three iPBS (inter-primer binding site)-retrotransposon markers data were compared among species of the *Crocus* genus, and their result showed *C. pallasii* subsp. *pallasii* and *C. cartwrightianus* were a close relative to *C. sativus* (Gedik et al. 2017). The same result was reported using IRAP (inter-retroelement amplified polymorphism) data (Alsayied et al. 2015). AFLP and MS-AFLP (methyl-sensitive amplified fragment length polymorphism) markers are used to analyse saffron to look for genetic (DNA sequence) and epigenetic (cytosine methylation) variations. Study concluded that genetic variations were low, while there was high epigenetic variability (Busconi et al. 2015). Further, a study investigated the variability and the presence of markers in the molecular pattern of different *Crocus* spp. and the distinction and variability of *C. sativus* from several geographic areas (Italy, Iran, Greece, and Spain) using 38 RAPD markers on dry stigmas as plant material (Pardo et al. 2003). From a saffron stigma cDNA library, 6603 high-quality ESTs (expressed sequence tags) were generated (D'Agostino et al. 2007). The RAPD primers developed 95 robust loci, and 75 SRAP (sequence-related amplified polymorphism) markers were also studied (Keify and Beiki 2012). Twelve novel polymorphic loci were developed and characterized from a repeat-enriched genomic library of *Crocus sativus* to study population and conservation genetics of this economically and medically important species (Nemati et al. 2012). Multilocus and SNP (single-nucleotide polymorphisms) data from GBS (genotyping by sequencing) revealed that *C. sativus* occupied a unique position within *C. cartwrightianus*, with no allies (Nemati et al. 2018). Nemati et al. (2019) looked at nuclear single-copy loci, GBS results, chloroplast genomes, and genome size in a more advanced analysis. Their findings revealed and confirmed that the wild *C. cartwrightianus* population in the south of Athens (Greece) is very similar to *C. sativus*, the saffron's single progenitor.

3.12 Genetic Improvement in Saffron Crop

Saffron is very particular in development as compared to most cultivated flowering plants. In general flora develop first vegetatively and then reproductive level. But saffron flowering is prior to vegetative development and is not immediately required for the production of stigmas. From an agronomic point of view, due to its agrological and eco-physiological features, saffron is a completely unknown plant. It is incapable of producing seeds and reproduces by means of an underground part, corm. Higher production and superiority of saffron can be accomplished, through a greater number of flowers per plant, a higher number of stigmas per flower, and enlarged stigma size or stigmas with a higher quantity of colour and odour (Caiola et al. 2004; Fernández 2004). Its sterility limits the implementation of breeding enhancements. Biotechnological methods such as micropropagation may be an alternate way of increasing the production of corm and improving the value of spices. The lack of genetic diversity inhibits the selection-based use of conventional plant breeding. Recent developments in sequencing and next-generation sequencing (NGS) offer successful methods such as transcriptome sequencing along with metabolome and proteome knowledge to harness the economic characteristics of saffron in functional genomics for genetic engineering.

Mutagenesis is another method used to generate genetic variants for the attribution of yield characteristics. Some of the most common mutagenic agents are UV light, X-rays, gamma rays, reactive oxygen species (ROS), alkylating agents, and base analogues. Many studies have reported improvement in different traits through mutagenesis. It was observed that when saffron corms were irradiated with gamma rays at a dose of 0.25, 0.50, 0.75, and 1 Kr from a Co60 source lower doses, radiation (0.25 Kr) enhances the plant growth, while higher doses reduced the plant growth (Dar et al. 2006). Colchicine too causes saffron heterogeneity by reducing the number of leaves and flowers per plant, resulting in flowers with oddly shaped reduced number of tepals, lobed and dentate tepals, flowers with deep coloration in stigmas extending to stylar regions, and flowers with orange-red pigmented anther. Studies on stomata found a decrease in stomata's number but an increase in their size (Zaffar et al. 2003). With the application of various mutagens, it was discovered that colchicine (0.05%) increased the length, quantity of pistill and leaves. EMS (0.1%) and Ethidium bromide (0.2%) also showed positive impacts saffron morphology however radiation dose had a negative impact on corm development. Morphological and anatomical variants caused by mutagens show increased numbers of stomata and thicker and wider leaves (Nehvi et al. 2009).

Breeding with wild ancestral species and the collection of random or triggered mutations should primarily discuss genetic improvement in saffron. New research focuses on enhancing stigma yield and efficiency. The development of the spice could be accelerated through extensive propagation of chosen pathogen-free ecotypes using tissue culture procedures employed in the propagation of saffron. The recovery of novel genotypes via somaclonal or in vitro mutagenesis variation, as well as the employment of genetic engineering to this crop tissue culture, can be highly useful and effective approaches for promoting the development of the spice saffron. To make somatic hybrids, protoplast culture is used, and stigma-like structure culture is used to make stigmas. The development of secondary metabolites is carried out by in vitro indirect somatic embryogenesis. The results indicated that optimum callus growth and useful development of secondary metabolites in in vitro saffron cultures could be achieved at moderate PGR concentrations (Firoozi et al. 2019). Hydroponics and aquaponics could be a better option for avoiding negative environmental indicators, preventing stigma yield reduction due to changing climate, and controlling weeds, diseases, and pests (Cardone et al. 2020).

3.13 Epigenetics and Saffron

Epigenetic marks are described as a collection of chemical modifications of both histones and DNA. Histone modification is the main area of concern because of the existence of inheritance mechanisms (Seymour and Becker 2017). Transgenerational epigenetic transmission occurs when heritable epigenetic marks sustain meiosis and are stably transferred to the next generation. Heritable epigenetic variation may result in heritable phenotypic variation when epigenetic changes occur near coding genes, influencing their transcriptional state. DNA methylation, among the most prevalent and well-studied epigenetic modifications in plant genomes, is essential for higher plants growth and development (Walker et al. 2018). There are many anomalies during gametophyte production due to irregular meiosis and sporogenesis thus developing abnormal pollen, while saffron ovules remain viable, which shows that infertility is mainly due to male gametophyte in this plant (Caiola 2003). As a result, although genetic variability in saffron is low, phenotypic variation is common in the region, and epigenetics may be a source of such substitute phenotypes. Alternative phenotypes in saffron could have genetic and epigenetic roots (Busconi et al. 2021). Remarkably phenotypic differences in the field have been found regularly plants with varying numbers of stigmas or various forms of tepals by researchers and saffron producers. Surprisingly, these morphological variations are always inconsistent, changing from one growth period to the next. Epigenetic variations and epigenetic-based phenotypes may have proved beneficial for crop improvement, but in order to use epigenetic knowledge, it is important to better understand the stability of epigenomic patterns in species, particularly under natural environmental conditions. If epigenetic trends remain constant during development and years, the epigenome may be used to classify an organism's epigenetic profile and predict traits.

3.14 Conclusions and Future Perspectives

Based on above-mentioned studies, it is concluded that *C. sativus* is an autotriploid with *C. cartwrightianus* as its single origin. Since it is a sterile crop and only propagated through vegetative means, micropropagation can be used for the large-scale production of corms. Further saffron has many medicinal properties (especially due to carotenoids), so sequencing and next-generation sequencing (NGS) methods such as transcriptome along with metabolome and proteome knowledge can be used to harness the economic and medicinal characteristics of saffron in functional genomics for genetic engineering. It has fewer genetic variations; therefore specific genome editing tools ZFN, TALENs, and CRISPR/Cas can be used for crop improvement. Phenotype is influenced by genetic as well as epigenetic variations, so a better understanding of the stability of epigenomic patterns will be crucial for this crop improvement.

References

Abdel-Rahman RF, El Awdan SA, Hegazy RR, Mansour DF, Ogaly HA, Abdelbaset M (2020) Neuroprotective effect of *Crocus sativus* against cerebral ischemia in rats. Metab Brain Dis 35:427–439

Abdullaev FI, Frenkel GD (1992) Effect of saffron on cell colony formation and cellular nucleic acid and protein synthesis. BioFactors (Oxford, England) 3(3):201–204

Abdullaev FI (2003) Antitumor effect of saffron (Crocus sativus L.): overview and perspectives. Acta Hortic 650:491–499

Agayev YM (2002) New features in karyotype structure and origin of saffron. Crocus sativus L Cytologia 67(3):245–252

Agayev YM, Zarifi E (2009) Peculiar evolution of saffron (*Crocus sativus* L.): prosperity and decline. Acta Hortic 850(850):29–34

Ahmad AS, Ansari MA, Ahmad M, Saleem S, Yousuf S, Hoda MN, Islam F (2005) Neuroprotection by crocetin in a hemi-parkinsonian rat model. Pharmacol Biochem Behav 81(4):805–813

Ahmad M, Zaffar G, Mir SD, Razvi SM, Rather MA, Mir MR (2011) Saffron (*Crocus sativus* L.) strategies for enhancing productivity. Res J Med Plant 5(6):630–649

Ahrazem O, Rubio-Moraga A, Berman J, Capell T, Christou P, Zhu C, Gómez-Gómez L (2016) The carotenoid cleavage dioxygenase CCD 2 catalysing the synthesis of crocetin in spring crocuses and saffron is a plastidial enzyme. New Phytol 209(2):650–663

Ahrazem O, Argandoña J, Fiore A, Aguado C, Luján R, Rubio-Moraga Á, Marro M, Araujo-Andrade C, Loza-Alvarez P, Diretto G, Gómez-Gómez L (2018) Transcriptome analysis in tissue sectors with contrasting crocins accumulation provides novel insights into apocarotenoid biosynthesis and regulation during chromoplast biogenesis. Sci Rep 8(1):2843

Ahrazem O, Diretto G, Argandoña Picazo J, Fiore A, Rubio-Moraga Á, Rial C, Varela RM, Macías FA, Castillo R, Romano E, Gómez-Gómez L (2019) The specialized roles in carotenogenesis and apocarotenogenesis of the phytoene synthase gene family in saffron. Front Plant Sci 10:249

Akhondzadeh S, Sabet MS, Harirchian MH, Togha M, Cheraghmakani H, Razeghi S, Hejazi SS, Yousefi MH, Alimardani R, Jamshidi A, Zare F (2010) Saffron in the treatment of patients with mild to moderate Alzheimer's disease: a 16-week, randomized and placebo-controlled trial. J Clin Pharm Ther 35(5):581–588

Alavi-Kia SS, Mohammadi SA, Aharizad S, Moghaddam M (2008) Analysis of genetic diversity and phylogenetic relationships in crocus genus of Iran using inter-retrotransposon amplified polymorphism. Biotechnol Biotechnol Equip 22(3):795–800

Alsayied NF, Fernández JA, Schwarzacher T, Heslop-Harrison JS (2015) Diversity and relationships of *Crocus sativus* and its relatives analysed by inter-retro element amplified polymorphism (IRAP). Ann Bot 116(3):359–368

Amin B, Hosseinzadeh H (2012) Evaluation of aqueous and ethanolic extracts of saffron, *Crocus sativus* L., and its constituents, safranal and crocin in allodynia and hyperalgesia induced by chronic constriction injury model of neuropathic pain in rats. Fitoterapia 83(5):888–895

Anonymous (2019). http://www.theplantlist.org/browse/A/Iridaceae/Crocus/

Ashktorab H, Soleimani A, Singh G, Amin A, Tabtabaei S, Latella G, Stein U, Akhondzadeh S, Solanki N, Gondré-Lewis MC, Habtezion A (2019) Saffron: the golden spice with therapeutic properties on digestive diseases. Nutrients 11(5):943

Ayatollahi H, Javan AO, Khajedaluee M, Shahroodian M, Hosseinzadeh H (2014) Effect of *Crocus sativus* L.(saffron) on coagulation and anticoagulation systems in healthy volunteers. Phytother Res 28(4):539–543

Baba SA, Mohiuddin T, Basu S, Swarnkar MK, Malik AH, Wani ZA, Abbas N, Singh AK, Ashraf N (2015) Comprehensive transcriptome analysis of *Crocus sativus* for discovery and expression of genes involved in apocarotenoid biosynthesis. BMC Genomics 16:1–14

Bathaie SZ, Mousavi SZ (2010) New applications and mechanisms of action of saffron and its important ingredients. Crit Rev Food Sci Nutr 50(8):761–786

Beiki AH, Keifi F, Mozafari J (2010) Genetic differentiation of crocus species by random amplified polymorphic DNA. Genet Eng Biotechnol J 18:1–10

Bhatia S (2015) Application of plant biotechnology. Modern applications of plant biotechnology in pharmaceutical sciences, vol 5, pp 157–207

Bolhasani A, Bathaie SZ, Yavari I, Moosavi-Movahedi AA, Ghaffari M (2005) Separation and purification of some components of iranian saffron. Asian J Chem 17:725–729

Boskabady MH, Seyedhosseini Tamijani SM, Rafatpanah H, Rezaei A, Alavinejad A (2011) The effect of *Crocus sativus* extract on human lymphocytes' cytokines and T helper 2/T helper 1 balance. J Med Food 14(12):1538–1545

Bouvier F, Suire C, Mutterer J, Camara B (2003) Oxidative remodeling of chromoplast carotenoids: identification of the carotenoid dioxygenase CsCCD and CsZCD genes involved in crocus secondary metabolite biogenesis. Plant Cell 15(1):47–62

Broadhead GK, Grigg JR, McCluskey P, Hong T, Schlub TE, Chang AA (2019) Saffron therapy for the treatment of mild/moderate age-related macular degeneration: a randomised clinical trial. Graefes Arch Clin Exp Ophthalmol 257:31–40

Buggs RJ, Chamala S, Wu WE, Gao LU, May GD, Schnable PS, Soltis DE, Soltis PS, Barbazuk WB (2010) Characterization of duplicate gene evolution in the recent natural allopolyploid Tragopogon miscellus by next-generation sequencing and Sequenom iPLEX MassARRAY genotyping. Mol Ecol 19:132–146

Busconi M, Colli L, Sánchez RA, Santaella M, De-Los-Mozos Pascual M, Santana O, Roldan M, Fernandez JA (2015) AFLP and MS-AFLP analysis of the variation within saffron crocus (*Crocus sativus* L.) germplasm. PLoS One 10(4):e0123434

Busconi M, Soffritti G, Stagnati L, Marocco A, Martínez JM, Pascual MDLM, Fernandez JA (2018) Epigenetic stability in Saffron (*Crocus sativus* L.) accessions during four consecutive years of cultivation and vegetative propagation under open field conditions. Plant Sci 277:1–10

Busconi M, Wischnitzki E, Del Corvo M, Colli L, Soffritti G, Stagnati L, Fluch S, Sehr EM, de los Mozos Pascual M, Fernández JA (2021) Epigenetic variability among saffron crocus (*Crocus sativus* L.) accessions characterized by different phenotypes. Front Plant Sci 12:642631

Caiola MG (1999) Reproduction biology of saffron and its allies. In: Negbi M (ed) Saffron: *Crocus sativus* L. CRC Press, London, pp 40–53

Caiola MG (2003) Saffron reproductive biology. In: International symposium on saffron biology and biotechnology, pp 25–37

Caiola MG, Caputo P, Zanier R (2004) RAPD analysis in crocus sativus L. accessions and related crocus species. Biol Plant 48:375–380

Caiola MG, Canini A (2010) Looking for saffron's (*Crocus sativus* L.) parents. Funct Plant Sci Biotechnol 4(2):1–14

Cardone L, Castronuovo D, Perniola M, Cicco N, Candido V (2020) Saffron (*Crocus sativus L.*) the king of spices: an overview. Sci Hortic 272:109560
Carmona M, Sánchez AM, Ferreres F, Zalacain A, Tomás-Barberán F, Alonso GL (2007) Identification of the flavonoid fraction in saffron spice by LC/DAD/MS/MS: comparative study of samples from different geographical origins. Food Chem 100(2):445–450
Chichiriccò G (1984) Karyotype and meiotic behaviour of the triploid *Crocus sativus L.* Caryologia 37(3):233–239
Cossignani L, Urbani E, Simonetti MS, Maurizi A, Chiesi C, Blasi F (2014) Characterisation of secondary metabolites in saffron from Central Italy (Cascia, Umbria). Food Chem 143:446–451
D'Agostino N, Pizzichini D, Chiusano ML, Giuliano G (2007) An EST database from saffron stigmas. BMC Plant Biol 7(1):1–8
Dar SA, Makhdoomi MI, Allie BA, Mir ZA, Nehvi FA, Wani SA (2006) Biological interventions for enhancing saffron productivity in Kashmir. Acta Hortic 739:25–32
Demurtas OC, Frusciante S, Ferrante P, Diretto G, Azad NH, Pietrella M, Aprea G, Taddei AR, Romano E, Mi J, Al-Babili S (2018) Candidate enzymes for saffron crocin biosynthesis are localized in multiple cellular compartments. Plant Physiol 177(3):990–1006
Deslauriers AM, Afkhami-Goli A, Paul AM, Bhat RK, Acharjee S, Ellestad KK, Noorbakhsh F, Michalak M, Power C (2011) Neuroinflammation and endoplasmic reticulum stress are coregulated by crocin to prevent demyelination and neurodegeneration. J Immunol 187(9):4788–4799
Ehsanzadeh P, Yadollahi AA, Maibodi AM (2003) Productivity, growth and quality attributes of 10 Iranian saffron accessions under climatic conditions of Chahar-Mahal Bakhtiari, Central Iran. Acta Hortic 650:183–188
Estep MC, DeBarry JD, Bennetzen JL (2013) The dynamics of LTR retrotransposon accumulation across 25 million years of panicoid grass evolution. Heredity 110(2):194–204
Fernández JA (2004) Biology, biotechnology and biomedicine of saffron. Recent Res Dev Plant Sci 2:127–159
Fernàndez JA, Horvat O, Vurdu H, Argento S, Sramko G, Balazs L, Constantinidis T, Krigas N, Pastor T, Sanchis E, Rodriguez MF (2011) The world saffron and crocus collection. Aristotle University of Thessaloniki
Firoozi B, Nasser ZA, Sofalian O, Sheikhzade-Mosadegh P (2019) In vitro indirect somatic embryogenesis and secondary metabolites production in the saffron: emphasis on ultrasound and plant growth regulators. J Agric Sci 25(1):1–10
Frello S, Ørgaard M, Jacobsen N, Heslop-Harrison JS (2004) The genomic organization and evolutionary distribution of a tandemly repeated DNA sequence family in the genus *crocus* (Iridaceae). Hereditas 141(1):81–88
Frizzi G, Miranda M, Pantani C, Tammaro F (2007) Allozyme differentiation in four species of the *Crocus cartwrightianus* group and in cultivated saffron (*Crocus sativus*). Biochem Syst Ecol 35(12):859–868
Frusciante S, Diretto G, Bruno M, Ferrante P, Pietrella M, Prado-Cabrero A, Rubio-Moraga A, Beyer P, Gomez-Gomez L, Al-Babili S, Giuliano G (2014) Novel carotenoid cleavage dioxygenase catalyzes the first dedicated step in saffron crocin biosynthesis. Proc Natl Acad Sci 111(33):12246–12251
Fukui H, Toyoshima K, Komaki R (2011) Psychological and neuroendocrinological effects of odor of saffron (*Crocus sativus*). Phytomedicine 18(8–9):726–730
Gao WY, Li YM, Zhu DY (1999) New anthraquinones from the sprout of Crocus sativus. Acta Bot Sin 41:531–533
Gedik A, Duygu AT, Erdogmus S, Comertpay G, Tanyolac MB, Ozkan H (2017) Genetic diversity of *Crocus sativus* and its close relative species analyzed by iPBS-retrotransposons. Turkish J Field Crop 22(2):243–252
Geromichalos GD, Lamari FN, Papandreou MA, Trafalis DT, Margarity M, Papageorgiou A, Sinakos Z (2012) Saffron as a source of novel acetylcholinesterase inhibitors: molecular docking and in vitro enzymatic studies. J Agric Food Chem 60(24):6131–6138
Ghaffari SM (1986) Cytogenetic studies of cultivated Crocus sativus (Iridaceae). Plant Syst Evol 153:199–204

Ghaffari S, Hatami H, Dehghan G (2015) Saffron ethanolic extract attenuates oxidative stress, spatial learning, and memory impairments induced by local injection of ethidium bromide. Res Pharm Sci 10(3):222–232

Ghaffari S, Roshanravan N (2019) Saffron: an updated review on biological properties with special focus on cardiovascular effects. Biomed Pharmacother 109:21–27

Gismondi A, Fanali F, Labarga JM, Caiola MG, Canini A (2013) *Crocus sativus* L. genomics and different DNA barcode applications. Plant Syst Evol 299:1859–1863

Golmohammadzadeh S, Jaafari MR, Hosseinzadeh H (2010) Does saffron have antisolar and moisturizing effects? IJPR 9(2):133

Gómez-Gómez L, Rubio-Moraga Á, Ahrazem O (2010) Understanding carotenoid metabolism in saffron stigmas: unravelling aroma and colour formation. Func Plant Sci Biotech 4:56–63

Gómez-Gómez L, Parra-Vega V, Rivas-Sendra A, Seguí-Simarro JM, Molina RV, Pallotti C, Rubio-Moraga Á, Diretto G, Prieto A, Ahrazem O (2017) Unraveling massive crocins transport and accumulation through proteome and microscopy tools during the development of saffron stigma. Int J Mol Sci 18(1):76

Gregory MJ, Menary RC, Davies NW (2005) Effect of drying temperature and air flow on the production and retention of secondary metabolites in saffron. J Agric Food Chem 53(15):5969–5975

Güner A, Özhatay N, Ekim T, Başer HC (2000) Flora of Turkey and the East Aegean Islands. Edinburgh Univ Pres Edinburgh, p 11

Harpke D, Meng S, Rutten T, Kerndorff H, Blattner FR (2013) Phylogeny of *Crocus* (Iridaceae) based on one chloroplast and two nuclear loci: ancient hybridization and chromosome number evolution. Mol Phylo genet Evol 66(3):617–627

Hashtjini MM, Jahromi GP, Meftahi GH, Javidnazar D (2018) Aqueous extract of saffron administration along with amygdala deep brain stimulation promoted alleviation of symptoms in post-traumatic stress disorder (PTSD) in rats. Avicenna J Phyto med 8(4):358

Hatziagapiou K, Kakouri E, Lambrou GI, Bethanis K, Tarantilis PA (2019) Antioxidant properties of *Crocus sativus* L. and its constituents and relevance to neurodegenerative diseases; focus on Alzheimer's and Parkinson's disease. Curr Neuro Pharmacol 17(4):377–402

Hausenblas HA, Saha D, Dubyak PJ, Anton SD (2013) Saffron (*Crocus sativus* L.) and major depressive disorder: a meta-analysis of randomized clinical trials. J Integr Med 11(6):377–383

Hooshmandi Z, Rohani AH, Eidi A, Fatahi Z, Golmanesh L, Sahraei H (2011) Reduction of metabolic and behavioral signs of acute stress in male Wistar rats by saffron water extract and its constituent safranal. Pharmaceut Biol 49(9):947–954

Hosseinzadeh H, Ghenaati J (2006) Evaluation of the antitussive effect of stigma and petals of saffron (*Crocus sativus*) and its components, safranal and crocin in guinea pigs. Fitoterapia 77(6):446–448

Hosseinzadeh H, Younesi HM (2002) Antinociceptive and anti-inflammatory effects of *Crocus sativus* L. stigma and petal extracts in mice. BMC pharmacol 2:1–8

Hosseinzadeh H, Sadeghnia HR, Ziaee T, Danaee A (2005) Protective effect of aqueous saffron extract *(Crocus sativus L.)* and crocin, its active constituent, on renal ischemia-reperfusion-induced oxidative damage in rats. J Pharm Pharm Sci 8(3):387–393

Hosseinzadeh H, Ziaee T, Sadeghi A (2008) The effect of saffron, Crocus sativus stigma, extract and its constituents, safranal and crocin on sexual behaviors in normal male rats. Phytomedicine 15(6–7):491–495

Hosseinzadeh H, Modaghegh MH, Saffari Z (2009) *Crocus sativus* L(Saffron) extract and its active constituents (crocin and safranal) on ischemia-reperfusion in rat skeletal muscle. Evid Based Complementary Alter Med 6:343–350

Ingram JS (1969) Saffron (*crocus-sativus* L). Trop sci 11(3):177

Jain M, Srivastava PL, Verma M, Ghangal R, Garg R (2016) De novo transcriptome assembly and comprehensive expression profiling in *Crocus sativus* to gain insights into apocarotenoid biosynthesis. Sci Rep 6(1):22456

Javadi B, Sahebkar A, Emami SA (2013) A survey on saffron in major Islamic traditional medicine books. Iran J Basic Med Sci 16(1):1

Joukar S, Najafipour H, Khaksari M, Sepehri G, Shahrokhi N, Dabiri S, Gholamhoseinian A, Hasanzadeh S (2010) The effect of saffron consumption on biochemical and histopathological heart indices of rats with myocardial infarction. Cardiovasc Toxicol 10:66–71

Kalendar R, Flavell AJ, Ellis TH, Sjakste T, Moisy C, Schulman AH (2011) Analysis of plant diversity with retrotransposon-based molecular markers. Heredity 106(4):520–530

Kalia D, Jose-Santhi J, Kumar R, Singh RK (2023) Analysis of PEBP genes in saffron identifies a flowering locus T homologue involved in flowering regulation. J Plant Growth Regul 42(4):2486–2505

Karasawa K (1932) Karyological studies of some flowering Crocus. Bot Mag, Tokyo, p 46

Karasawa K (1933) On the triploidy of Crocus Sativus L. and its high sterility. Jpn J Genet 9(1):6–8

Keify F, Beiki AH (2012) Exploitation of random amplified polymorphic DNA (RAPD) and sequence-related amplified polymorphism (SRAP) markers for genetic diversity of saffron collection. J Med Plant Res 6(14):2761–2768

Khorasani G, Hosseinimehr SJ, Zamani P, Ghasemi M, Ahmadi A (2008) The effect of saffron (*Crocus sativus*) extract for healing of second-degree burn wounds in rats. Keio J Med 57(4):190–195

Khorasany AR, Hosseinzadeh H (2016) Therapeutic effects of saffron (*Crocus sativus* L.) in digestive disorders: a review. Iran J Basic. Med Sci 19(5):455

Khoulati A, Ouahhoud S, Mamri S, Alaoui K, Lahmass I, Choukri M, Kharmach EZ, Asehraou A, Saalaoui E (2019) Saffron extract stimulates growth, improves the antioxidant components of Solanum lycopersicum L., and has an antifungal effect. Ann Agric Sci 64(2):138–150

Kianbakht S, Ghazavi A (2011) Immunomodulatory effects of saffron: a randomized double-blind placebo-controlled clinical trial. Phytother Res 25(12):1801–1805

Koç H (2012) Farklıekolojikşartlarınsafranın. *Crocus sativus* pp 13–15

Li CY, Lee EJ, Wu TS (2004) Antityrosinase Principles and Constituents of the Petals of *Crocus sativus*. J Nat Prod 67(3):437–440

Lopresti AL, Drummond PD (2017) Efficacy of curcumin, and a saffron/curcumin combination for the treatment of major depression: a randomised, double-blind, placebo-controlled study. J Affect Disord 207:188–196

Madlung A (2013) Polyploidy and its effect on evolutionary success: old questions revisited with new tools. Heredity 110(2):99–104

Mardani H, Maninang J, Appiah KS, Oikawa Y, Azizi M, Fujii Y (2019) Evaluation of biological response of lettuce (*Lactuca sativa* L.) and weeds to safranal allelochemical of saffron (*Crocus sativus*) by using static exposure method. Molecules 24(9):1788

Masaki M, Aritake K, Tanaka H, Shoyama Y, Huang ZL, Urade Y (2012) Crocin promotes non-rapid eye movement sleep in mice. Mol Nutr Food Res 56(2):304–308

Mathew BR (1999) Botany, taxonomy and cytology of *C. sativus* L. and its allies. Saffron: *Crocus sativus L* 1730

Mehraj SS, Kamili AN, Bhat GA, Nazir R (2005) Resource allocation in dormant corms of *Crocus sativus* L. and their heavy metal analysis. Int J Adv Res 10:27–33

Melnyk JP, Wang S, Marcone MF (2010) Chemical and biological properties of the world's most expensive spice: Saffron. Food Res Int 43(8):1981–1989

Mohammad R, Daryoush M, Ali R, Yousef D, Mehrdad N (2011) Attenuation of oxidative stress of hepatic tissue by ethanolic extract of saffron (dried stigmas of *Crocus sativus* L.) in streptozotocin (STZ)-induced diabetic rats. Afr. J Pharm Pharmacol 5(19):2166–2173

Molina RV, Valero M, Navarro Y, Guardiola JL, Garcia-Luis AJ (2005) Temperature effects on flower formation in saffron (*Crocus sativus* L.). Sci Hortic 103(3):361–379

Mollazadeh H, Emami SA, Hosseinzadeh H (2015) Razi's Al-Hawi and saffron (Crocus sativus): a review. Iran J Basic Med Sci 18(12):1153–1166

Moradzadeh M, Sadeghnia HR, Tabarraei A, Sahebkar A (2018) Anti-tumor effects of crocetin and related molecular targets. J Cell Physiol 233(3):2170–2182

Moraga AR, Trapero-Mozos A, Gómez-Gómez L, Ahrazem O (2010) Intersimple sequence repeat markers for molecular characterization of *Crocus cartwrightianus* cv. albus. Ind Crops. Prod 32(2):147–151

Mousavi SZ, Bathaie SZ (2011) Historical uses of saffron: identifying potential new avenues for modern research. Avicenna J Phyto med 1(2):57–66

Mzabri I, Addi M, Berrichi A (2019) Traditional and modern uses of saffron (*Crocus sativus*). Cosmetics 6(4):63

Nagatoshi M, Terasaka K, Owaki M, Sota M, Inukai T, Nagatsu A, Mizukami H (2012) UGT75L6 and UGT94E5 mediate sequential glucosylation of crocetin to crocin in *Gardenia jasminoides*. FEBS Lett 586(7):1055–1061

Negbi M (1999) Saffron cultivation: past, present and future prospects. Saffron *Crocus sativus* L 154:1–17

Nehvi FA, Khan MA, Lone AA, Maqhdoomi MI, Wani SA, Yousuf V, Yasmin S (2009) Effect of radiation and chemical mutagen on variability in saffron *(Crocus sativus* L.). Acta Hortic. 850:67–74

Nemati Z, Zeinalabedini M, Mardi M, Pirseyediand SM, Marashi SH, Khayam Nekoui SM (2012) Isolation and characterization of a first set of polymorphic microsatellite markers in saffron, *Crocus sativus* (Iridaceae). Am J Bot 99(9):e340–e343

Nemati Z, Zeinalabedini M, Majidian P, Eftekharian Jahromi A, Kiani D (2014) Phylogenetic relationships among Iranian and Spanish date palms (*Phoenix dactylifera* L.) revealed by microsatellite markers. J.hortic sci biotechnol 89(2):115–120

Nemati Z, Blattner FR, Kerndorff H, Erol O, Harpke D (2018) Phylogeny of the saffron-crocus species group, Crocus series Crocus (Iridaceae). Mol Phylogenet Evol 127:891–897

Nemati Z, Harpke D, Gemicioglu A, Kerndorff H, Blattner FR (2019) Saffron (*Crocus sativus*) is an autotriploid that evolved in Attica (Greece) from wild *Crocus cartwrightianus*. Mol Phylogenet Evol136:14–20

Nescatelli R, Carradori S, Marini F, Caponigro V, Bucci R, De Monte C, Mollica A, Mannina L, Ceruso M, Supuran CT, Secci D (2017) Geographical characterization by MAE-HPLC and NIR methodologies and carbonic anhydrase inhibition of Saffron components. Food Chem 221:855–863

Ochiai T, Shimeno H, Mishima KI, Iwasaki K, Fujiwara M, Tanaka H, Shoyama Y, Toda A, Eyanagi R, Soeda S (2007) Protective effects of carotenoids from saffron on neuronal injury in vitro and in vivo. Biochim Biophys Acta Gen Subj 1770(4):578–584

Pardo J, Fernández JA, Gomez LG (2003) Development of molecular markers for origin determination in saffron. Acta Hortic. 650:95–98

Petersen G, Seberg O, Thorsøe S, Jørgensen T, Mathew B (2008) A phylogeny of the genus *Crocus* (Iridaceae) based on sequence data from five plastid regions. Taxon 57(2):487–499

Pfander H, Schurteberger H (1982) Biosynthesis of C20-carotenoids in Crocus sativus. Phytochemistry 21:1039–1042

Pintado C, de Miguel A, Acevedo O, Nozal L, Novella JL, Rotger R (2011) Bactericidal effect of saffron (Crocus sativus L.) on Salmonella enterica during storage. Food Control 22(3–4):638–642

Plessner O, Negbi M, Ziv M, Basker D (1989) Effects of temperature on the flowering of the saffron *crocus* (*Crocus sativus L.*): induction of hysteranthy. Isr J Plant Sci 38(1):1–7

Poma A, Fontecchio G, Carlucci G, Chichiricco G (2012) Anti-inflammatory properties of drugs from saffron crocus. Antiinflamm Antiallergy Agents Med Chem 11:37–51

Ramsey J, Schemske DW (1998) Pathways, mechanisms, and rates of polyploid formation in flowering plants. Annu Rev. Ecol Evol Syst 29(1):467–501

Rezaee R, Hosseinzadeh H (2013) Safranal: from an aromatic natural product to a rewarding pharmacological agent. Iran J Basic 16(1):12

Rubio-Moraga A, Trapero A, Ahrazem O, Gómez-Gómez L (2010) Crocins transport in *Crocus sativus*: the long road from a senescent stigma to a newborn corm. Phytochemistry 71(13):1506–1513

Sadeghnia H, Cortez M, Liu D, Hosseinzadeh H, Snead OC (2008) Antiabsence effects of safranal in acute experimental seizure models: EEG and autoradiography. J Pharm Pharm Sci 11(3):1–4

Sadraei H, Ghannadi A, Takei-bavani M (2003) Effects of *Zataria multiflora* and *Carum carvi* essential oils and hydroalcoholic extracts of *Passiflora incarnata*, *Berberis integerrima* and *Crocus sativus* on rat isolated uterus contractions. Int J Aromather 13:121–127

Saito N, Mitsui S, Hayashi K (1960) Delphin, the Anthocyanin of Medicinal Saffron and its Identity with Hyacin as Shown by Paper Chromatography of Partial Hydrolysates Studies on Anthocyanins.P. JPN ACAD 36(6):340–345

Sampathu SR, Shivashankar S, Lewis YS (1984) Saffron (*Crocus sativus* L.) cultivation, processing, chemistry and standardization. Crit Rev Food Sci Nutr 20:123–157

Santhosh MS, Thushara RM, Hemshekhar M, Sunitha K, Devaraja S, Kemparaju K, Girish KS (2013) Alleviation of viper venom induced platelet apoptosis by crocin (*Crocus sativus)*: implications for thrombocytopenia in viper bites. J thrombo thrombolysis 36:424–432

Sarris J, McIntyre E, Camfield DA (2013) Plant-based medicines for anxiety disorders, part 1: a review of preclinical studies. CNS drugs 27:207–219

Sarshoori JR, Asadi MH, Mohammadi MT (2014) Neuroprotective effects of crocin on the histopathological alterations following brain ischemia-reperfusion injury in rat. Iran J Basic Med Sci 17(11):895

Schmidt T, Heitkam T, Liedtke S, Schubert V, Menzel G (2019) Adding color to a century-old enigma: multi-color chromosome identification unravels the autotriploid nature of saffron (Crocus sativus) as a hybrid of wild Crocus cartwrightianus cytotypes. New Phytol 4:1965–1980

Seymour DK, Becker C (2017) The causes and consequences of DNA methylome variation in plants. Curr Opin Plant Biol 36:56–63

Shahi T, Assadpour E, Jafari SM (2016) Main chemical compounds and pharmacological activities of stigmas and tepals of 'red gold'; saffron. Trends Food Sci Technol 58:69–78

Sheidai M, Tabasi M, Mehrabian MR, Koohdar F, Ghasemzadeh-Baraki S, Noormohammadi Z (2018) Species delimitation and relationship in *Crocus L* (Iridaceae). Acta Bot Croat 77(1):10–17

Sik L, Candan F, Soya S, Karamenderes C, Kesercioglu T, Tanyolac B (2008) Genetic variation among Crocus sativus L. species from western Turkey as revealed by RAPD and ISSR markers. J Appl Biol Sci 2:73–78

Srivastava RP (1963) Cultivation of saffron in India. Fertil News 8(9):7–11

Stebbins GL (1971) Chromosomal evolution in higher plants. Edward Arnold LTD, London, pp 87–89

Tammaro F (1990) *Crocus sativus* L.–cv. Piano di Navelli (L'Aquila saffron): environment, cultivation, morphometric characteristics, active principles, uses. In: Proceedings of the international conference on saffron, pp 47–57

Tavana S, Eimani H, Azarnia M, Shahverdi A, Eftekhari-Yazdi P (2012) Effects of saffron (*Crocus sativus* L.) aqueous extract on in vitro maturation, fertilization and embryo development of mouse oocytes. Cell Journal (Yakhteh) 13(4):259

Tsaftaris A, Pasentsis K, Makris A, Darzentas N, Polidoros A, Kalivas A, Argiriou A (2011) The study of the E-class SEPALLATA3-like MADS-box genes in wild-type and mutant flowers of cultivated saffron crocus (*Crocus sativus* L.) and its putative progenitors. J Plant Physiol 168(14):1675–1684

Vavilov NI (1951) The origin, variation, immunity and breeding of cultivated plants. Soil Sci 72(6):482

Walker J, Gao H, Zhang J, Aldridge B, Vickers M, Higgins JD, Feng X (2018) Sexual-lineage-specific DNA methylation regulates meiosis in Arabidopsis. Nat Genet 50(1):130–137

Winterhalter P, Straubinger M (2000) Saffron—renewed interest in an ancient spice. Food Rev Int 16(1):39–59

Zaffar G, Wani SA, Anjum T, Zeerak NA (2003) Colchicine induced variability in saffron. Acta Horti:277–280

Zheng YQ, Liu JX, Wang JN, Xu L (2007) Effects of crocin on reperfusion-induced oxidative/nitrative injury to cerebral micro vessels after global cerebral ischemia. Brain Res 1138:86–94

Zubor Á, Surányi G, Győri Z, Borbély G, Prokisch J (2004) Molecular Biological Approach of *Crocus sativus* L. and its Allies. Acta Agrar Debr 4(13):112–115

Insight into the Genetics and Genomics Studies of the *Fritillaria* Species

4

Vinay Kumar, Shagun Sharma, and Pankaj Kumar

4.1 Introduction

For as long as there have been human cultures, people have relied on plants for both food and medicine. It is believed that medicinal plants have ready access to the ingredients required for the synthesis and manufacture of drugs (Jan and Abbas 2018). Native peoples of the Himalayas and the trans-Himalayas have long relied on the region's flora as an important source of medicinal ingredients (Kapoor et al. 2021). Ayurveda, Unani, Siddha, Tibetan, and other forms of traditional informal medicine systems have been in use for millennia. Jammu and Kashmir, Ladakh, Himachal Pradesh, and Uttarakhand are just a few of the states in India that constitute high-altitude Indian Himalayas (Kala 2006; Jan and Abbas 2018). The Indian Himalayan Region (IHR), which has long been recognized as a key location for the collection of medicinal plants, is home to an exceptionally rich and diverse flora as a result of its extreme elevational and climatic variations (Kumar et al. 2021). High-altitude medicinal plants are a growing concern in the Himalayas because of their relevance in conventional medicine and commercial harvesting. Inaccessible areas and protected habitats are currently the only places where many important medicinal plant species including species of *Aconitum*, *Taxus*, *Ephedra*, *Dactylorhiza*, *Fritillaria*, *Polygonatum*, *Picrorrhiza*, *Allium*, etc. are present (Gairola et al. 2010; Kumar et al. 2021, Kumar et al. 2023; Kapoor et al. 2021).

The perennial bulbous plants known as *Fritillaria* are classified under the family Liliaceae. Approximately 140 species are widely dispersed across the temperate

---

Vinay Kumar and Shagun Sharma contributed equally to this work.

V. Kumar · S. Sharma · P. Kumar (✉)
Department of Biotechnology, Dr. Y.S. Parmar University of Horticulture and Forestry, Solan, Himachal Pradesh, India
e-mail: pksharmabiotech@yspuniversity.ac.in

V. Gahlaut, V. Jaiswal (eds.), *Genetics and Genomics of High-Altitude Crops*,
https://doi.org/10.1007/978-981-99-9175-4_4

regions of the Northern Hemisphere. The altitude range meters above sea level vary from 2500 to 4600 m (Wang et al. 2017). The acidic, well-drained light sand or medium loam soils found on alpine slopes are ideal for their growth. *Fritillaria* species are known to be indigenous to the temperate zone of the Northern Hemisphere. However, recent studies suggest that the primary evolutionary hotspot for this genus is located in the East Mediterranean region, specifically in Iran. Majority of *Fritillaria* taxa are predominantly distributed in Turkey, with a total of 33 taxa. China and Greece followed closely behind with 30 and 24 taxa, respectively. Meanwhile, California and India had 18 and 6 taxa, respectively (Kumar et al. 2021). At altitudes between 2800 and 4000 meters above sea level, *Fritillaria* has been observed in the western Himalayas of India (Kala 2006; Singh and Rawat 2011; Chauhan et al. 2011; Dad and Reshi 2015). To treat tuberculosis, burns, asthma, lung congestion, and stomach issues, many pharmaceuticals and dietary supplements include *Fritillaria* bulbs as an active ingredient (Dasgupta and Deb 1986; Naito and Otsuki 2005; Kiani et al. 2017; Singh et al. 2020; Samaropoulou et al. 2021; Marković et al. 2021). It is also believed to have beneficial antioxidant, anti-aging, anti-stress, life-promoting, and rejuvenating properties (Kumar et al. 2020). Interest in *Fritillaria* species has risen significantly due to their commercial value as a source of materials used in traditional medicine and their aesthetic appeal as ornamental plants (Day et al. 2014). Numerous Ayurvedic medicines, including Astavarga, Chyavanprash, Mahatraiphala Ghritham, Jeevanthyadi Ghrutham, and Danwantharam Thailam, are formulated from this perennial plant with a bulb that grows underground (Kaul 2010). A total of 35 species of *Fritillaria* have been studied, and so far about 140 compounds have been isolated from them. Imperialine, verticine, verticinone, isoverticine, ebeiedinone, and ebeienine are all examples of isosteroidal alkaloids; the next most common group is non-alkaloids, which includes saponin, terpenoids, steroids, succinic acid, thymidine, and adenosine (72.7%) (Chang et al. 2020). Plants from this genus are well-known for their medicinal properties and serve as important building blocks in many medications used in the ISM (Indian System of Medicine) due to the presence of peimine ($C_{27}H_{45}O_3N$), peiminine ($C_{27}H_{43}O_3N$), fritimine, fritillarin, verticin, and verticillin in the bulbs of some other species of this genus (Chauhan et al. 2011). Due to the high concentration of alkaloids (valuable secondary metabolites) in the bulbs of many fritillaries, they have been used in traditional medicine for many years in China, Turkey, and Japan. The name "beimu" (Petriü et al. 2012) comes from traditional Chinese medicine and alludes to the emergence of fritillary plant bulbs. Considering the high starch content of *Fritillaria* species, they are also useful as food industry starch sources (Wang et al. 2017).

It is now clear that molecular biology plays a critical role in the identification of medical plant drugs, particularly in the selection and use of appropriate screening assays targeted at physiologically relevant molecular targets (Al-Snafi 2019). Certain species of *Fritillaria* exhibit a remarkable characteristic of possessing genomes of exceptionally large size. The genomes of all species examined thus far have been found to exceed 1C, equivalent to 30 gigabases (Gb). This is approximately 190 times larger than the genome of the model plant *Arabidopsis thaliana*.

In addition, the formation of such massive genomes is partly attributable to the lack of recent whole-genome duplication. Of all the diploid individuals of the genus *Fritillaria*, the Japanese endemic species in the subgenus *Japonica* have the largest genomes, at over 85 Gb (Day et al. 2014). DNA markers have been shown to be an effective scientific technique for conserving the *Fritillaria* population by revealing its genetic diversity and elucidating its unique structure. Studies have demonstrated that the phylogenetic evidence provided by the nuclear internal transcribed spacer (ITS) and various sections of the plastid genome (trnL-trnF, matK, rbcL, and rpl16) is relatively weak. Nonetheless, despite their limitations, these markers continue to be utilized for species classification. The chloroplast (CP) genome has been studied extensively for its potential to reveal molecular markers that are more sensitive and useful in tracking evolutionary relationships. These markers have been used as universal plant DNA barcodes; examples include trnH-psbA, matK, and rpl16. Presently, the GenBank repository harbours a collection of 23 chloroplast (CP) genomes belonging to the *Fritillaria* genus. These CP genomes hold promise to better understand the evolutionary connections between these plants and to discover molecular markers between them (Zhang et al. 2021).

4.2 Genetics

The utilization of medicinal plants by humans for various purposes such as food, medicine, and other applications has been a common practice throughout history. The utilization of conventional biotechnological methodologies has been implemented in the cultivation of medicinal flora. It is now time to put these valuable plants to use to expedite biotechnology-based breeding methods (BBBMs). The major strategies through which genetics and biotechnology may help improve medicinal plants more rapidly are genetic diversity evaluation, conservation, proliferation, and overproduction (Jan and Abbas 2018). It is critical to employ plant tissue culture (PTC) as a basis for future BBBM applications in medicinal plants. Artificial polyploidy induction and *Agrobacterium*-mediated gene transformation are the two basic BBBMs that depend on PTC directly. The targeted modulation of secondary metabolite pathways in medicinal plants can be achieved through the controllable regulation of endogenous and/or transferred genes using tailored zinc-finger proteins or transcription activator-like effectors. The utilization of next-generation sequencing techniques holds great promise for the investigation of genetic diversity in medicinal plants. The restriction site-associated DNA sequencing (RAD-seq) method can be employed to this end, while transcriptome profiling (RNA-seq) can be utilized to accurately identify the genes and enzymes involved in the biosynthesis of secondary metabolites (Kapoor et al. 2021). The development of custom-designed medicinal plants is a promising area of research in the field of genome editing. Various approaches have been explored for this purpose, including Clustered Regularly Interspaced Short Palindromic Repeats-Associated (Cas) technology, Zinc-Finger Nucleases, and Transcription Activator-Like Effector Nucleases (TALENs). These techniques offer the potential to precisely modify the genetic makeup of plants, resulting in the creation of novel and highly targeted modifications. The utilization of advanced targeted genome editing techniques holds great promise in the field of plant synthetic biology and may offer novel opportunities for the integration of medicinal plants into relevant industries (Canter et al. 2005). *F. cirrhosa*, *Fritillaria thunbergii* Miq., *F. imperialis* L., *F. pallidiflora* Schrenk, *F. walujewii*, *F. ussuriensis*, *F. unibracteata*, *F. tubiformis*, *F. hupensis*, and *F. verticillate* are among the important species in the genus *Fritillaria* (Li et al. 2018). The chromosomal counts of more than 50 *Fritillaria* species have been reported. Despite reports of $n = 9$ (3 species), $n = 11$ (2 species), and $n = 13$ (2 species), the majority of species have $n = 12$ as their basic chromosomal number (Ahmadi-Roshan et al. 2016). It is made up of ten rod-shaped (I) (subtelocentric and acrocentric) chromosomes and two V-shaped (metacentric and submetacentric) chromosomes. That was another element leading to the genus's scarcity of taxonomic studies so far (Zaharof 1989).

In the eighteenth century, the conception of morphology- or phenotype-based genetic markers started, which may visually distinguish traits such as seed form, flower colour, and other key agronomic aspects without the need for professional molecular approaches. The number of advantages of morphological markers is limited and influenced by plant development stages and other environmental variables. The morphological indicators also have disadvantages such as dominance, low polymorphism, epistasis, pleiotropy, and a significant environmental influence, making it difficult to identify the best-connected gene for selection (Collard et al. 2005). Biochemical markers, commonly referred to as isozymes, are a group of enzymes that exist in multiple molecular forms. These forms are encoded by distinct genes, yet they share identical functions. PCR-based molecular markers have been effectively employed for the identification of genetic diversity, population structure, gene flow, etc. However, these markers possess certain limitations such as the absence of appropriate enzyme loci, limited ability to detect polymorphism, challenges associated with diverse extraction techniques, and expression patterns that are dependent on the season and stage. As a consequence, molecular or DNA-based markers were developed to overcome the limitations of morphological and biochemical markers. These markers indicate nucleotide differences at the genomic level. These polymorphisms are characterized by insertions, deletions, point mutations, duplications, and translocations. These markers are limitless in number and unaffected by environmental factors or developmental stages and may be detected in any tissues independent of their growth, differentiation, development, or state of defence (Collard et al. 2005; Mondini et al. 2009). SSRs (simple sequence repeats) markers are 1–6 nucleotide tandem repeat patterns found in the genomes of many different species. They have become increasingly popular markers for investigating genetic relatedness across accessions and evaluating genetic diversity within a collection because of their abundance in eukaryotic genomes, co-dominant inheritance, high rate of polymorphism, repeatability, and relative ease of analysis (Schlotter 2004). SSRs are widely recognized for their potential in various scientific fields such as genome mapping, population genetics, conservation biology, and marker-assisted selection. They offer advantages such as automation potential and codominant inheritance, as reported by Zheng et al. (2012). However, the population genetics and molecular phylogeny of *Fritillaria* species remain largely understudied. To address this, different molecular markers have been developed, including RFLP (Idrees and Irshad 2014), RAPD (Kibria et al. 2009), ISSR (Sofalian et al. 2008), AFLP (Johnson et al. 2007), SSR (Tigano et al. 2010), and SCAR (Tigano et al. 2010). These markers have been employed to assess the genetic diversity and population structure of *Fritillaria* species, such as *F. thunbergii* Miq., revealing significant genetic differences among populations (Liu et al. 2010; Xu et al. 2010; Li et al. 2011; Da-Cheng et al. 2013; Mucciarelli et al. 2014). Similarly, AFLP markers were utilized to study nine *F. cirrhosa* populations, highlighting notable genetic diversity and minimal genetic differentiation (Wu et al. 2020). Furthermore, these molecular approaches have facilitated investigations into the phylogenetic relationships and species evolution of *Fritillaria* species, as demonstrated by Türktaş et al. (2012) (Table 4.1 and Fig. 4.1).

A comprehensive understanding of a plant species' evolutionary history relies on the assessment of its genetic diversity and its spatial distribution. This knowledge is crucial for the development of efficient conservation and management strategies (Ge et al. 1997; Francisco-Ortega et al. 2000; Shah et al. 2008). Thorough investigations in population genetics are essential for addressing specific concerns in conservation biology, such as the loss of genetic diversity and the restoration of vulnerable populations (Hamrick 1996). The ability of a species to adapt to changing environments and population changes may worsen in the short and long term as a consequence of genetic variation loss (Milligan et al. 1994; Reisch et al. 2003). Geographically wide species often have greater genetic diversity than species with limited distribution, while rare plants frequently have low levels of genetic variation (Vrijenhoek 1985; Hattemer 1991; Karron 1991; Hamrick 1996; Li et al. 2002). Yet, even despite their very limited habitats, numerous endangered species demonstrated considerable genetic diversity (Kang and Chung 2000; Nakagawa 2004; Ellis et al. 2006). The AFLP data population genetic analysis revealed that *F. cirrhosa* from China had significant genetic diversity at the population and species levels. Overharvesting of natural resources, in general, may result in a considerable decrease in population sizes and numbers, with isolated patches of extinction, which may damage an organism's ability to adapt to changing environmental circumstances and evolutionary potential. It has been shown that genetic diversity and its distribution within and across populations are inextricably linked to breeding techniques, seed dispersion, and geographic distribution (Hamrick 1996; Rajeswara Rao 2016). Seed dispersion by wind and water is critical for sustaining gene flow in *F. cirrhosa*, a perennial outcrossing plant. RAPD markers were employed to evaluate the genetic diversity of 16 populations of *F. thunbergii* Miq. (Xiaye, Kuanye, and Duozi), three of which were distributed in the Zhangshui, Yinjiang, Longguan, and Yinzhou Districts of Zhejiang Province (Yizhe et al. 2012). Madosa et al. (2016) evaluated the genetic diversity of Romania's *F. meleagris* species. Although the population is diminishing and the geographic range is narrowing, plants were classified using the average cluster technique. The plants studied were classified into four primary clusters, with an average similarity of 61%. The first cluster was made up of plants that had an average degree of similarity of 68%. The second cluster was

Table 4.1 Application of molecular markers to evaluate genetic diversity of *Fritillaria* species

Sr. no.	Species	Region	Molecular techniques	Outcome	References
1.	*F. thunbergii* Miq. (16 populations of 3 varieties Xiaye, Kuanye, and Duozi)	Ningbo of Zhejiang Provinces, Zhangshui, Yinjiang, Longguan, and Yinzhou District (China)	RAPD markers	Gene flow between populations were very low	Yizhe et al. (2012)
2.	*F. meleagris*	"Meadow Pogonici" Romania	RAPD markers	Population divided into four major clusters based on the similarity	Madosa et al. (2016)
3.	Five *Fritillaria* species (*F. thunbergia* (4 cultivars), *F. cirrhosa*, *F. anhuiensis*, *F. thunbergii* var. *chekiangensis*, and *F. pallidiflora*)	China	RAPD markers	*F. cirrhosa* was more closely related to *F. thunbergii* than to *F. pallidiflora*, while *F. thunbergii* var. *chekiangensis* and *F. thunbergii* cultivars had a close relationship	Lu et al. (2009)
4.	19 *Fritillaria* populations	Sichuan Province (China)	ISSR markers	Four original species of Bulhus *F. cirrhosae* listed in the Chinese Pharmacopoeia and other species of the *Fritillaria* from Sichuan province showed interspecific identification and connections between *Fritillaria* genus cluster and geographic distribution	Li et al. (2009)
5.	Twenty-two ornamental *Fritillaria* taxa	Mediterranean and East Asia	ISSR markers	Population divided into four major clusters based on the similarity	Wu et al. (2010)
6.	*F. cirrhosa* and *F. anhuiensis* and four landraces of *F. thunbergia*	Nine different locations in China	ISSR markers	High amount of genetic variety in *F. thunbergii* at the species level but a low level of genetic diversity at the landrace level	Li et al. (2011)

Table 4.1 (continued)

Sr. no.	Species	Region	Molecular techniques	Outcome	References
7.	*F. thunbergii* var. *chekiangensis* and horticultural species (Daye Duozi), *F. przewalskii*, and *F. anhuiensis*	Different locations in China and Zhejiang Province	ISSR markers	Due to artificial selection made during the cultivation of herbs, *F. thunbergii* var. chekiangensis was varied from *F. thunbergii* whereas *F. thunbergii* (Daye) and *F. thunbergii* (Duozi) were almost identical	Liu et al. (2010)
8	6 representative populations of *F. thunbergia*	Zhejiang Province (China)	AFLP markers	Due to the lack of considerable genetic difference across the populations, *F. thunbergii* was shown to have high intra-population diversity	Xu et al. (2010)
9	12 *Fritillaria* species	Turkey	AFLP markers	Population divided into three major clusters based on the similarity	Metin et al. (2013)
10	140 genotypes from seven different populations within and across populations of *F. imperialis*	Iran's mountainous Zagros area	AFLP, ISSR, and RAPD markers	AFLP is most useful marker for examining genetic diversity and determining how *F. imperialis* populations from various geographical areas (Iran)	Badfar-Chaleshtori et al. (2012)
11	Evolutionary connections of *F. cirrhosa* and closely related species	China	AFLP markers	More endemic species typically exhibited lower genetic diversity than widely dispersed one, and *F. cirrhosa* populations in the centre Hengduan Mountains were probably better protected	Wu et al. (2020)

made up of three subclusters with a total diversity of 32%. These findings show that the population had genetic variety, which may remain even if the population declines. *Fritillaria thunbergii* encompasses five species and four cultivars, namely, Duozi (abundant seed), broad-leaf, narrow-leaf, and Lingxia (*F. thunbergii*, *F. cirrhosa*, *F. anhuiensis*, *F. thunbergii* var. *chekiangensis*, and *F. pallidiflora*). Subsequently, nine polymorphic primers were employed for RAPD amplification.

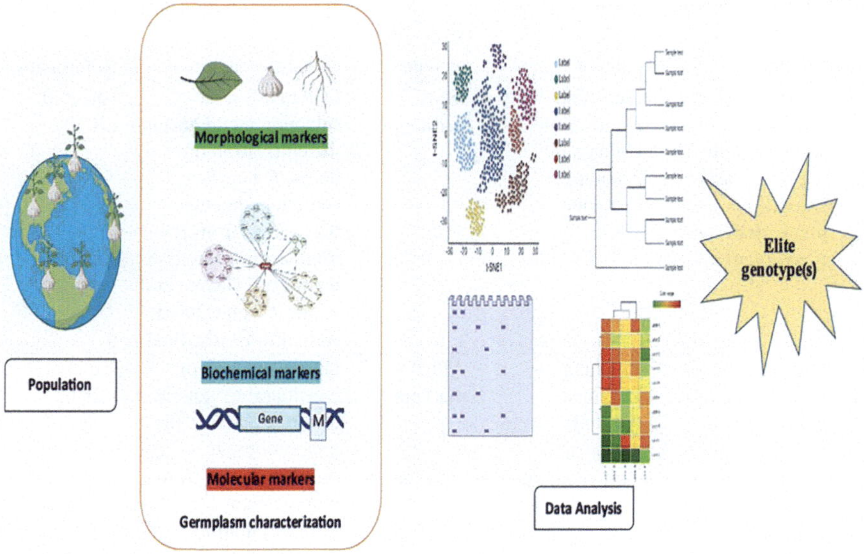

Fig. 4.1 Overview of markers process to evaluate genetic diversity of *Fritillaria* species

This yielded 48 polymorphic bands or 55.7 amplicons, constituting 84.21% of all bands. Notably, Duozi, the fourth cultivar of *F. thunbergii*, exhibited distinct characteristics from the previous three cultivars. Lingxia displayed greater similarity to the broad-leaf variant. *F. cirrhosa* demonstrated a closer relationship to *F. thunbergii* than to *F. pallidiflora*, whereas *F. thunbergii* var. *chekiangensis* exhibited a close association with *F. thunbergii*. The screened RAPD markers hold potential for discriminating between various *F. thunbergii* cultivars and *Fritillaria* species. Additionally, RAPD analysis was employed to explore the possible parentage of *F. wabuensis* in relation to Chuan Bei Mu. The taxa were classified into four groups, with one group comprising *F. taipaiensis*; *F. delavayi*, one sample of *F. cirrhosa* var. *ecirrhosa*; and one sample of *F. cirrhosa*, which is sister to the group consisting of *F. wabuensis*. In Turkey, the taxonomic status of 42 *Fritillaria* species was examined using RAPD-PCR and seed protein analyses. The study focused on the sister plants *F. imperialis* and *F. persica*, as well as the closely related *F. acmopetala* subsp. *acmopetala* and *F. sororum*, which displayed morphological, RAPD, and protein similarities. Consequently, these two taxa could be considered synonymous. Furthermore, *F. zagrica*, *F. caucasica*, *F. baskilensis*, *F. armena*, and *F. pinardii* were clustered together, suggesting their potential synonymy based on morphological and genetic data (Lu et al. 2009; Li et al. 2010; Celebi et al. 2008).

In Sichuan Province, the application of inter-simple sequence repeat (ISSR) analysis was employed to investigate genetic variation and interconnections among 19 distinct populations of *Fritillaria* (Li et al. 2009). Out of the 35 ISSR primers initially tested, 11 were selected, resulting in the amplification of 179 DNA fragments from the 19 populations. The collected samples were classified into four groups. Additionally, ISSR analysis was utilized to classify 22 ornamental *Fritillaria* taxa, encompassing species originating from the Mediterranean and East Asia (Wu et al. 2010). Thirteen primers were used to amplify DNA, yielding a total of 160 polymorphic bands. Subsequently, the samples were categorized into two groups. Notably, within the smaller group, *F. davidii* of subgenus *Davidii* exhibited a closer genetic resemblance to *F. anhuiensis* than to *F. camtschatcensis* of subgenus *Liliorhiza*. Within the larger group, three subdivisions were observed: (1) a subgroup consisting of Mediterranean taxa such as *F. acmopetala* ssp. *wendelboi*, *F. acmopetala*, *F. elwesii*, and *F. uvavulpis* and (2) a subgroup comprising the source plants for Yi Bei Mu, Zhe Bei Mu, and Chuan Bei Mu, all belonging to the subgenus *Theresia*. Based on studies utilizing random amplified polymorphic DNA (RAPD) analysis (Celebi et al. 2008) and phylogenetic data obtained from the internal transcribed spacer (ITS) and matK sequences, it was determined that *Fritillaria* subgroup 1 shared a closer relationship with subgroup 2 than with subgroup 3. The genetic diversity and differentiation of two closely related species, *F. cirrhosa* and *F. anhuiensis*, as well as four *F. thunbergii* landraces, were investigated using ISSR markers (Li et al. 2011). *F. thunbergii* displayed considerable genetic variation at the species level but exhibited a lower level of variation among landraces. Previous ISSR (Liu et al. 2010) and RAPD (Lu et al. 2009) analyses indicated that *F. anhuiensis* showed a closer genetic affinity to *F. cirrhosa* than to *F. thunbergii*. Interestingly, *F. thunbergii* cultivar Duozi, along with *F. thunbergii* vars. *chekiangensis*, *anhuiensis*, and *przewalskii*, occupied basal positions in the UPGMA tree, suggesting the possibility of Duozi being a hybrid taxon that emerged after extensive cultivation. The genetic diversity of *F. thunbergii* populations from Pan'an, Zhejiang Province, a primary production area, was comparatively lower than that of fritillary populations from other regions (Liu et al. 2010). However, ISSR markers revealed significant genetic diversity and heterogeneity within *F. thunbergii* (Zhou and Wang 2012). An analysis combining ISSR molecular markers and 32 morphological traits was conducted to assess the genetic diversity of native Crown Imperial (*F. imperialis* L.) populations in the Zagross region. A total of 160 samples from eight Zagross communities across six regions were collected and analysed. Six primers were utilized, resulting in the generation of 57 bands, of which 49 were polymorphic. The average polymorphism percentage was 85.6%, with an average of 8.17 polymorphic bands per test unit. The polymorphism information content (PIC) value averaged at 0.31 (Momenei et al. 2013). Furthermore, the genetic diversity and differentiation of *F. cirrhosa* and *F. anhuiensis*, as well as four landraces of *F. thunbergii*, were investigated using ISSR markers. The results revealed significant genetic diversity at the species level for *F. thunbergii* but relatively limited diversity at the landrace level. A notable degree of genetic differentiation among the four *F. thunbergii* landraces was observed using gene differentiation coefficient and analysis of molecular variance (AMOVA), indicating minimal gene flow between landraces. MS-tree analysis revealed that the four *F. thunbergii* landraces clustered closely, whereas Kuanye (Ft-KY) exhibited distinct genetic characteristics compared to other landraces. Principal coordinate analysis (PCoA) supported the MS-tree analysis, confirming the existence of two groups and three subgroups within the three *Fritillaria* species, exhibiting substantial genetic diversity between them. *F. thunbergii* demonstrated significant genetic diversity according to ISSR markers, with high genetic divergence observed among landraces. *F. cirrhosa* and *F. anhuiensis* displayed a close genetic relationship (Li et al. 2011). Notably, the ISSR markers failed to differentiate between the four original species of Bulbous *F. cirrhosa* mentioned in the Chinese Pharmacopoeia and other *Fritillaria* species from the Sichuan region. Additionally, associations were found between the genetic clustering of *Fritillaria* species and their regional distributions. Amplified fragment length polymorphism (AFLP) analysis involves digesting genomic DNA with restriction enzymes, attaching adaptors to the resulting fragments, and selectively amplifying a subset of fragments using primers specific to the adaptors and some of the restriction site fragments. The amplified fragments can be visualized through autoradiography (Zhang et al. 2010) or fluorescence methods using denaturing polyacrylamide gels (Xu et al. 2010). AFLP analysis was employed to assess the genetic diversity of *F. thunbergii* using 32 individuals from 6 representative populations (Xu et al. 2010). In contrast to the absence of significant genetic differences across populations, *F. thunbergii* was shown to exhibit substantial intra-population genetic diversity. Moreover, the genomic structure and diversity of *F. cirrhosa* in southwest China were studied using AFLP (Zhang et al. 2010). Using two primer combinations, 148 reproducible, clear, and polymorphic bands with diameters ranging from 120 to 750 bp were produced. Populations originating from the Hengduan Mountains exhibit greater genetic diversity compared to populations found at the eastern edge of the distribution range. The majority of variety (72.29%) was discovered within populations, but gene flow across groups is often fast. For thousands of years, wild *F. cirrhosa* has been harvested as an important medicinal plant, but the past 30 years have witnessed the greatest pressure. Overharvesting may result in a loss of genetic diversity in wild populations. The amplified fragment length polymorphism (AFLP) method was employed to investigate the genetic relationships among 12 *Fritillaria* species collected from diverse locations in Turkey (Metin et al. 2013). The study evaluated seven different primer pair combinations. Taxonomic identification of the species was accomplished through the application of neighbour-joining and principal coordinate analysis techniques. The results obtained from these analyses were consistent both with each other and with previous studies. While the neighbour-joining analysis categorized the species into three distinct groups, the principal coordinate analysis did not support the separation of the third group. Overall, the data strongly indicated that the subgenera *Fritillaria* and *Petilium* had diverged. Utilizing the neighbour-joining tree, the members of the subgenus *Fritillaria* were further classified into two subgroups. Statistical analysis of the AFLP data using NTSYS 2.1 yielded an r-value of 0.91, indicating a highly accurate match between the data matrix and the cophenetic matrix. This study successfully demonstrated the potential of AFLP for investigating genetic links among *Fritillaria* species by generating a comprehensive polymorphic band profile. Consequently, AFLP analysis offers a valuable tool for obtaining a deeper understanding of the genetic relationships between *Fritillaria* species. *Fritillaria imperialis* thrives at high altitudes in Iran's mountainous Zagros region and is prized for its medicinal and decorative properties. Badfar-Chaleshtori et al. (2012) carried out the study using AFLP, ISSR, and RAPD markers to identify genetic diversity within and across *F. imperialis* populations. 140 genotypes from 7 different populations were tested for a genetic organization using 6 randomly chosen primers, 6 ISSR primers, and 5 different AFLP primer combinations. Nei's genetic diversity index and Shannon's information index were used to estimate the percentage of polymorphic loci in multiple populations of *F. imperialis*. The analysis revealed moderate levels of genetic variation within these populations. The coefficient of gene differentiation (Gst) between populations was calculated and found to be 0.415, 0.47, and 0.63, indicating that 58.5%, 52.3%, and 36.1% of the genetic diversity was observed, respectively. The AMOVA analysis confirmed that there was significant variation between populations of *F. imperialis*. Assessing gene flow using RAPD, ISSR, and AFLP markers indicated low levels of gene flow, with values of 0.70, 0.54, and 0.29, respectively. Overall, the AFLP marker proved to be the most effective in evaluating genetic diversity and distinguishing between *F. imperialis* populations from different geographic regions within Zagros, Iran. These findings suggest that several management techniques, including maintaining adequate population numbers, creating in situ conservation areas, preserving seeds ex situ, and domesticating this wild plant species, are required to preserve the natural population of *F. imperialis* (Badfar-Chaleshtori et al. 2012). In Wu et al. (2020) study, amplified fragment-length polymorphism markers were employed to analyse the genetic diversity, population dynamics, and evolutionary relationships of *F. cirrhosa* and closely related species. The findings revealed that *F. cirrhosa* exhibited the highest genetic diversity at the species level (HSP = 0.2960), whereas *F. dajinensis* displayed the lowest (HSP = 0.1785). Regarding population-level analysis, *F. unibracteata* demonstrated the greatest average genetic diversity (HPOP = 0.2609), while *F. dajinensis* exhibited the lowest value (HPOP = 0.1785).

In contrast to the fringe populations (YD and QJ), the *F. cirrhosa* centre populations (HLS, DDS, and DQ) produced a high level of genetic diversity. In comparison to molecular variance analysis, which discovered only 19.15% variation between groups, genetic divergence between populations was relatively minor (0.0622–0.1561). The 30 original populations were classified into 6 groups using a UPGMA dendrogram. *F. cirrhosa* was a member of a multi-lineage cluster that was relatively complicated in nature. The intricate evolutionary links between species were also validated by STRUCTURE and main component analysis. They discovered that endemic species have lower genetic diversity than widely dispersed species. Moreover, due to their significantly higher genetic diversity in comparison to marginal populations, the populations of *F. cirrhosa* in the central Hengduan Mountains likely experienced enhanced protection. This study offers valuable insights into the conservation and sustainable utilization of *F. cirrhosa* and its closely related species.

Due to rapid DNA sequencing technologies, *Fritillaria* taxonomic and phylogenetic research has used several nuclear and chloroplast DNA markers. PCR primers were used to selectively amplify 5s rDNA sequences from *F. przewalskii* and *F. unibracteata*, two traditional Chuan Bei Mu source plants, which resulted in no product amplification. ITS1 regions of nine *Fritillaria* species and one variation were sequenced (Wang et al. 2007). PCR-RFLP was employed for discrimination of four distinct Chuan Bei Mu species, as documented in the 2005 Chinese Pharmacopoeia, in contrast to other medicinal *Fritillaria* species. SmaI restriction enzyme was utilized to detect a specific mutation site located within the ITS1 region, which is shared among *F. cirrhosa* and other Chuan Bei Mu source species. This method could also be utilized with *F. wabuensis* and *F. taipaiensis*, which the 2010 Chinese Pharmacopoeia designates as Chuan Bei Mu source plants (Xu et al. 2010). ITS2 sequencing was used as a DNA barcode to distinguish Chuan Bei Mu raw plants from adulterants (Luo et al. 2012). The ITS2 evolutionary tree differentiated Chuan Bei Mu plants from their adulterants. Intriguingly, *F. cirrhosa* and other Chuan Bei Mu source species samples didn't cluster on the neighbour-joining (NJ) tree. These taxa may be young and experiencing fast diversification and species radiation. Macromorphology, microscopic morphology, and floral features have diverged to separate species, although molecular sequences and chemical profiles have not. Based on the maximum parsimony (MP) phylogenetic tree analysis using ITS sequences (ITS1 + 5.8s rDNA + ITS2), the species that serves as the source for Chuan Bei Mu is found to be more closely related to *F. thunbergii* than to *F. yuminensis*, which is the parent species for Yi Bei Mu (Gao et al. 2012). This study utilized a limited set of ten *Fritillaria* ITS sequences. In a related study, Rønsted et al. (2005) conducted phylogenetic analyses involving 37 *Fritillaria* taxa, 15 *Lilium* species, and 6 Liliaceae outgroup taxa to establish the boundaries of the *Fritillaria* genus. Chemosystematics, a field that contributes to the discovery of pharmaceutical resources, employs biochemical differences and commonalities to classify and identify plants (Hao et al. 2012). Chemotaxonomy must be compared to genetic phylogeny and traditional morphology-based classification to rationalize quality control and verify the validity of plant materials used in therapeutic environments, research labs, and pharmaceuticals. Molecular phylogeny research is necessary to preserve and perpetuate *Fritillaria* resources. They help researchers learn about therapeutic *Fritillaria* species. The regulation of *Fritillaria* biological processes at the genomic, epigenomic, transcriptional, post-transcriptional, translational, and post-translational levels is largely unknown (Kumar et al. 2021), despite being crucial for sustainable development. The development can be found in the biological and chemical investigations of *Taxus* (Hao et al. 2011) and *Polygonum* (Hao et al. 2012) using systems biology and hence omics will drive *Fritillaria* medicinal research (Sharma et al. 2023). Nuclear and plastid DNA markers were employed to investigate the origin and evolutionary status of *Fritillaria* species (Protopopova et al. 2023). This study represents an original endeavour to elucidate the origins of *Fritillaria sonnikovae* and its evolutionary relationships with other *Fritillaria* species, utilizing nuclear (ITS) and plastid (*matK* + *rps16* + *trnH-psbA*) DNA markers. *F. sonnikovae* Freyn, along with *F. dagana* and *F. maximowiczii*, belongs to the North Asian lineage within the *Liliorhiza* subgenus. However, there is no evidence linking *F. sonnikovae* with *F. roylei*. Despite both phylogenetic analysis and morphology indicating a close relationship between *F. sonnikovae* and *F. maximowiczii*, the species' monophyly could not be definitively established. The study reveals that

F. dagana is associated with the *F. maximowiczii* + *F. sonnikovae* group. It is concluded that *F. sonnikovae* represents a light-perianth morph of *F. maximowiczii*, a species that originated in the Western Sayan region and has persisted there as a tertiary relict (Protopopova et al. 2023).

4.3 Genomics Studies

Even though angiosperms have only been evolving for a short period of time, their genome sizes have shown more variation than those of any other major group of organisms. Ecological and evolutionary factors may both act as constraints on how vast a genome may become. Larger genomes seem to be associated with plant growth and more diverse ecosystems. While the correlation between genome size and natural selection has been hypothesized, it has not been thoroughly investigated. The wide-ranging cellular and physiological impacts of large genomes could be attributed, at least in part, to selective pressures acting on crucial processes involved in genome evolution, including transposable elements and gene duplication. To unravel the intricacies of selection operating on these genomic elements, a comprehensive approach combining population genetic and comparative methods is essential (Gaut and Ross-Ibarra 2008). Genome sizes in higher plants have expanded with many folds as 2000, with the smallest being 64 megabases (Mb) in *Genlisea* (corkscrew plants) (Greilhuber et al. 2006) and the largest being 124 Gb in *Fritillaria*. Plant species characterized by large genomes exhibit correspondingly large cells and seeds, imparting significant implications for various life history traits. However, these plants with extensive genomes also exhibit reduced growth rates and lower photosynthetic efficiency and are less prevalent in challenging environments (Knight et al. 2005). It is noteworthy that a clear association exists between genome size and its corresponding evolutionary and external manifestations. In plants, the largest genome sizes tend to correspond with the lowest species diversity, suggesting that genome size has an effect on speciation rates. Genome sizes vary across members of the same species, and this diversity has been linked to natural selection. Comparing genome sizes among individuals with the same chromosome number may reveal a 40% range (Rayburn et al. 1985). Its variation among species is linked to environmental gradients and phenotypic growth characteristics, and it may be impacted by indirect selection on other aspects (Meagher and Vassiliadis 2005). There is currently a lack of information about the correlation between genome size and phenotype. If DNA content affects cell volume and replication, it may help explain why growth rates are often lower. DNA accumulation may also impact cells via mechanisms such as gene regulation and copy number variation (Meagher and Vassiliadis 2005). Nonetheless, direct selection on either the genome size itself or on the fundamental mechanisms influencing genome size may occur because of the multiple cellular and physiological advantages of having a larger genome. The majority of the increase in genome size is due to the presence of repetitive DNA, primarily transposable elements (TEs). Genomic sequence data has been used extensively to investigate several questions about TEs, including their prevalence, where they accumulate in genomes, how they are removed from genomes, and how quickly they replicate. Certain transposon insertions, especially those within the genetic code, may be very deleterious and quickly eliminate offending genes from the population (Naito et al. 2006). To fully appreciate how natural selection influences transposon diversity, however, a population genetics perspective is required. Gene duplication is also a major contributor to the expansion of the plant genome. This duplication is mostly attributable to polyploid occurrences, which result in the duplication of every gene in the genome and therefore entirely remove genetic redundancy. This process is not random, even if many duplicated genes are lost due to mutations and deletions (Blanc and Wolfe 2004). Genes implicated in transcription, signal transduction, and development display a higher tendency for duplicate retention compared to genes involved in other functional gene groups. Certain genes are more or less susceptible to dosage effects, and selection acts to maintain healthy stochiometric ratios, both of which might lead to this biased retention. Tandem duplication is an efficient alternative to conventional gene duplication. Uncertainties remain about the extent of copy number variation in plants, the role of copy number variations in local adaptation, and the contribution of copy number variants to individual genome size variation. By analysing the relative strengths of refining, balancing, and directing selection, genomic data from a population may be utilized to pinpoint genes that have been the focus of selective pressure. Eventually, explicit multi-population sampling will need to be a part of these diversity assessments so that patterns of variety can be analysed and indicators of local adaptation may be discovered (Blanc and Wolfe 2004). Genome sizes in the genus *Fritillaria* vary widely, from 30.15 to 85.38 Gbp. The nuclear genome of *Fritillaria assyriaca* is about 125 Gb in size (Suwabe et al. 2006). Unlike the nuclear and mitochondrial genomes, the chloroplast genome (CP) exhibits remarkable conservation across a wide range of angiosperms. Notably, the chloroplast genome of *Fritillaria* has garnered significant attention in recent years due to comprehensive investigations. It has been claimed that by utilizing just a few molecules, a complete CP genome may be swiftly generated (Li et al. 2016). This approach has been used to investigate the CP genomes of three different *Fritillaria* species: *F. cirrhosa*, *F. taipaiensis*, and *F. hupehensis* (Kumar et al. 2021). Researchers employed next-generation sequencing, specifically Illumina sequencing, to ascertain the complete chloroplast genome sequences of *Fritillaria ussuriensis* and *Fritillaria cirrhosa*. These genomes have respective sizes of 151,524 and 151,083 base pairs (bp) (Park et al. 2017). Yet, there has been a little in-depth study of the genetic makeup and evolutionary linkages of *F. cirrhosa* and closely related species (Wu et al. 2020) (Fig. 4.2).

A case study revealed that the evolutionary position and heterochromatin content of the majority of North American species (subgenus *Liliorhiza*) differ from those of their European counterparts. Two *Fritillaria* species, *F. affinis* (1C = 45.6 pg) from North America and *F. imperialis* (1C = 43.0 pg) from Eurasia, were chosen from each phylogeographical group based on their comparable genome sizes. Fosmid libraries were created from their genomic DNAs and employed for highly repetitive sequence clones for chromosomal localization, sequence

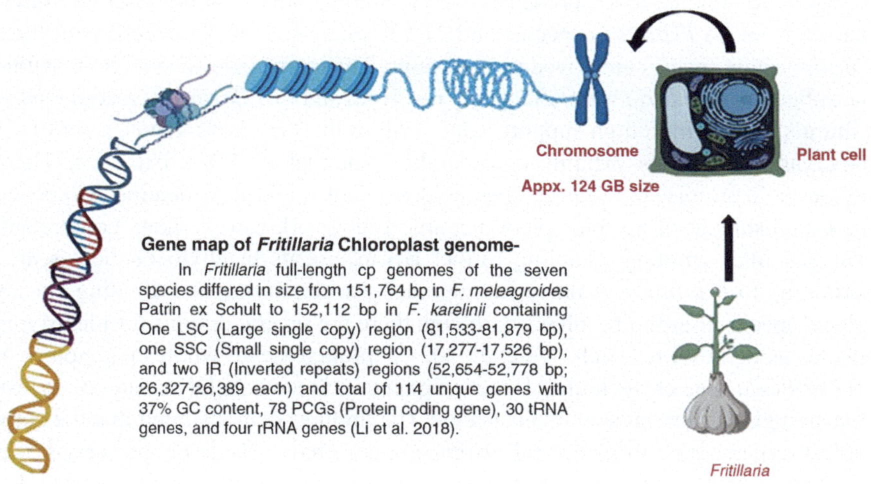

Fig. 4.2 Genome in plant cell of *Fritillaria*

characterization, and identification. Repetitive sequences, accounting for 6.7% and 4.7% of the genomes of *F. affinis* and *F. imperialis*, respectively, were detected. Within these genomes, the dominant repetitive fractions were found to consist mainly of the *Tat lineage of Ty3/gypsy* group long terminal repeat retrotransposons and chromoviruses in *F. affinis* and *F. imperialis*, respectively. Nevertheless, there was no evidence linking heterochromatin concentration to variance in genome size, even though no dominating repeats that mirrored the rising/falling patterns in the evolution of *Fritillaria's* genome size could be found in the analysis. On the contrary, the vast genomes of *Fritillaria* exhibit a diverse assemblage of transposable element families. Hence, the bigger genome may be caused by the removal processes' failure to adequately counteract the proliferation of retrotransposons (Ambrožová et al. 2011).

Huang et al. (2020) conducted a comprehensive study utilizing complete chloroplast (CP) genome sequences to investigate the phylogenetic relationships within the *Fritillaria* genus. The researchers carefully selected markers based on *Fritillaria* clades and employed these markers in their phylogenetic analysis. They scrutinized 11 recently acquired whole-plastome sequences from various *Fritillaria* species, which exhibited remarkable similarity in terms of their overall size (ranging from 151,652 to 152,434 bp), genome structure, gene richness, and gene order. Although sharing six out of ten families with other Liliales species, these *Fritillaria* plastomes exhibited structural variations resulting from the expansion or contraction of the inverted repeat (IR) regions. Among the various types of repetitive elements, A/T mononucleotides, palindromic repeats, and forward repetitions were the most abundant. Additionally, the researchers identified six highly variable regions within the 26 *Fritillaria* whole-plastomes, namely, *rps16-trnQ*, *rbcL-accD*, *accD-psaI*, *psaJ-rpl33*, *petD-rpoA*, and *rpl32-trnL*, which could serve as potential molecular markers. To conduct a comprehensive phylogenomic analysis, the plastome information from 26 *Fritillaria* species and 21 Lilium species was utilized, with three *Cardiocrinum* species employed as outgroups. The interspecies connections within the subgenus *Fritillaria* were effectively resolved, and *Fritillaria* was recognized as Lilium's sister with a high support value. Within the *Fritillaria* genus, a set of six hypervariable regions exhibits considerable potential as DNA barcodes. These regions hold promise for accurate species identification and classification. Moreover, the establishment of a robust phylogenomic framework can facilitate comprehensive genomic sampling, enabling further advancements in phylogenetic research pertaining to *Fritillaria.* A universal approach for classifying and dividing closely related species is still difficult, even though there are several options for identifying species using DNA research. Wu et al. 2021 exploited the 26 whole cp genomes of 10 *Fritillaria* species, including 18 new sequences, as a whole DNA barcode to test. Repeat regions were present in all species. The findings demonstrated that the chloroplast (cp) genomes of medicinal *Fritillaria* plants have effectively preserved their genome structure, gene types, and gene content. Comparison analysis revealed that the intergenic spacer regions of the *Fritillaria* cp genomes were more divergent than other regions. Individuals of each *Fritillaria* species formed a monophyletic clade when the phylogenetic tree was created using the maximum likelihood and maximum parsimony techniques, demonstrating the full cp genome's high degree of discriminating for these species (Wu et al. 2021). The People's Republic of China Pharmacopoeia states that Chinese herbal remedies use the bulbs of 11 distinct *Fritillaria* species. Yet, using standard morphological classification techniques, it is difficult to discern between closely related species of *Fritillaria.* As a result, broad molecular markers were used, although there wasn't enough phylogenetic data to support them. Comparisons were conducted on the complete chloroplast genomes of eight distinct *Fritillaria* species in relation to others. The sizes of these *Fritillaria* chloroplast genomes ranged from 151,009 to 152,224 base pairs. Through this analysis, a total of 136 SSR loci were identified, with 124 of them exhibiting polymorphisms. Furthermore, 108 repeat loci encompassing 4 distinct repeat types were detected, along with 10 highly variable regions that possess potential as molecular markers. The utilization of these SSRs, large repeat sequences, and highly variable regions holds great promise for genetic marker development and DNA fingerprinting purposes. Phylogenetic investigations revealed that all datasets displayed well-resolved and congruent topological structures, except for the IR regions (Bi et al. 2018).

Xinjiang, one of China's two biodiversity hotspots, is home to eight *Fritillaria* species, two of which are indigenous. The evolutionary positioning of Xinjiang *Fritillaria* species, including *F. yuminensis*, within the genus's evolutionary tree remains unclear. To address this, the chloroplast (cp) genomes of seven *Fritillaria* species in Xinjiang were analysed using Illumina HiSeq technology. The primary objectives of this analysis were to identify highly variable DNA sequences and assess global structural changes within the genomes. Phylogenetic research and stigma development examined the Xinjiang species' relationships. A total of 151,764–152,112 bp were the sizes of the seven quadripartite cp genomes. Gene composition and order were the same in the seven cp genomes. The 13 cp genomes showed highly conserved structure. Ten strongly dissimilar locations may be beneficial for phylogenetic and population genetic studies. The 13 *Fritillaria* species evolutionary connections were well-defined by protein-coding genes, major single-copy, small single-copy, and inverted repeat regions. As the genus is non-monophyletic, species with undivided stigmas may have evolved several times (Li et al. 2018). Similar to this, the southwest Chinese alpine *Fritillaria cirrhosa* D. Don species undergoes complex morphological variations depending on the distribution. Many new taxa were found via cryptic physical changes. However, the conventional taxonomy system presents challenges in determining and elucidating the evolutionary connections of these species. In the case of *F. cirrhosa* and its closely related species, the utilization of next-generation sequencing (NGS) has enabled the identification of eight plastid genomes ranging from 151,058 to 152,064 base pairs, each containing 115 genes. The GC concentration, gene content, gene order, and architectures of the IR/SC boundaries were found to be similar across these genomes. The counts of five large repeat sequences and three SSRs varied from 63 to 75 and 73 to 79, respectively. *Fritillaria* genetic markers were identified in six quite different locations. Three clades supported by phylogenetic research revealed eight *Fritillaria* species. *F. cirrhosa*, *F. przewalskii*, and *F. sinica* are closely related (Chen et al. 2019). The Illumina sequencing technique was employed to sequence and assemble the complete chloroplast (CP) genomes of four *Fritillaria* species: *F. unibracteata*, *F. przewalskii*, *F. delavayi*, and *F. sinica*. The CP genomes of *F. unibracteata* and *F. przewalskii* were found to be 151,076 and 152,043 base pairs in length, respectively. These CP genomes exhibited a quadripartite structure consisting of two inverted repeats (26,078–26,355 bp) and large and small single-copy sections (17,537–17,569 bp) separating them. The species *F. przewalskii*, *F. delavayi*, and *F. sinica* shared 133 unique genes, including 38 transfer RNA genes, 8 ribosomal RNA genes, and 87 protein-coding genes. However, *F. unibracteata* had only 132 genes as it lacked the rps16 gene. Subsequent comparative analysis of the complete CP genomes revealed that certain molecular markers, namely, *ycf1*, *trnL*, *trnF*, *ndhD*, *trnN-trnR*, *trnE-trnT*, *trnN*, *psbM-trnD*, *atpI*, and *rps19*, exhibited significant interspecies variations, making them valuable tools for taxonomic research. Three *Cardiocrinum* and five Amana species were employed as outgroups in a phylogenomic investigation of 53 *Fritillaria* and *Lilium* species' comprehensive CP genome data. The phylogenetic investigation showed that *Lilium* and *Fritillaria* were sisters, and the interspecies linkages within subgenus *Fritillaria* were well characterized (Zhang et al. 2021). Four plastid areas (*ycf1*, *matK-trnG-GCC*, *rpoC1*, and *matK*) have higher resolution than others; however only matK may be relevant for *Fritillaria* species identification due to its length. Phylogenomic analysis divided China's *Fritillaria* subgenus into four primary clades with the obvious regional organization. Clade I, a complex and young group from southwest China, the Middle to Late Miocene geological and climatic variables affected the formation and divergence of the subgenus *Fritillaria* and its four core clades, according to molecular dating (Chen et al. 2022). The plastid genome of 117 *Fritillaria* individuals from 92 species, which accounts for 66% of the genus, and 9 other Liliaceae genera have been subjected to sequencing for three regions. A pooled plastid dataset phylogenetic research using maximum parsimony and Bayesian inference validated the monophyly of most subgenera. Japan's unique subgenus *Japonica* is sister to Rhinopetalum, which is largely found in the Middle East and Central Asia and includes the diploid plant species with the largest genome size (Day et al. 2014). When polyploidy is absent, genome size differences across species are mostly due to transposable elements and tandem repeats. Yet, the relative importance of repetitive DNA deletion vs. amplification in genome size regulation is still debated. Using 454 sequencing data, an investigation into the genomic expansion of large diploid plant species, including representatives from the *Fritillaria* genus, reveals that their substantial genome size is not solely driven by recent and extensive amplification of a limited number of repeat families, as observed in species with smaller genomes. Instead, these vast genomes predominantly consist of diverse DNA sequences with relatively low repeat frequencies. This finding supports the hypothesis that repetitive DNA accumulates over time due to infrequent removal processes. The study's results indicate that the extraordinary size of these genomes cannot be solely attributed to a small subset of highly abundant repeat families, emphasizing the significance of limited deletion and inadequate turnover of repetitive DNA as principal factors contributing to their expansion (Kelly et al. 2015). A circular consensus sequencing (CCS) method based on single molecule, real-time (SMRT) DNA sequencing was utilized to de novo build chloroplast genomes and identify SNPs (SNPs). To establish a species shotgun library, chloroplast DNA was recovered from enriched chloroplasts of pooled individuals. Sequencing reactions were performed on PacBio RS. The chloroplast genome sequence was generated through a meticulous process involving the iterative alignment of each read against the draft sequence. This method, known as full-chain, PCR-free approach, effectively mitigates any biases associated with library construction and sequencing, ensuring accurate representation of the chloroplast genome. Utilizing information from a single SMRT Cell, the chloroplast genome was efficiently and comprehensively assembled. Additionally, 34.1 kb of Sanger sequencing data was obtained, which exhibited a perfect match of 100% with the three *Fritillaria* genomes, thereby providing validation. Furthermore, the analysis revealed the presence of intraspecies single-nucleotide polymorphisms (SNPs) with a minimum frequency variation of 15%. This simple approach produces high-quality chloroplast genomes that can be parallel sequenced and analysed for SNPs (Li et al. 2014). In a case study, eight intergenic spacer regions were revealed that might be universal markers. *Fritillaria* and *Lilium* have nine and seven variable intergenic spacer regions, respectively. Phylogenetic, phylogeographic, and population genetic studies in *Fritillaria* may benefit from intergenic spacer regions *rpoB-trnC*, *trnS-trnG*, *trnTpsbD*, and *trnT*-trnL (Lu et al. 2021). The complete chloroplast (cp) genome of *Fritillaria taipaiensis* was compared to NCBI Genbank's complete cp genomes of Liliaceae, Nartheciaceae, Amaryllidaceae, and Asparagaceae species to determine its characteristics, sequence divergence, and phylogenetic relationships. A total of nine genes, namely, *atpB*, *atpE*, *ndhF*, *ndhH*, *petB*, *rpl2*, *rpl20*, *rpl22*, and *yc*, have been identified to possess positive selection sites. This finding was based on a comprehensive examination of the entire chloroplast (cp) genomes and protein-coding sequences, which established *Fritillaria's* closest phylogenetic relationship to *Lilium* (She et al. 2020). The complete cp genome of *Fritillaria thunbergii*, spanning 152,155 base pairs, conforms to the typical quadripartite structure observed in other plant species. Moreover, *F. thunbergii* cp genome exhibits remarkable similarities in gene content, order, orientation, and GC content when compared to other Fritillaria cp genomes. Notably, a comparative analysis of the cp genomes of *F. thunbergii* and *F. hupehensis* highlighted the potential of two specific genes, *matK* and *rps16*, to serve as effective markers for distinguishing between different herbal medicines and five distinct *Fritillaria* species (Moon et al. 2018). *F. unibracteata* var. *wabuensis* chloroplast genome was rebuilt as a circular sequence of 151,009 base pairs using PacBio RS. 88 proteins, 37 tRNAs, and 8 rRNAs are encoded by the assembled genome's 133 genes. This genome sequencing would help identify *Fritillaria* plants and adulterants and study the genus evolution. *F. yuminensis* chloroplast genome was sequenced using Illumina Hiseq. 26,377 inverted repeats (IRa and IRb), 81,533 large single copies (LSC), and 17,526 short single copies (SSC) make up the cp genome's 151,813 base pairs (bp) (SSC). 36.9% GC was present. 30 tRNA, 4 rRNA, and 80 protein-coding genes were annotated. *F. yuminensis* and *F. verticillata* were closely related in maximum-likelihood phylogenetic research (Li et al. 2017).

Scientific harvesting in *Fritillaria* also provides new opportunities in research. In case of *F. hupehensis*, there is limited knowledge related to dynamic variations of steroidal alkaloid metabolites and their concentrations throughout bulb growth and their molecular regulation mechanisms. To systematically explore the variations in steroidal alkaloid metabolite levels and identify the genes modulating their accumulation and associated regulatory mechanisms, bioactive chemical investigations through metabolome and transcriptome profiles were carried out. At IM03 (post-withering stage, early July), the weight, size, and total alkaloid content of the regenerated bulbs peaked, whereas peiminine content peaked at IM02 (withering stage, early June). IM02 and IM03 did not differ significantly from one another, indicating that regenerated bulbs might be harvested successfully in early June or July. In comparison to IM01 (vigorous development stage, early April), the levels of peiminine, peimine, tortifoline, hupehenine, korseveramine, delafrine, hericenone N-oxide, korseveridine, puqiedinone, pingbeinone, puqienine B, puqienine E, pingbeimine A, jervine, and ussuriedine were elevated. The Kyoto Encyclopedia of Genes and Genomes enrichment analysis revealed that the accumulation of steroidal alkaloid metabolites mainly occurred before IM02. HMGR1, DXR, CAS1, CYP 90A1, and DET2 may have a positive impact on the synthesis of peiminine, peimine, hupehenine, korseveramine, korseveridine, hericenone N-oxide, puqiedinone, delafrine, jervine, and ussuriedine biosynthesis, while the downregulation in FPS1, SQE, and 17-DHCR gene may lead to a reduction in peimisine levels (Duan et al. 2023).

An endogenous pararetrovirus (EPRV) repeat was found in *Fritillaria imperialis* using the RepeatExplorer pipeline. This repeat is located in a very intricate region of the genome that is rather close to the centromere in the majority of chromosomes (Becher et al. 2014). Three linked loops were formed when the repeat was dissected into its constituent consensus sequences; one of these loops contained sequence patterns unique to EPRV (including the gag and pol domains). The phylogenetic analysis of *FriEPRV's* relationship to other pararetroviruses indicates that it is most closely related to *Petuvirus* (Becher et al. 2014). Studies have shown that EPRVs are more common than the *Arabidopsis* genome; however, it is very unlikely that a complete EPRV sequence exists. An increased frequency of C/T and G/A transitions, as measured by single-nucleotide polymorphism, is indicative of methylated cytosine deamination. Methylation levels as high as one hundred percent were found in CG and CHG motifs and averaged between 15 and 20% in CHH motifs, as determined by bisulfite sequencing. The centromeric localization of *FriEPRV* suggests a mechanism for targeted insertion associated with the meiotic drive. Several short RNAs (24 nt) were discovered that specifically target *FriEPRV*, suggesting that it may serve as a marker for RNA-dependent DNA methylation. Indications from epigenetic control markers suggest that *F. imperialis* large genome did not arise from a catastrophic breakdown in repeat amplification regulation (Becher et al. 2014). The chloroplast genome sequences of six *Fritillaria* medicinal plants were analysed using bioinformatics methods such as multiple sequence alignment, SNP screening, and restriction enzyme analysis. Chloroplast genomic DNA from *Fritillaria* showed a homology of 98.38%. Through analysis, a total of 4508 SNP loci were found; of these, 71 were potential candidates for identifying *F. cirrhosa*, 25 for identifying *F. unibracteata* var. *wabuensis*, 120 for identifying *F. taipaiensis*, 79 for identifying *F. ussuriensis*, and 61 for identifying *F. hupehensis*. The chloroplast genome SNP molecular marker has a broad variety of uses due to its high density, outstanding discriminatory power, and ease of analysis (Qimgkuo et al. 2018).

4.4 Conclusion

The escalating interest in the medicinal potential of *Fritillaria* species has led to the identification of various compounds, including alkaloids, terpenoids, and saponins. Additionally, research efforts have focused on investigating their genetic diversity, molecular phylogeny, and genomes. Despite these endeavours, a mere 80% of the total 165 *Fritillaria* species have undergone comprehensive evaluation employing phytochemical analysis, molecular biology techniques, and pharmacological investigations. This makes it difficult to regulate the standards of *Fritillaria*-related products and utilize them therapeutically. It is quite conceivable that other *Fritillaria* species will be explored that can be used in medicine as well as the identification of lead compounds for drug development. Future pharmaceutical investigations utilizing *Fritillaria* bioactive components will increasingly rely on systems biology and omics technologies.

References

Ahmadi-Roshan M, Karimzadeh G, Babaei A, Jafari H (2016) Karyological studies of *Fritillaria* (Liliaceae) species from Iran. Cytologia 81(2):133–141

Al-Snafi AE (2019) *Fritillaria imperialis*-A review. IOSR J Pharm 9(3):47–51

Ambrožová K, Mandáková T, Bureš P, Neumann P, Leitch IJ, Koblížková A, Macas J, Lysak MA (2011) Diverse retrotransposon families and an AT-rich satellite DNA revealed in giant genomes of Fritillaria lilies. Ann Bot 107(2):255–268

Badfar-Chaleshtori S, Shiran B, Kohgard M, Mommeni H, Hafizi A, Khodambashi M, Sorkheh K (2012) Assessment of genetic diversity and structure of Imperial crown (*Fritillaria* imperialis L.) populations in the Zagros region of Iran using AFLP, ISSR and RAPD markers and implications for its conservation. Biochem Syst Ecol 42:35–48

Becher H, Ma L, Kelly LJ, Kovarik A, Leitch IJ, Leitch AR (2014) Endogenous pararetrovirus sequences associated with 24 nt small RNA s at the centromeres of Fritillaria imperialis L. (L iliaceae), a species with a giant genome. Plant J 80(5):823–833

Bi Y, Zhang MF, Xue J, Dong R, Du YP, Zhang XH (2018) Chloroplast genomic resources for phylogeny and DNA barcoding: a case study on Fritillaria. Sci Rep 8(1):1184

Blanc G, Wolfe KH (2004) Functional divergence of duplicated genes formed by polyploidy during Arabidopsis evolution. Plant Cell 16(7):1679–1691

Canter PH, Thomas H, Ernst E (2005) Bringing medicinal plants into cultivation: opportunities and challenges for biotechnology. Trends Biotechnol 23:180–185

Celebi A, Teksen M, Acik L (2008) Taxonomic relationships in genus Fritillaria (Liliaceae): evidence from RAPD-PCR and SDS-PAGE of seed proteins [J]. Acta Bot Hun 50(3–4):325–343

Chang HC, Xie HM, Lee MR, Lin CY, Yip MK, Agrawal DC, Tsay HS (2020) In vitro propagation of bulblets and LC–MS/MS analysis of isosteroidal alkaloids in tissue culture derived materials of Chinese medicinal herb *Fritillaria* cirrhosa D. Don. Botan Stud 61(1):1–9

Chauhan RS, Nautiyal MC, da Silva JAT, Prasad P, Purohit H (2011) Habitat preference, ecological parameters and conservation of *Fritillaria* roylei hook. An endangered medicinal herb of the Astavarga group. Bioremed Biod Bioavailab 5(1):73–76

Chen Q, Wu X, Zhang D (2019) Phylogenetic analysis of *Fritillaria cirrhosa* D. Don and its closely related species based on complete chloroplast genomes. PeerJ 7:e7480

Chen Q, Hu H, Zhang D (2022) DNA barcoding and Phylogenomic analysis of the genus *Fritillaria* in China based on complete chloroplast genomes. Front Plant Sci 13:13

Collard BCY, Jahufer MZZ, Brouwer JB, Pang ECK (2005) An introduction to markers, quantitative trait loci (QTL) mapping and marker-assisted selection for crop improvement: the basic concepts. Euphytica 142:169–196

Da-Cheng HAO, Xiao-Jie GU, Pei-Gen XIAO, Yong PENG (2013) Phytochemical and biological research of *Fritillaria* medicine resources. Chin J Nat Med 11(4):330–344

Dad JM, Reshi ZA (2015) Influence of environmental and anthropogenic factors on the species distribution in alpine rangelands of Figurez valley, Kashmir, India. Trop Ecol 56(3):335–346

Dasgupta S, Deb DB (1986). Taxonomic revision of the genus *Fritillaria* L.(Liliaceae) in India and adjoining regions

Day PD, Berger M, Hill L, Fay MF, Leitch AR, Leitch IJ, Kelly LJ (2014) Evolutionary relationships in the medicinally important genus Fritillaria L.(Liliaceae). Mol Phylogen Evol 80:11–19

Duan Y, Wu J, Wang F, Zhang K, Guo X, Tang T, Mu S, You J, Guo J (2023) Transcriptomic and metabolomic analyses provide new insights into the appropriate harvest period in regenerated bulbs of Fritillaria hupehensis. Front Plant Sci 14:1132936. https://doi.org/10.3389/fpls.2023.1132936

Ellis JR, Pashley CH, Burke JM, McCauley DE (2006) Mol Ecol 15:2345

Francisco-Ortega J, Santos-Guerra A, Kim SC, Crawford DJ (2000) Plant genetic diversity in the Canary Islands: a conservation perspective. Am J Bot 87(7):909–919

Gairola S, Shariff NM, Bhatt A, Kala CP (2010) Influence of climate change on production of secondary chemicals in high altitude medicinal plants: issues needs immediate attention. J Med Plant Res 4(18):1825–1829

Gao YD, Zhou SD, He XJ, Wan J (2012) Chromosome diversity and evolution in tribe Lilieae (Liliaceae) with emphasis on Chinese species. J Plant Res 125:55–69

Gaut BS, Ross-Ibarra J (2008) Selection on major components of angiosperm genomes. Science 320(5875):484–486

Ge S, Zhang DM, Wang HQ, Rao GY (1997) Allozyme variation in *Ophiopogon xylorrhizus*, an extreme endemic species of Yunnan, China. Conserv Biol 11(2):562–565

Greilhuber J, Borsch T, Müller K, Worberg A, Porembski S, Barthlott W (2006) Smallest angiosperm genomes found in Lentibulariaceae, with chromosomes of bacterial size. Plant Biol 8(06):770–777

Hamrick JL (1996). Conservation genetics of endemic plant species. In: Conservation genetics: case histories from nature, pp 281–304

Hao DC, Yang L, Xiao PG (2011) The first insight into the *Taxus* genome via fosmid library construction and end sequencing [J]. Mol Gen Genomics 285(3):197–205

Hao DC, Ma P, Mu J (2012) De novo characterization of the root transcriptome of a traditional Chinese medicinal plant *Polygonum cuspidatum* [J]. Sci China Life Sci 55(5):452–466

Hattemer HH (1991) Measuring genetic variation. In: Müller-Starck G, Ziehe M (eds) Genetic variation in European populations of forest trees. Sauerlander's Verlag, Frankfurt, MA, pp 2–19

Huang J, Yu Y, Liu YM, Xie DF, He XJ, Zhou SD (2020) Comparative chloroplast genomics of Fritillaria (Liliaceae), inferences for phylogenetic relationships between *Fritillaria* and *Lilium* and plastome evolution. Plan Theory 9(2):133

Idrees M, Irshad M (2014) Molecular markers in plants for analysis of genetic diversity: a review. Eur Acad Res 2(1):1513–1540

Jan S, Abbas N (2018) Himalayan phytochemicals: sustainable options for sourcing and developing bioactive compounds. Elsevier

Johnson RC, Kisha TJ, Evans MA (2007) Characterizing safflower germplasm with AFLP molecular markers. Crop Sci 47(4):1728–1736

Kala CP (2006) Medicinal plants of the high-altitude cold desert in India: diversity, distribution and traditional uses. Int J Biodivers Sci Ecosyst Serv Manag 2(1):43–56

Kang SS, Chung MG (2000) High levels of allozyme variation within populations and low allozyme divergence within and among species of *Hemerocallis* (Liliaceae). Am J Bot 87(11):1634–1646

Kapoor B, Kumar A, Kumar P (2021) Transcriptome repository of North-Western Himalayan endangered medicinal herbs: a paramount approach illuminating molecular perspective of phytoactive molecules and secondary metabolism. Mol Gen Genomics 296:1177–1202

Karron JD (1991) Patterns of genetic variation and breeding systems in rare plant species. In: Falk DA, Holsinger (eds) Genetics and conservation of rare plants. Oxford University Press, New York, pp 87–98

Kaul MK (2010) High altitude botanicals in integrative medicine-case studies from northwest Himalaya. pp 1825

Kelly LJ, Renny-Byfield S, Pellicer J, Macas J, Novák P, Neumann P, Lysak MA, Day PD, Berger M, Fay MF, Nichols RA (2015) Analysis of the giant genomes of *Fritillaria* (L iliaceae) indicates that a lack of DNA removal characterizes extreme expansions in genome size. New Phytol 208(2):596–607

Kiani M, Mohammadi S, Babaei A, Sefidkon F, Naghavi MR, Ranjbar M, Potter D (2017) Iran supports a great share of biodiversity and floristic endemism for *Fritillaria* spp.(Liliaceae): a review. Plant Diversity 39(5):245–262

Kibria K, Nur F, Begum SN, Islam MM, Paul SK, Rahman KS, Azam SMM (2009) Molecular marker based genetic diversity analysis in aromatic rice genotypes using SSR and RAPD markers. Int J Sustain Crop Prod 4(1):23–34

Knight CA, Molinari NA, Petrov DA (2005) The large genome constraint hypothesis: evolution, ecology and phenotype. Ann Bot 95(1):177–190

Kumar P, Partap M, Ashrita Rana D, Kumar P, Warghat AR (2020) Metabolite and expression profiling of steroidal alkaloids in wild tissues compared to bulb derived in vitro cultures of *Fritillaria roylei*–high value critically endangered Himalayan medicinal herb. Ind Crop Prod 145:111945

Kumar P, Partap M, Warghat AR (2021) *Fritillaria* roylei. In: Himalayan medicinal plants. Academic Press, pp 57–66

Kumar V, Kumar P, Bhargava B, Sharma R, Irfan M, Chandora R (2023) Transcriptomic and metabolomic reprogramming to explore the high-altitude adaptation of medicinal plants: a review. J Plant Growth Regul 13:1–5

Li Q, Xu Z, He T (2002) Ex situ genetic conservation of endangered *Vatica guangxiensis* (Dipterocarpaceae) in China. Biol Conserv 106(2):151–156

Li KQ, Wu W, Zheng Y (2009) Genetic diversity of Fritillaria from Sichuan Province based on ISSR [J]. China J Chin Mater Med 34(17):2149–2154

Li Q, Chen X, Wang S (2010) Study on the relative in molecular biology among *Fritillariae* of Sichuan species [J]. West China J Pharm Sci 25(2):140–1433u

Li S, Hu K, Guo J (2011) Genetic diversity and relationship of *Fritillaria thunbergii* Miq. landraces and related taxa [J]. Biochem Syst Ecol 39(4–6):725–731

Li Q, Li Y, Song J, Xu H, Xu J, Zhu Y, Li X, Gao H, Dong L, Qian J, Sun C (2014) High-accuracy de novo assembly and SNP detection of chloroplast genomes using a SMRT circular consensus sequencing strategy. New Phytol 204(4):1041–1049

Li Y, Li Q, Li X, Song J, Sun C (2016) Complete chloroplast genome sequence of *Fritillaria* unibracteata var. wabuensis based on SMRT sequencing technology. Mitochondrial DNA Part A 27(5):3757–3758

Li Y, Zhang Z, Lv G (2017) The complete chloroplast genome of *Fritillaria yuminensis*, a rare and endangered species endemic to China. Mitochondrial DNA Part B 2(2):913–914

Li Y, Zhang Z, Yang J, Lv G (2018) Complete chloroplast genome of seven Fritillaria species, variable DNA markers identification and phylogenetic relationships within the genus. PLoS One 13(3):e0194613

Liu X, Chen C, Pan L, Chen Y, Fu C (2010) Genetic diversity of bulbus *Fritillaria*e thunbergii Miq. varieties based on ISSR markers. J Zhejiang Univ (Agriculture and Life Sciences) 36(3):246–254

Lu H, Zhu SH, Zhou SJ (2009) Specific RAPD screening of breeds in *Fritillaria thunbergii* and genetic diversity analysis of five species of *Fritillaria* [J]. J Ningbo Univ 22(1):44–47

Lu RS, Yang T, Chen Y, Wang SY, Cai MQ, Cameron KM, Li P, Fu CX (2021) Comparative plastome genomics and phylogenetic analyses of Liliaceae. Bot J Linn 196(3):279–293

Luo R, Liu B, Xie Y, Li Z, Huang W, Yuan J, He G, Chen Y, Pan Q (2012) SOAPdenovo2: an empirically improved memory-efficient short-read de novo assembler. Gigasci 1:18

Madosa E, Velicevici G, Sasu L, Ciulca A, Ciulca S (2016) Studies on genetic variability of the species *Fritillaria meleagris* in western Romania. Int Multidiscip Sci GeoConference 1:761–766

Marković M, Trifunović Momčilov M, Uzelac B, Jevremović S, Subotić A (2021) Bulb dormancy in vitro—*Fritillaria meleagris*: initiation, release and physiological parameters. Plan Theory 10(5):902

Meagher TR, Vassiliadis C (2005) Phenotypic impacts of repetitive DNA in flowering plants. New Phytol 168(1):71–80

Metin ÖK, Türktaş M, Aslay M, Kaya E (2013) Evaluation of the genetic relationship between Fritillaria species from Turkey's flora using fluorescent-based AFLP. Turk J Biol 37(3):273–279

Milligan BG, Leebens-Mack J, Strand AE (1994) Conservation genetics: beyond the maintenance of marker diversity. Mol Ecol 3(4):423–435

Momenei H, Shiran B, Khodambashi M, Cheghamirza K (2013) An assessment of the genetic diversity of Imperial crown (*Fritillaria imperialis* L.) populations from Zagross region of Iran through ISSR markers and morphological traits. Iran. J Plant Pathol 44(1):61–72

Mondini L, Noorani A, Pagnotta MA (2009) Assessing plant genetic diversity by molecular tools. Diversity 1:19–35

Moon BC, Park I, Kim WJ, Yang S, Kang YM (2018) The complete chloroplast genome sequence of Fritillaria thunbergii Miq., an important medicinal plant, and identification of DNA markers to authenticate Fritillariae Bulbus. Hortic Environ Biotechnol 59:71–80

Mucciarelli M, Ferrazzini D, Belletti P (2014) Genetic variability and population divergence in the rare *Fritillaria* tubiformis subsp. moggridgei Rix (Liliaceae) as revealed by RAPD analysis. PLoS One 9(7):e101967

Naito T, Otsuki Y (2005) The endemic *Fritillaria* species of Japan. Curtis's Botanical Magazine 22(3):189–196

Naito K, Cho E, Yang G, Campbell MA, Yano K, Okumoto Y, Tanisaka T, Wessler SR (2006) Eukaryotic transposable elements and genome evolution special feature: dramatic amplification of a rice transposable element during recent domestication. Proc Natl Acad Sci 103:17620–17625

Nakagawa M (2004) Genetic diversity of fragmented populations of *Polygala reinii* (Polygalaceae), a perennial herb endemic to Japan. J Plant Res 117:355–361

Park I, Kim WJ, Yeo SM, Choi G, Kang YM, Piao R, Moon BC (2017) The complete chloroplast genome sequences of *Fritillaria* ussuriensis maxim. And *Fritillaria* cirrhosa D. Don, and comparative analysis with other *Fritillaria* species. Molecules 22(6):982

Petriü M, Subotiü A, Trifunoviü M, Jevremoviü S (2012). Morphogenesis in vitro of *Fritillaria* spp.

Protopopova M, Sandanov D, Pavlichenko V, Selyutina I, Stepanov N (2023) The curious case of *Fritillaria sonnikovae* (Liliaceae) in South Siberia: new insights into its origin and phylogeny. Diversity 15:193

Qimgkuo L, Chen R, Zhao X, Wang C, Chen X, Wang X, Shen X, Liu N, Zhu Z, Liu Z, Wang Y (2018) Bioinformatics analysis for single nucleotide polymorphism sites of chloroplast genome of Fritillaria. J Food Saf Qual 9(17):4527–4533

Rajeswara Rao BR (2016) Genetic diversity, genetic erosion, conservation of genetic resources, and cultivation of medicinal plants. In: Genet divers eros plants. Springer, Cham, pp 357–407

Rayburn AL, Price HJ, Smith JD, Gold JR (1985) C-band heterochromatin and DNA content in Zea mays. Am J Bot 72(10):1610–1617

Reisch C, Poschlod P, Wingender R (2003) Genetic variation of *Saxifraga paniculata* mill. (Saxifragaceae): molecular evidence for glacial relict endemism in Central Europe. Biol J Linn Soc 80(1):11–21

Rønsted N, Law S, Thornton H, Fay MF, Chase MW (2005) Molecular phylogenetic evidence for the monophyly of Fritillaria and Lilium (Liliaceae; Liliales) and the infrageneric classification of Fritillaria. Mol Phylogenet Evol 35(3):509–527

Samaropoulou S, Bareka P, Bouranis DL, Kamari G (2021) Studies on the pollen morphology of *Fritillaria* (Liliaceae) taxa from Greece. Plant Biol:1–12

Schlotter C (2004) The evolution of molecular markers—a matter of fashion. Nat Rev Genet 5:63–69

Shah A, Li DZ, Gao LM, Li HT, Möller M (2008) Genetic diversity within and among populations of the endangered species *Taxus fuana* (Taxaceae) from Pakistan and implications for its conservation. Biochem Syst Ecol 36(3):183–193

Sharma S, Kumar P, Sharma R, Warghat AR (2023) In vitro propagation and omics breakthroughs for understanding specialized metabolite production in high-value Himalayan Fritillaria species. Ind Crop Prod 205:117541

She R, Zhao P, Zhou H, Yue M, Yan F, Hu G, Gao X, Zhang S (2020) Complete chloroplast genomes of Liliaceae (sl) species: comparative genomic and phylogenetic analyses. Nordic J Bot 38:1

Singh G, Rawat GS (2011) Ethnomedicinal survey of Kedarnath wildlife sanctuary in Western Himalaya, India. Indian J Fundam Appl Life Sci 1(1):35–46

Singh K, Kumar P, Kumar B, Lone JF, Sharma PR, Gairola S (2020) Morpho-anatomical and palynological standardization and DNA barcoding of *Fritillaria cirrhosa* D. Don (Syn. *Fritillaria* roylei Hook.). Plant Arch 20(2):1304–1313

Sofalian O, Chaparzadeh N, Javanmard A, Hejazi MS (2008) Study the genetic diversity of wheat landraces from northwest of Iran based on ISSR molecular markers. Int J Agric Biol 10:466–468

Suwabe K, Tsukazaki H, Iketani H, Hatakeyama K, Kondo M, Fujimura M, Matsumoto S (2006) Simple sequence repeat-based comparative genomics between Brassica rapa and Arabidopsis thaliana: the genetic origin of clubroot resistance. Genetics 173(1):309–319

Tigano M, De Siqueira K, Castagnone-Sereno P, Mulet K, Queiroz P, Dos Santos M, Carneiro RMDG (2010) Genetic diversity of the root-knot nematode Meloidogyne enterolobii and development of a SCAR marker for this guava-damaging species. Plant Pathol 59(6):1054–1061

Türktaş M, Aslay M, Kaya E, Ertuğrul F (2012) Molecular characterization of phylogeneticrelationships in *Fritillaria* species inferred from chloroplast trnL-trnF sequences. Turkish J Biol 36(5):552–560

Vrijenhoek RC (1985) Animal population genetics and disturbance: the effects of local extinctions and recolonizations on heterozygosity and fitness. In: Pickett STA, White PS (eds) The ecology of natural disturbance and path dynamics. Academic Press, London, MA, pp 265–285

Wang CZ, Li P, Ding JY, Peng X, Yuan CS (2007) Simultaneous identification of Bulbus Fritillariae cirrhosae using PCR-RFLP analysis. Phytomed 14(9):628–632

Wang D, Chen X, Atanasov AG, Yi X, Wang S (2017) Plant resource availability of medicinal *Fritillaria* species in traditional producing regions in Qinghai-Tibet plateau. Front Pharmacol 8:502

Wu XQ, Cheng HZ, Zhou YZ (2010) Analysis of genetic relationship ornamental Fritillaria germplasm resources by ISSR markers [J]. North Horti 18:157–160

Wu X, Duan L, Chen Q, Zhang D (2020) Genetic diversity, population structure, and evolutionary relationships within a taxonomically complex group revealed by AFLP markers: a case study on *Fritillaria* cirrhosa D. Don and closely related species. Global Ecol Conserv 24:e01323

Wu L, Wu M, Cui N, Xiang L, Li Y, Li X, Chen S (2021) Plant super-barcode: a case study on genome-based identification for closely related species of *Fritillaria*. Chin Med 16(1):1–11

Xu J, Zhang H, Ma X, Cai J, Dong J (2010). AFLP analysis on genetic diversity for germplasm resources of *Fritillaria* thunbergii cultivated in Zhejiang Province. Chin Traditional and Herbal Drugs

Yizhe J, Qian Z, Qin X, Zhou S, Zhang L, Zhu S (2012) Molecular characterization of genetic diversity of *Fritillaria thunbergii* Miq. varieties in native habitat. Acta Botan Boreali-Occiden Sin 32(6):1126–1132

Zaharof E (1989) Karyological studies of twelve *Fritillaria* species from Greece. Caryologia 42(2):91–102

Zhang DQ, Gao LM, Yang YP (2010) Genetic diversity and structure of a traditional Chinese medicinal plant species, *Fritillaria cirrhosa* (Liliaceae) in Southwest China and implications for its conservation [J]. Biochem Syst Ecol 38(2):236–242

Zhang T, Huang S, Song S, Zou M, Yang T, Wang W, Zhou J, Liao H (2021) Identification of evolutionary relationships and DNA markers in the medicinally important genus Fritillaria based on chloroplast genomics. PeerJ 9:e12612

Zheng JY, Wang H, Chen XX, Wang P, Gao P, Li XN, Zhu GP (2012) Microsatellite markers for assessing genetic diversity of the medicinal plant *Paris polyphylla* var. *chinensis* (Trilliaceae). Genet Mol Res 11:1975–1980

Zhou J, Wang ZH (2012) ISSR analysis of genetic diversity of *F. thunbergii* [J]. J Zhejiang Agri Sci 2:156–159

Genetic and Genomic Resources of *Bunium persicum* (Boiss.) Fedtsch

5

Sapna, Satakshi Sharma, Ramesh Chauhan, and Satbeer Singh

5.1 Introduction

Bunium persicum (Boiss.) Fedtsch is a spice that grows wild in the dry temperate parts of Central Asia, such as Egypt, Iran, Afghanistan, Kazakhstan, Pakistan, and the Northern Himalayan portions of India. It comes under the family Umbelliferae (now called Apiaceae). It is well-known for its therapeutic properties all throughout the world and goes by several names, including black cumin, kala jeera, shahi zeera, and siah zeera. There are three species of the *Bunium* genus found in India: *Bunium persicum (B. persicum)* Boiss. *B. Fedtsch*, *Bunium cylindricum* Boiss. and Hohen. Drude, and *Bunium nothum* CB Clarke., PK Mukh. Among them, *B. persicum* is naturally found in the grassy slopes of Himachal Pradesh, Uttarakhand, Jammu, and Kashmir at a wide range of altitude (1800–3500 m), while species *B. cylindricum* is mostly found in Himachal Pradesh. The other species *B. nothum* is endemic to India and occurs in Nilgiri hills, Tamil Nadu (Singh et al. 2021).

B. persicum is a high-value medicinal spice plant that is economically important. It is used for various culinary and novel food applications, such as forming bioactive films/coatings that extend the shelf life of food products (Bansal et al. 2022). In a variety of cuisine, it is mostly used as a spice and condiment. However, it is industrially important because of its strong aroma and richness in essential oils. The percentage of oil in seeds ranges from 5 to 14. As stimulants and carminatives, seeds are used for medicinal purposes in the pharmaceutical industry, for treating diarrhea, dyspepsia, fever, flatulence, stomachic hemorrhoids, hiccoughs, and antihistaminic.

Sapna and Satakshi Sharmaa contributed equally to this work.

Sapna · S. Sharma · R. Chauhan · S. Singh (✉)
Division of Agrotechnology, Council of Scientific and Industrial Research - Institute of Himalayan Bioresource Technology, Palampur, Himachal Pradesh, India
e-mail: satbeer@ihbt.res.in

V. Gahlaut, V. Jaiswal (eds.), *Genetics and Genomics of High-Altitude Crops*,
https://doi.org/10.1007/978-981-99-9175-4_5

In addition to being a source of essential oils rich in terpenoids, polyene, phenylpropanoids, and phototoxic furanocoumarins are typical of this family. Its phytochemical bioactivity and pharmacological actions have been thoroughly studied (Chahota et al. 2017). It has the potential to be used in various food systems as a natural antioxidant and preservative.

There is a drastic decline in its population in India due to overexploitation of this important spice plant making it a plant of special conservation concern in the northwestern Himalayas. The propagation of *B. persicum* through seeds takes 4–6 months for germination in its natural habitats, and the crop can only be harvested 3 years after it has been sown. A major drawback to this plant's economic production is its long life cycle. Farmers are least interested in cultivating it because of its slow growth and long crop duration. These factors have caused an alarming rate of depletion of this economically and medicinally important spice plant in its natural habitats, and the plant is now listed as an endangered species. Himachal Pradesh state in India has protected it by awarding geographical indication (GI) under GI number 432 due to its unique nature and properties (Bansal et al. 2022). Although some recent efforts have been made to domesticate it in some areas of the northwestern Himalayas, it has remained relatively limited due to the lack of available genotypes. There is only one location where it is being grown by people and that is Shong in the Kinnaur district of Himachal Pradesh. However, the overall wild and domesticated resources of *B. persicum* are currently insufficient to satisfy the needs of various consumers. As a result, this crop requires considerably greater attention at this time to ensure that farmers can utilize it sustainably and that it can be improved using scientific techniques. Therefore, in order to maximize its production and preserve the various resources pertinent to this plant species found in the western parts of the Himalayas, farmers must establish scientific and sustainable ways. The desired elite genetic stocks for commercial cultivation can be developed through breeding programs.

Recognizing the appropriate genotypes in a plant that may be used more for the beginning of breeding and other enhancement programs requires exploring and evaluating genetic diversity. An essential tool for the accurate assessment of genetic range is DNA-based markers; they are not influenced by environmental conditions. Many plant species are difficult to distinguish from one another morphologically, and in that case, using molecular markers, especially DNA markers, is necessary. These markers can be used to identify accessions, eliminate duplicates, determine the amount of genetic range, and indicate the path of domestication. Many people have started using various types of markers to study the genetic range of *B. persicum* in various parts of the globe (Hashemi et al. 2010). In Iran, cultivars of *B. persicum* with RAPD and AFLP markers, 86% and 75% polymorphism have been reported (Azimzadeh et al. 2012). On the molecule level, TIBMBA-06 and OPR-16 primers exhibited high polymorphic characteristics w.r.t. polymorphism information content and marker index values (Chahota et al. 2017). There is a severe deficit of information concerning the genetic range within the diverse populations of *B. persicum*, particularly in the northwestern Indian mountain ranges. So, this chapter aims to provide comprehensive information on important traits, selection strategy, genetic diversity, molecular markers, genetic improvements, and biotechnological interventions of *B. persicum.*

5.2 Genetic Improvement via Breeding Techniques.

The *B. persicum* has many issues: a longer juvenile period, latent seeds, a dormant vegetative phase, and a low yielding (Fig. 5.1). Classical and modern breeding techniques for genetic improvement could be a key strategy for resolving these issues and maximizing its use through extensive cultivation. Future breeding plans in *B. persicum* will thus focus on breaking seed dormancy, reducing the vegetative period, and developing high-yielding cultivars with improved nutritional attributes. Any crop improvement program must have access to genetic diversity in order to offer superior selections at the outset. Domestication and cultivation of better plant genotypes are required due to their great therapeutic potential and demand in food and pharmacology (Pezhmanmehr et al. 2009). In Iran, a few attempts to domesticate this highly prized species were made during the start of this century (Khosravi 2005). Similar research was also conducted in the northwest Himalayas; however, relatively few agricultural methods are available due to a lack of suitable agrotechnology and better genotypes. To comprehend domestication, management, and development of novel varieties, studies on variation among various populations and

Fig. 5.1 Life cycle and growth phases of *B. persicum*

genetic variation of germplasm are required (Govindaraj et al. 2015). The high-altitude cold desert regions of the Indian Western Himalayas, Iran, and Pakistan have huge natural variability of *B. persicum* (Dar et al. 2011). The first step in developing better planting material or cultivars of *B. persicum* may be to choose plants that perform better than the base population. According to some studies on *B. persicum*, seed yield is directly and strongly correlated with the number of umbels, number of umblets, number of branches, size of umbel in diameter, and tuber weight. Hence these traits under direct selection could enhance genetic gain for seed yield in *B. persicum* (Devi 2004). An investigation on 105 accessions of *B. persicum* from Himachal Pradesh, Jammu, and Kashmir's western Himalayan regions identified 19 best genotypes for further large-scale cultivation for the region (Singh et al. 2021).

5.2.1 Genetic Resources

Since ruthless collection to meet increasing demand leads depletion of genetic resources in natural habitats, the significance of genetic resources is now being acknowledged as a specific topic. The accessibility and evaluation of genetic variation are essential for improving any crop species. Finding the relevant cultivars in a crop that may be employed further for the start of breeding and other quality measures requires the exploration and evaluation of genetic diversity. Targeting apex genetic stocks for commercial production may be developed through breeding programs. The genetic diversity among the numerous populations of *B. persicum* is quite little understood, especially in the northwest Himalayan region. Similarly to this, this crop lacks data on morphological and biochemical diversity. A list of populations characterized by various chemical constituents is given in Table 5.1.

A herbarium database of diverse collections of *B. persicum* was developed at Ferdowsi University of Mashhad, Iran (Gincarlo et al. 2006). A collection of 252 different accessions from the northwestern Himalayan region showed huge morphological variation (Khan et al. 2022). A collection of different accessions from Himachal Pradesh was also conserved at Center of High Altitude Biology (CeHAB) of CSIR-Institute of Himalayan Bioresource Technology (Fig. 5.2). Also, a collection of more than 31 accessions from the Westmost parts of Himalayan ranges locales (Ladakh, J&K, and upper parts of Himachal Pradesh) were evaluated and maintained (Fayaz et al. 2022). Only 5 of the 14 populations—Tunaar, Purthi, Saptal, Chamrat, Tindi, Kukumseri, Kalpa, Rispa, Moorang, Jangi, Shong, Batseri, Chansu, and Sangla—were genetically diverse enough to be examined for phytochemical diversity. Inflorescence, seed, and branch features were the morphological traits that were most effective at spotting variation (Chahota et al. 2017). The following important morphological and yield-related traits were estimated using an analysis of variance: plant ht. (centimeters), the quantity of umblets secondary-1 umbel, number of primary branches plant-1, number of umblets primary-1 umbel, number of umbels plant-1, number of umblets tertiary-1 umbel, number of seeds primary-1 umbel, number of seeds tertiary-1 umbel, number of seeds secondary-1

Table 5.1 List of diverse populations characterized by chemical constituents in *Bunium persicum*

Sr. no.	Population	Origin	Constituents	Quantity (%)	References
1	Shong	India	γ-Terpene	21.96	Chahota et al. (2017)
			α-Terpinolene	0.24	
2	Kalpa	India	γ-Terpene	41.26	Chahota et al. (2017)
			α-Terpinolene	0.68	
3	Moorang	India	γ-Terpene	26.65	Chahota et al. (2017)
			α-Terpinolene	0.61	
4	Saptal	India	γ-Terpene	41.27	Chahota et al. (2017)
			α-Terpinolene	0.76	
			3-Menthene	0.04	
5	Chansu	India	γ-Terpene	–	Chahota et al. (2017)
			α-Terpinolene	0.27	
6	Wild	Iran	Terpinolene	0.9–0.6	Azizi and Davareenejad 2011
			Cumin aldehyde	14.1–11.8	
			γ-Terpinene	40.8–36.8	
			p-Cymene	9.5–9.4	
			Limonene	2.5–2.4	
7	Wild	Italy	Carvone	23.3	Iacobellis et al. (2005)
			Limonene	18.2	
8	Cultivated	India	Cumin aldehyde	27–34	Sofi et al. (2009)
9	Kashmir	India	p-Cymene	24–27	Sofi et al. (2009)
			γ-Terpene	25–42	
			Carvone	44.0	
10	Wild	Iran	Carvone	23.3	Hassanzadazar et al. (2018)
11	Wild	Iran	Cumin aldehyde	9.1–18.9	Talebi et al. (2018)
			Safranal	3.4–7.9	
			Cuminic alcohol	16.4–24.4	
			Limonene	3.7–6.4	
			p-Cymene	9.4–15.6	
			γ-Terpinene	29.2–49.1	
12	Wild	India	Cumin aldehyde	34.1	Thappa et al. (1991)
			γ-Terpinene	25.6–42.9	
			p-Cymene	24.0–27.8	
13	Cultivated	Iran	Cumin aldehyde	6.0–27.8	Haghirogsadat et al. (2010)
14	Wild	Iran	Cumin aldehyde	27.8	Chizzola et al. (2014)
			γ-Terpinene	9.9–46.1	

(continued)

Table 5.1 (continued)

Sr. no.	Population	Origin	Constituents	Quantity (%)	References
			α-Terpinen-7-al	19.2	
			p-Cymene	13.5	
			Limonene	5.8	
15	Breeders' population	Iran	Cumin aldehyde	16.9	Oroojalian et al. (2010)
			γ-Terpinene	46.1	
			γ-Terpinen-7-al	10.6–18.7	
			α-Terpinen-7-al	0.7–0.5	
			p-Cymene	8	
			Lidmonene	2.0	
			Linalool	0.1	
16	Cultivated	Serbia	Limonene	10.1	Samojlik et al. (2010)
17	Mysore	India	Linalool	0.06	Padmashree et al. (2007)

Fig. 5.2 Field view of (**a**) plantation and (**b**) flowered plant of *B. persicum*

umbel, and seed yield m-2 (grams). Findings show that morphological, yield, and yield-attributing traits have significant genotypic diversity. Nearly all the factors had highly significant genotypic variance. It was evident across all the examined variables that cultivar generated had significantly greater values than the other sources of variation (Dar et al. 2011). Thirty-one Kala zeera germplasm plants growing in diverse places were examined for different seed traits like length (millimeter), length/breadth ratio (millimeter), breadth (millimeter), and 100 seed weight (gram). Ten plants from each location were chosen for statistical analysis to characterize the seed trait parameters.

There was significant and widespread morphological variation among the Kala zeera germplasm. The mean length of the seed was calculated and observed to be 2.46 ± 0.24 millimeter lying in the range of 2.21 ± 0.07 millimeter (RMZ31) and 2.83 ± 0.00 millimeter (RMZ17). The range of the seed breadth turned out to be

0.46 ± 0.05 millimeter (RMZ-12) to 0.93 ± 0.04 millimeter (RMZ-5), making an average of 0.73 ± 0.18 millimeter. Length and breadth (millimeter) fell in the range of 2.44 ± 0.00 (RMZ-5) and 5.59 ± 2.32 (RMZ-31) averaging 3.72 ± 0.23 millimeter, and the average weight (gram) of 100 seeds turned out to be in the range of 0.05 ± 0.005 (RMZ 9) to 0.35 ± 0.005 (RMZ5) averaging 0.11 ± 0.005 g. Fifteen ecotypes were divided into five main groups using a dendrogram of clustering using the UPGMA method and Euclidean distance. Ecotypes in each group had the same geographic and habitat conditions, and the biplot of PC1 and PC2 confirmed this. Two ecotypes from the Yazd region (Mehriz Yazd and Masih Yazd) and one ecotype from the region of Semnan are included in the first group and have the same climatic and geographic conditions. The second group consists of ecotypes from the provinces of Isfahan, Ghazvin, and Kerman (Joopar, Alamoot, Ashaee 350, Khoro Biabanak, and Sirch), which share a similar climate and geographic habitat specification. The third group consists of samples from the Hormozgan province (Hormozgan, 351 and Hormozgan, 390) that are roughly equivalent. The next group consisted of ecotypes from Khorasan province from Mashhad University, i.e., Kalat naderi as well as one from Isfahan province. The finest ecotypes that were produced ultimately made up the final group. Saghafi and Natanz ecotypes from the province of Isfahan are included in this category. While doing the cluster analysis, the gap between the ecotypes of Mehriz and Masih was the smallest, whereas Natanz and Mashhad ecotypes had the maximum distance. Principle components analysis (PCA) decreased the 16 variables (which were originally used) to 2 main variables, which resulted for 95.83% of the total variance (Azimzadeh et al. 2012).

5.2.2 Association Among the Traits

Seed yield and its chemical constituent profile are the economic traits of this plant. Genetic gain of these target traits in a breeding program depends on the direct and indirect selection of their component traits. Association among phenotypic traits and chemical constituents leads to an effective selection strategy in any genetic improvement program. The results among different traits showed that limonene had the largest coefficients of variation (59.5), and plant height had the lowest (15.05). Fifteen character combinations out of a total of 120 revealed significant correlations.

The secondary umbels (+0.61) were the only ones, which linked positively and substantially with the number of primary umbels at the 5% probability level. At the 5% probability level, there was a significant positive correlation (+0. 51) between the number of plant height and secondary umbels. It was discovered that the number of stems significantly correlated with the γ-terpinene-7-al percentage (−0.58) and the cuminaldehyde percentage (+0.52) at a 5% probability level. The diameter of the stem was found to be negatively and highly correlated with the earliness (−0.67) and the percentage of γ-terpinene-7-al (−0.67), as well as positively and highly correlated with the plant height (+0.59) and the first stem height surface (+0.32) at the 5% probability level (−0.67). Plant height (−0.64) and stem diameter (−0.67) had a significant negative correlation with earliness at the 1% probability level. The thousand-grain weight (+0.60) and the percentage of α-pinene (+0.52) were found to be positively and significantly correlated with the oil yield at a 5% probability level. The % of cuminaldehyde exhibited a highly positive correlation (+0.52) with the number of stems at the 5% probability level and a significant negative correlation with the percentage of γ-terpinene (−0.71) at the 1% probability level. At a 5% probability level, there was a major positive correlation between the percentages of limonene and α-pinene (0.45). Both the percentage of ρ-cymene and the α-terpinene-7-al had no significant correlation with any character. At a 1% probability level, it was discovered that the proportion of γ-terpinene-7-al had a negative and significant correlation with the diameter of the stem (−0.67) (Azimzadeh et al. 2012).

According to Pearson's correlation analysis, there is a significant positive correlation between phenol and flavonoids ($r = 0.385$, $p \leq 0.05$) as well as between FRAP and flavonoids ($r = 0.606$, $p \leq 0.01$), while there is a significant negative correlation within IC50 DPPH and flavonoids ($r = 0.180$, $p \leq 0.05$) as well as between 100 seed wt. and length: breadth ($r = -0.356$, $p \leq 0.05$). Breadth and length: breadth was shown to have a highly significant negative connection ($r = 0.932$, p 0.01). Additionally, a positive but not significant correlation was discovered between length, breadth, and 100 seed wt. ($r = 0.322$ and 0.334), and a negative correlation was found, which was not statistically significant between IC50 DPPH and phenol ($r = -0.064$) (Fayaz et al. 2022).

5.2.3 Heritability of the Important Traits

Heritability estimates vary among traits by more than 23% on average. The first stem's height above the field surface had the highest heredity rate (98.7%), whereas the number of stems had the lowest heritability rate (75.4%). In addition, the heritability rate of primary umbels was 90.7%; secondary umbels was 88.7%; for the height of the plant, it was 92.9%; for the diameter of the stem, it was 95.1%; and for the weight of the grain, it was 98.4% (Azimzadeh et al. 2012). The outcomes showed that the maximum of its characteristics had heritability lying between 74.07 and 92.98%. The highest heritability evaluations were for the number of umblets in the tertiary-1 umbel (92.98%), after that the secondary-1 umbel (91.99%), tertiary-1 umbel (88.32%), secondary-1 umbel (85.30%), and the weight of the 1000 seeds (84.37%). The heredity of seed output m-2 was equally high (74.07%), but the heritability of the number of primary-1 umbel umblets was the lowest (19.12%) (Dar et al. 2011).

5.3 Genomic Resources

The first 10 years of the twenty-first century saw the quick discovery, revolutionary technological advancement, and decreasing cost of genomics technologies. In the second decade, the focus shifted to making sense of the enormous volume of genomic data, which led to advancements in reliably predicting gene-to-phenotype connections. Since *B. persicum* has numerous challenges like a longer juvenile phase, a dormant vegetative phase, and a low-yielding potential, molecular markers and advanced genomic tools could effectively be utilized in the genetic improvement of this plant. Related species share morphological characteristics and are difficult to distinguish from one another; it is challenging to characterize their morphology (Martinez et al. 2007). It resulted in the adoption of molecular markers, in particular DNA markers, to eliminate copies, distinguish accessions, DNA fingerprint vague genotypes, gauge the level of genetic range, and denote the route of breeding (Chahota et al. 2017).

5.3.1 Molecular Marker-Based Genotyping

The fact that DNA-based markers are unaffected by environmental factors makes them an important tool for the accurate estimate of genetic variation. A list of available molecular markers genotyped in *B. persicum* is given in Table 5.2. Using 32 polymorphic RAPD primers, a molecular diversity study produced 164 bands with sizes ranging from 100 to 2800 bp, and 120 of them (73.1%) were discovered to be polymorphic in the examined populations. Between all the primers, there were 3.75 on average polymorphic bands. Most of the polymorphic information content (PIC) and marker index (MI) were identified by primers OPR-16 and TIBMBA-06, respectively, on the other hand primer OPR-16 amplified a maximum of 9 bands. They discovered median polymorphic information content and marker index values of 0.36 and 1.51, respectively, indicating there is enough polymorphism in all groups examined. Other researchers used other types of markers to examine the genetic variance of different varieties of *B. persicum* in various areas in the world (Hashemi et al. 2010). Using RAPD and AFLP markers, Iranian populations of

Table 5.2 List of molecular markers genotyped in *Bunium persicum*

Sr. no.	Type of molecular marker	Number	Number of alleles detected	Size of reference set	Reference
1	AFLP (amplified fragment length polymorphism)	17	228	20	Pezhmanmehr et al. (2009)
2	RAPD (random amplified polymorphic DNA)	32	164	14	Chahota et al. (2017)
3	RAPD	36	–	15	Majeed et al. (2009)
4	RAPD	15	192	20	Pezhmanmehr et al. (2009)
5	RAPD	26	–	20	Hashemi et al. (2010)
6	SSR (simple sequence repeats)	68	101	25	Bansal et al. (2022)

B. persicum have been found to have 86% and 75% polymorphism, respectively. However, they found 73.1% overall polymorphism, which is similar to these previous findings.

By utilizing Jaccard's dissimilarity method coefficient, the combined populations of *B. persicum* were predominantly divided into two clusters in the dendrogram based on RAPD polymorphism. The Saptal, Purthi, and Thunaar populations made constituted the Saptal out-group group I, which was separate from the Shong population. In contrast, group II was made up of ten populations, including Kukumseri, Jangi, Moorang, Chamrat, Kalpa, and Rispa. The two-directional factor analysis showed that different populations were primarily scattered into two clusters, even while specimens from the same/nearby place were grouped together to a large extent. This showed consistency with the previous studies on this species from various parts of the world (Majeed and Sharma 2006; Pezhmanmehr et al. 2007; Jahansooz et al. 2007). This could be a reason because of its extended cultivation, which might have caused some beneficial allelic combinations to accumulate in the population as a result of constant selection pressure. Furthermore, grouping of the RAPD-based dendrogram differed from morphological grouping yet demonstrated greater precision than the former. There may be a variety of explanations for the discrepancies a genetic diversity details produced by morphological features and molecular markers. Environmental factors significantly impact morphological diversity, and genetic differences or similarities in morphology are not always represented (John et al. 1997). Furthermore, regardless of the genetic origins of the populations, values were adapted for particular features, which may have accrued in particular physical conditions of their environment (Steiner and los Santos 2001). Four clusters were formed by 14 populations using Bayesian clustering. With seven different specimen clusters, including Jangi, Rispa, Moorang, Kalpa, Chamrat, Kukumseri, and Sangla, cluster I was the largest. Except for the Sangla community, every population in this cluster displayed a membership rate of at least 80%. Populations from Saptal, Thunaar, Purthi, and Tindi were present in cluster II. Except for the Tindi population, all three populations displayed membership rates of at least 95%. In the Himalayan region, cluster III had the sole agricultural population. Chansu and Batseri, two populations in cluster IV, displayed 100% similarities in this population sample. It was intriguing to see that the one cluster from Shong that was cultivated developed a distinct cluster and had no genetic resemblance to the other populations. Geographic proximity and the plant's habits impact the four genetic clusters. Cluster II stood out from the other three clusters because it was completely pure. It represented the Shong population under cultivation, a tradition that dates back to ancient times. Under selection pressure and recombination events within the same population, completely new allelic combinations may have developed. When the genetic structure was analyzed geographically, it concluded that the two gene clusters (clusters I and IV) were scattered throughout the whole region under study and appeared to represent ancestral or parent gene pools from which two other gene pools (clusters II and III) had developed through means of isolation and self-hybridization occasions. Admixture between clusters (Tindi and Sangla) was noted; this could result from cross-breeding behavior between populations. It is, therefore, clear that the plant's geographic topography and dietary preferences have a significant role in determining its genetic makeup (Chahota et al. 2017).

This critically endangered medicinal plant was gathered from 15 sites in Himachal Pradesh and Jammu and Kashmir, India, and its genetic variance was checked using random amplified polymorphic DNA (RAPD) markers. There was a significant amount of genetic diversity in the accessions that were gathered, and 36 random 10-mer primers produced a total of 173 bands, 168 of which were found to be polymorphic, demonstrating a high level (97.11%) of polymorphism among the accessions studied. Dendrogram analysis and similarity measurements using UPGMA led to forming two significant clusters. All the accessions gathered from Jammu and Kashmir (Gurez, Khrew, and Harwan) were grouped in cluster I, while accessions gathered from Himachal Pradesh (Sangla, and Kalpa) were placed in the second cluster. The cluster I member with the highest divergence was accession KW-4. According to the preliminary findings of this study, there is significant genetic diversity among wild accessions of this threatened species of medicinal plant, which needs to be identified, preserved, and used for its improvement (Majeed et al. 2009).

The main determinants of any genetic improvement breeding program's success are the genetic diversity currently available, its knowledge, management, and further utilization (Akbulut et al. 2008). It is crucial to research how to measure genetic diversity at various molecular and phenotypic levels. However, there are few publications and few investigations on the genetic diversity evaluation at the DNA level in *B. persicum*. Most research on genetic diversity has relied heavily on RAPD molecular markers. High polymorphism was found across 20 Iranian populations of *B. persicum* in research by Pezhmanmehr et al. (2009) on the assessment of genetic polymorphism by 15 RAPDs (random amplified polymorphic DNA) and 17 AFLPs (amplified fragment length polymorphism). The study also noted a wide range of similarity coefficients between populations (0.4–0.82 and 0.39–0.96 for RAPDs and AFLPs, respectively). Majority of populations were grouped independently, each representing a distinct genetic background. The clustering pattern and genetic divergence showed that RAPD and AFLP, two different marker systems, were not correlated. Additionally, there was no relationship between the geographic distribution and genetic distinctness of groups (Pezhmanmehr et al. 2009; Hashemi et al. 2010) found a significant amount of genetic variability in 20 populations of *B. persicum* using 26 RAPD markers which were collected from all over the world (15 from Iran, 2 from India, 2 from Afghanistan, and 1 from Europe). The study showed that the populations from Europe and Iran were clustered separately, whereas those belong to India and Afghanistan formed a large cluster. Also, they discovered a strong relationship between morphological and molecular diversity, which supports the use of such RAPD markers in trait-focused research. Chahota et al. (2017) used 32 RAPD primers to study genetic diversity within 14 populations from the north-western Himalayan areas and found significant genetic variation within the examined accessions. According to PIC (polymorphism information content) values, TIBMBA-06 and OPR-16 primers were reported to be highly polymorphic. The study found that there are two primary populations of *B. persicum*. The STRUCTURE analysis supported the separation of the groups from the Shong region from other populations in this study. This proved that the Shong population's genetic background is distinct from the other studied populations. However, in the study mentioned above, STRUCTURE analysis was used to divide the population into four groups in addition to the dendrogram. Using 36 RAPD primers, Majeed et al. (2009) found 97.11% polymorphism across 15 accessions representing various locations in Himachal Pradesh and Jammu and Kashmir. Furthermore, 22 RAPD markers (Zehra et al. 2018) examined the genetic polymorphism between *B. persicum* and *Carum carvi*, the two distinct species, and found a wide range of genetic diversity (34–100%) among them (Singh et al. 2021).

5.3.2 Other Biotechnological Tools

Farmers are discouraged from producing *B. persicum* extensively because of its lengthy juvenile phase and need for 3–4 years to produce seeds. By employing tubers as planting material, the problem might also be handled by shortening the dormant vegetative phase. Consequently, the development of tubers by biotechnological methods opens up new possibilities. On Murashige-Skoog (MS) medium, Grewal (1996) successfully described the induction of in vitro tubers utilizing kinetin and sucrose. They stated that whereas a dose of 30 g/L sucrose created the most shoot plant conversion, a dose of 60 g/L sucrose allegedly produced the most tubers. Various tissue culture techniques utilizing plant growth regulators may also be used to help the plant get through its vegetative hibernation period. There have been various studies on *B. persicum* that focus on in vitro regenerations. A combination of BA and NAA, according to Grewal (1996), could rejuvenate *B. persicum*. They successfully attempted both embryogenesis and organogenesis on MS media that had been enriched with kinetin and 2,4-D. Explants including mericarp (Wakhlu et al. 1990), embryo (Ebrahimie et al. 2003), hypocotyl and cotyledons (Sharefi and Pouresmael 2006) embryo (Ebrahimie et al. 2003), and B5 mediums containing 2,4-D at a level of 2 mg/L and kinetin at a dose of 2–4 mg, were all effective. Likewise, Valizadeh et al. (2007) discovered that when callus from hypocotyl explants was used in B5 medium with NAA (1 mg/L), 2, 4-D (0.1 mg/L), and kinetin (2 mg/L), quick root and shoot regenerations took place in 20 days. As a result, it may be inferred from previous studies that NAA (1–2 mg/L) and kinetin (2–4 mg/L) may be beneficial for efficient root and shoot regeneration and that 2,4-D (2 mg/L) may also have positive effects on embryogenesis (Singh et al. 2021).

5.4 Conclusions

Overall, *B. persicum* showed substantial morphological, molecular, and phytochemical diversity, offering ample and wide genetic resources for future improvement programs. Adapting to shifting climatic conditions, the divergence of

B. persicum germplasm may aid in the continuation of this species. The conservation of Kala zeera (*B. persicum*) presents one of the biggest obstacles for scientific communities because this plant's seed-to-seed cycle takes almost 4 years, and its tuber does not produce daughter tubers as the saffron plant does. Furthermore, high phytochemical diversity refers to the germplasm of *B. persicum* as having a better potential for essential oil-based industrial products. Additionally, the different populations mentioned in the current study may be helpful in developing breeding programs for this crop. Diverse lines should be multiplied through cultivation and preserved under in situ conditions. The genomic resources and information aggregated in this chapter will offer up new opportunities for population structure research; genotype selection for superior traits; population mapping; breeding for new, improved varieties; and developing long-term conservation plans for this priceless spice. By minimizing anthropogenic disruptions, the sites with natural populations of *B. persicum* should be identified and cultivated as conservatories.

References

Akbulut M, Ercisli S, Orhan E et al (2008) Determination of genetic relationships among some cherry laurel (*Laurocerasus officinalis* Roem.) genotypes by using RAPD markers. Romanian. Biotechnol Lett 13(3):3698–3702

Azimzadeh M, Amiri R, Assareh MH et al (2012) Genetic diversity of Iranian *Bunium persicum* germplasm by morphological markers and essential oil components. J Med Plants Res 6(7):1119–1129. https://doi.org/10.5897/JMPR10.085

Azizi M, Davareenejad M (2011) Essential Oil Content and Constituents of Black Zira (Bunium persicum [Boiss.] B. Fedtsch.) from Iran During Field Cultivation (Domestication). J Essent Oil Res 21(1):78–82. https://doi.org/10.1080/10412905.2009.9700117

Bansal S, Kumar A, Lone AA et al (2022) Development of novel genome-wide simple sequence repeats (SSR) markers in *Bunium persicum*. Indust Crops Products 178:114625. https://doi.org/10.1016/j.indcrop

Chahota RK, Sharma V, Ghani M et al (2017) Genetic and phytochemical diversity analysis in *Bunium persicum* populations of north-western Himalaya. Physiol Mol Biol Plant 23(2):429–444. https://doi.org/10.1007/s12298-017-0428-9

Chizzola R, Saeidnejad AH, Azizi M et al (2014) *Bunium persicum*: variability in essential oil and antioxidants activity of fruits from different Iranian wild populations. Genet Resour Crop Evol 61(8):1621–1631. https://doi.org/10.1007/s10722-014-0158-6

Dar ZA, Zeerak N, Wani SA et al (2011) Morpho-characterization of different populations of black caraway (*Bunium persicum* Bioss. Fedts) with respect to yield and yield traits across important growing sites of Kashmir Valley. J Agric Biotechnol Sustain Dev 3:60–64. https://doi.org/10.5897/JABSD.9000028

Devi S (2004) Evaluation of *Bunium persicum* (Boiss.) Fedtsch (kala zeera) germplasm in Himachal Pradesh. PhD thesis, Dr. YS Parmar University of Horticulture and Forestry, Solan, Himachal Pradesh, India

Ebrahimie E, Habashi AA, Ghareyazie B et al (2003) A rapid and efficient method of regeneration of plantlets from embryo explants of cumin. Plant Cell Tissue Organ Cult 75:19–25

Fayaz S, Mahajan R, Hami A et al (2022) Polyphenolics, antioxidant characterization and DNA barcoding of kala zeera [*Bunium persicum* (Boiss.) Fedtsch] through multiple barcode analysis to unravel best barcode combination. Mol Biol Rep 49:7205–7217. https://doi.org/10.1007/s11033-022-07682-w

Gincarlo S, Rosa LM, Nadjafi F et al (2006) Hypoglycemic activity of two spices extracts: *Rhus Coriaria* L. and *Bunium persicum* Boiss. Nat Prod Res 20:882–886. https://doi.org/10.1080/14786410500520186

Govindaraj M, Vetriventhan M, Srinivasan M (2015) Importance of genetic diversity assessment in crop plants and its recent advances: an overview of its analytical perspectives. Genet Res Int:1–14. https://doi.org/10.1155/2015/431487

Grewal S (1996) Microtubers from somatic embryos of *Bunium persicum*. Indian J Exp Biol 34:813–815

Haghirogsadat F, Bernard F, Kalantar SM et al (2010) Survey of effective compounds and antioxidant properties of essential oil of black cumin in Yazd province (in Persian). J Shahid Sadoughi Univ Med Sci 18(4):284–291

Hashemi H, Safarnejad A, Bagheri A (2010) The use of RAPD marker for assessing the genetic diversity of *Bunium persicum* (Boiss.) B. Fedtsch populations. Int J Sci Nat 1:202–208

Hassanzadazar H, Taami B, Aminzare M et al (2018) *Bunium persicum* (Boiss) B Fedtsch: an overview on Phytochemistry, Therapeutic uses and its application in the food industry. J Appl Pharm Sci 8(10):150–158. https://doi.org/10.7324/JAPS.2018.81019

Iacobellis NS, Lo CP, Capasso F et al (2005) Antibacterial activity of *Cuminum cyminum* L. and *Carum carvi* L. essential oils. J Agric Food Chem 53(1):57–61

Jahansooz F, Najafi AA, Sefidkon F et al (2007) Assessment of genetic diversity in ecotypes of *Bunium persicum* using AFLP markers. In: Proceeding of the 5th national biotechnology congress of Iran, Tehran

John MA, Skroch PW, Nienhuis J (1997) Gene pool classification of common bean landraces from Chile based on RAPD and morphological data. Crop Sci 37:605–613

Khan M, Altaf S, Shafi S et al (2022) Exploration, collection and characterization of kala zeera (*Bunium persicum* Boiss. Fedtsch.) germplasm from northwestern Himalayas. Plant genetic. Resources 20(1):62–65. https://doi.org/10.1017/S1479262122000028

Khosravi M (2005) Intercropping black Zira (*Bunium Persicum*) with saffron and annual crops: agroecological and economic perspectives. PhD thesis, College of Agriculture, Ferdowsi University of Mashhad Iran

Majeed S, Sharma D (2006) Assessment of genetic divergence of Kalazeera of different geographical races from India. In: Proceedings of 1st international conference on biotechnological approaches for alleviating malnutrition and human health, Bangalore

Majeed S, Sayeed M, Kaur R et al (2009) Molecular analysis of genetic diversity in wild accessions of *Bunium persicum* (Boiss.) Fedtsch—a critically endangered medicinal plant of temperate Himalayas. Asian Australasian J Plant Sci Biotechnol 3(1):7–10

Martinez GP, Sanchez PR, Dicenta F et al (2007) Genome mapping and molecular breeding in plants. Springer, Berlin, pp 229–242

Oroojalian F, Kermanshahi RK, Azizi M et al (2010) Phytochemical composition of the essential oils from three Apiaceae species and their antibacterial effects on food-borne pathogens. Food Chem 120:765–770

Padmashree A, Roopa N, Semwal AD et al (2007) Star-anise (*Illicium verum*) and black caraway (*Carum nigrum*) as natural antioxidants. Food Chem 104:59–66

Pezhmanmehr M, Hassani ME, Fakhre et al (2007) Evaluation of genetic diversity in some ecotypes of black cumin (Bunium persicum) using RAPD markers. In: Proceeding of the 5th national biotechnology congress of Iran, Tehran

Pezhmanmehr M, Hassani ME, Jahansooz F et al (2009) Assessment of genetic diversity in some Iranian populations of *Bunium persicum* using RAPD and AFLP markers. Iran J Biotechnol 7(2):93–100

Samojlik I, Lakic N, Mimica-Dukic N et al (2010) Antioxidant and hepatoprotective potential of essential oils of coriander (*Coriandrum sativum L.*) and caraway (*Carum carvi* L.) (Apiaceae). J Agric Food Chem 58(15):8848–8853

Sharefi M, Pouresmael M (2006) Breaking seed dormancy in *Bunium persicum* by stratification and chemical substances. Asian J Plant Sci 5:695–699

Singh S, Kumar V, Ramesh (2021) Biology, genetic improvement and agronomy of *Bunium persicum* (Boiss.) Fedtsch.: a comprehensive review. J Appl Res Med Aromatic Plants 22:100304

Sofi PA, Zeerak NA, Singh P (2009) Kala zeera (*Bunium persicum* Bioss.): a Kashmirian high value crop. Turk J Biol 33:249–258. https://doi.org/10.3906/biy-0803-18

Steiner JJ, Los Santos GG (2001) Adaptive ecology of Lotus corniculatus L. genotypes: I. Plant morphology and RAPD marker characterization. Crop Sci 41:552–563

Talebi M, Moghaddam M, Pirbalouti AG (2018) Variability in essential oil content and composition of *Bunium persicum* Boiss. Populations growing wild in northeast of. Iran 30(4):258–264. https://doi.org/10.1080/10412905.2018.1441077

Thappa RK, Ghosh S, Agarwal G (1991) Comparative studies on major volatiles of Kalazira (*Bunium persicum* seed) of wild and cultivates sources. Food Chem 41:129–134

Valizadeh M, Kazemi S, Nematzadeh G (2007) A novel method for regeneration of plantlets from embryo explants of *B. persicum*. Int J Plant Breed Genet 1(12):13–17

Wakhlu A, Nagori S, Burna K (1990) Somatic embroyogenesis and plant regeneration from callus cultures of *B. persicum* Bioss. Plant Cell Rep 9:137–138

Zehra M, Razaq A, Khan IA (2018) Molecular analysis in medicinally important species *Carum carvi* and *Bunium persicum* (family Apiaceae) from district Astore. Pak J Bot 50(1):301–305

Genetic and Breeding Advancement in Buckwheat: A Pseudocereal of Himalaya

6

Vishal Kumar, Priya Kumari, Himanshi Gangwar, Vishek Choudhary, Vijay Gahlaut, and Vandana Jaiswal

6.1 Introduction

Buckwheat is an annual or biennial herb belonging to the Polygonaceae family. Despite having similar characteristics and chemical composition to cereals, it is categorized differently. The name "buckwheat" is derived from the Old English words "Boc," meaning beech (due to its beechnut-like seeds), and "Whoet," resembling "wheat" (Edwardson et al. 1995). There are nine distinct species of buckwheat with significance in agriculture and nutrition. Among these, *Fagopyrum esculentum* (common buckwheat) and *Fagopyrum tataricum* are the two most commonly cultivated species (Baljeet et al. 2010). The different developmental stages of *F. esculentum* and *F. tataricum* including vegetative, flowering, seeds at early stage, seeds at late stage and harvested seed have been shown in Fig. 6.1. Tartary buckwheat (*F. tataricum Gaertn*) is typically grown at much higher altitudes, around 3500 m, compared to common buckwheat (*F. esculentum L. Moench*), which grows at elevations of about 2500 m and is regarded to be more important in the Himalayan region (Rana 2004). Interestingly, in China and Nepal, common buckwheat is known as "sweet buckwheat," while Tartary buckwheat is referred to as "bitter buckwheat" (Zhou et al. 2018a). In the Japanese language, it is known as soba (Ikeda 2002). According to previous study by Ohnishi and Yasui (1998), common buckwheat is

V. Kumar · P. Kumari · H. Gangwar · V. Choudhary · V. Jaiswal (✉)
Biotechnology Division, CSIR-Institute of Himalayan Bioresource Technology, Palampur, Himachal Pradesh, India

Academy of Scientific and Innovative Research (AcSIR), Ghaziabad, Uttar Pradesh, India

V. Gahlaut
Biotechnology Division, CSIR-Institute of Himalayan Bioresource Technology, Palampur, Himachal Pradesh, India

Department of Biotechnology and University Center for Research and Development, Chandigarh University, Mohali, Punjab, India

V. Gahlaut, V. Jaiswal (eds.), *Genetics and Genomics of High-Altitude Crops*,
https://doi.org/10.1007/978-981-99-9175-4_6

Fig. 6.1 Different developmental stage of *F. tataricum* (**a–e**) and *F. esculentum* (**f–j**) including vegetative, flowering, seeds at early stage, seeds at late stage, and harvested seed, respectively

believed to have originated in western and central China. It was subsequently distributed to various Asian countries through both overland and maritime trade routes approximately 3000 years ago (Murai and Ohnishi 1996). The primary producers of common buckwheat are Kazakhstan, Russia, Ukraine, and China, while China, Brazil, Canada, the United States, and France are the leading exporters of this versatile crop (Campbell 1995).

Cultivating buckwheat offers several benefits, including its ease of growth, a rapid growth cycle of 2–2.5 months, extended storage capabilities without quality deterioration, and notable antioxidant properties (Ahmed et al. 2014). Buckwheat flour surpasses wheat flour in terms of lysine, iron, copper, and magnesium content, making it a superior choice (Ikeda and Yamashita 1994), and it also has a high vitamin content, particularly from the B group (Li and Zhang 2001). Buckwheat has also been linked to various health advantages, such as reducing cholesterol levels, aiding in blood sugar regulation, and lowering the risk of cancer (Fabjan et al. 2003; Kim et al. 2004). Selenium (Se) levels in buckwheat vary from 0.009 to 0.1208 mg/g (Shi et al. 2011). Selenium is a critical micronutrient for human health, enhancing resistance to diseases such as diabetes, AIDS, and cancer, as well as cardiovascular and cerebrovascular conditions. Importantly, buckwheat is a safe dietary option for individuals with celiac disease due to its absence of gluten protein (Skerritt 1986). A variety of food products derived from buckwheat, including bread, cakes, noodles, honey, tea, tarhana, alcoholic beverages such as beer and shochu, vinegar, and sprouts, have been created and are currently available in the food market (Giménez-Bastida et al. 2015). Additionally, it is worth noting that in northern and western India, buckwheat flour is recognized as "kuttu ka atta" (Zhou et al. 2018a, b).

Buckwheat production comes with several significant disadvantages, including lower grain yield, asynchronous seed ripening, rapid seed setting, susceptibility to lodging, self-incompatibility, and vulnerability to frost (Yasui et al. 2016).

Addressing these fundamental challenges requires a deep understanding of genetics, genomics, and molecular mechanisms to support breeding programs. Molecular methods have been successfully employed in numerous crops for genetic analysis of important traits and crop development (Jaiswal et al. 2019). In the case of buckwheat, extensive research has been conducted to date. Many molecular markers have been developed, such as RAPD, SSR, AFLP, ISSR (Deng et al. 2011; Gupta et al. 2012; Kishore et al. 2013; Hou et al. 2016). Additionally, various QTLs (Quantitative Trait Loci) associated with different morphological and physiological traits, such as photoperiod sensitivity, stem length, preharvest sprouting, grain length, and flowering time, have been identified (Yabe et al. 2014; Hara et al. 2020; Takeshima et al. 2021; Fang et al. 2022). To expedite molecular breeding programs, a draft assembly of the buckwheat genome (*F. esculentum* Moench; 2n = 2x = 16) was generated using short reads obtained from Illumina sequencers, encompassing 264.5 Gb of data, as part of the Buckwheat Genome DataBase (BGDB; http://buckwheat.kazusa.or.jp) (Yasui et al. 2016). This drafted genome served as a crucial reference for genotyping-by-sequencing (GBS) investigations in buckwheat. Additionally, a transcriptome map based on the newly assembled genome was constructed and made publicly available on the TraVA database (http://travadb.org/browse/Species=Fesc/). Notably, a recent development in buckwheat research involved the publication of a buckwheat genome assembly using single-molecule real-time (SMRT) read technology and the creation of a high-resolution expression atlas covering 46 organs and developmental stages (Penin et al. 2021).

Despite the progress made in genetic, genomic, and molecular methodologies, conventional techniques continue to be the predominant approach in the field of buckwheat breeding. A significant divide exists between molecular geneticists and buckwheat breeders. Within this book chapter, we have gathered the available genetic and genomic information pertaining to buckwheat and examined potential approaches to enhance our understanding and exploration of genetic and molecular mechanisms. These progressions in genomics and molecular breeding techniques are expected to develop numerous new buckwheat varieties possessing desirable characteristics in the near future.

6.2 Origin, Evolution, and Cultivation

In 1913, Gross reported the first taxonomic study on buckwheat, utilizing various morphological characteristics for identification (Gross 1913). This research laid the groundwork for classifying the gene pool (Singh et al. 2020). The *Fagopyrum* genus is categorized into two distinct monophyletic groups: cymosum, known as the group with larger achenes, and urophyllum, consisting of the group with smaller achenes. These groupings encompass a total of approximately 28 species. Among them, four species fall into the cymosum group in which two are cultivated and two are wild type, while the rest belong to the urophyllum group and are considered wild varieties (Table 6.1).

Table 6.1 List of species that are presently classified as member of the genus *Fagopyrum*

Group	Species	Type	References
Cymosum	*F. tataricum*	Cultivated	Gaertner (1791)
	F. esculentum	Cultivated	Moench (1794); Ohnishi and Yasui (1998)
	F. cymosum	Wild	Meisner (1857)
	F. homotropicum	Wild	Ohnishi and Yasui (1998)
Urophyllum	*F. gracilipes*	Wild	Diels (1901)
	F. statice, F. urophyllum	Wild	Gross (1913)
	F. gilesii, F. leptopodum	Wild	Hedberg (1946)
	F. lineare	Wild	Haraldson (1978)
	F. callianthum, F. capillatum, F. pleioremosum,	Wild	Ohnishi and Matsuoka (1996)
	F. megaspartanium, F. pilus, F. zuogongense	Wild	Chen (1999)
	F. macrocarpum, F. rubifolium	Wild	Ohsako and Ohnishi (1998)
	F. gracilipedoides, F. jinshaense	Wild	Ohsako et al. (2002)
	F densovillosum	Wild	Liu et al. (2008)
	F. crispatifolium	Wild	Liu et al. (2008)
	F. pugense	Wild	Tang et al. (2014)
	F. qiangcai, F. wenchuanense	Wild	Shao et al. (2011)
	F. luojishanense	Wild	Hou et al. (2015)
	F. hailuogouense	Wild	Zhou et al. (2015)
	F. longistylum	Wild	Zhang et al. (2021)

Buckwheat is likely to have its origins in the temperate regions of central and western China, as suggested by several studies (Yururi and Zhongging 1984; Ohnishi and Yasui 1998; Ahmad et al. 2018; Singh et al. 2020). The earliest evidence of cultivated buckwheat seeds dates back to the first and second centuries BC in China (Li 1992). According to Murai and Ohnishi (1996), buckwheat was introduced from both northern and southern China to other parts of Asia about 3000 years ago, eventually reaching Europe during medieval times. However, some archaeological assessments suggest that buckwheat may have appeared in Europe as early as the first and second centuries (Ohnishi 1993). Research involving allozyme analysis and DNA polymorphism indicates that *F. megaspartanium* and *F. pilus* may have served as ancestors for *F. esculentum* and *F. tataricum*, respectively (Chen et al. 2004; Li et al. 2013b). Campbell (1995) proposed that cultivated buckwheat is distantly related to *F. esculentum* ssp. *Ancestralis*, while Ohnishi considered it as the wild ancestor of common buckwheat and *F. tataricum* ssp. *Potanini Batalin* as the wild ancestor of Tartary buckwheat (Ohnishi 1988). Both *F. esculentum* and *F. tataricum* were domesticated in western Yunnan and Sichuan, China (Ahmad et al. 2018). In addition to southwestern China, common buckwheat is also grown in temperate mountainous regions of Nepal, India, China, Japan, and Korea. There are records of more than 2500 accessions of various buckwheat species from around the world, providing a diverse gene pool for the development of different quality traits (Dar et al. 2018).

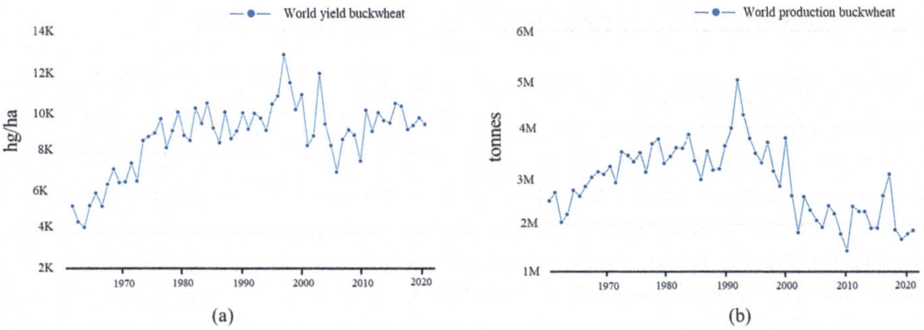

Fig. 6.2 The global buckwheat chart for the year 2020. (**a**) Yield graph measured in hectograms per hectare (hg/ha), and (**b**) the worldwide production of buckwheat in that same year, measured in metric tons. (Source: http://faostat.fao.org/)

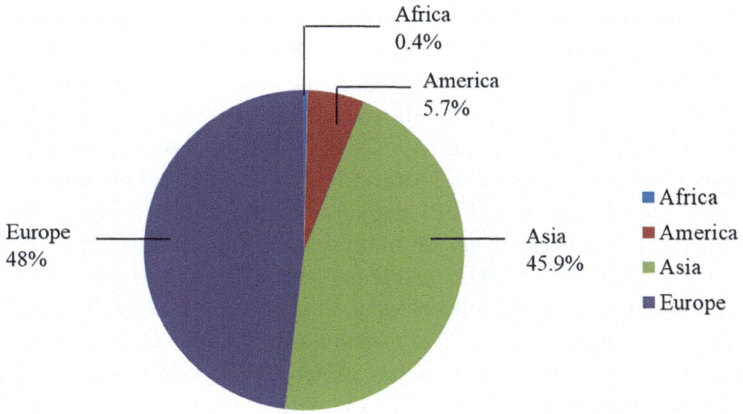

Fig. 6.3 The distribution of buckwheat production among different geographical regions

In 2014, the global production of buckwheat exceeded two million tons, as reported by FAOSTAT in 2016. The primary producers of buckwheat were China and Russia, which collectively yielded over 700,000 tonnes each year. Poland, Kazakhstan, Ukraine, and France followed closely in buckwheat production. The charts (Figs. 6.2 and 6.3) illustrate the buckwheat world yield graph in hectograms per hectare (hg/ha) and the worldwide buckwheat production measured in tonnes, according to FAOSTAT data from 2020.

Buckwheat has traditionally been cultivated in several countries, such as Japan, Bhutan, Nepal, Korea, Belarus, the Czech Republic, and Slovenia. However, it is still doubtful that buckwheat will become a dominant global food crop in the future. While some regions may expand their buckwheat cultivation areas, the primary focus is on improving existing buckwheat varieties. This is driven by the rising interest in using buckwheat as a nutritional supplement and functional food, leading to ongoing research and development efforts in this direction.

6.3 Genetic Diversity and Germplasm Resources

Genetic diversity plays pivotal role in the survival of any species, as it directly influences a population's ability to adapt to changing environmental conditions (Malysheva-Otto et al. 2007; Bashir et al. 2015). To assess the genetic diversity of buckwheat, a range of tools are available, including morphological characteristics, biochemical markers, and molecular markers (Govindaraj et al. 2015). Understanding the diversity in the morphological traits of buckwheat seeds is particularly important in the field of buckwheat breeding (Oshawa et al. 1998). A study conducted by Gao et al. (2010) examined the genetic diversity of buckwheat germplasms belonging to *F. esculentum* and *F. tataricum* using both morphological and molecular markers. This research aimed to provide essential insights for both the utilization and conservation of buckwheat germplasm. Several types of molecular markers, such as AFLP, allozymes, RFLP, RAPD, SSR, conserved sequences derived from chloroplast and mitochondrial DNA, SRAP, and STS, have been employed to assess relatedness and population structure among different buckwheat accessions (see Table 6.2). For example, a 1996 study utilizing random amplified polymorphic DNA (RAPD) markers classified 80 accessions of common buckwheat into European and Asian groups, along with some additional clusters and variations in populations from both regions. In 2001, an investigation based on amplified fragment length polymorphism (AFLP) markers suggested that Tartary buckwheat likely originated in eastern Tibet or northwestern Yunnan in China (Tsuji and Ohnishi 2001). Similarly, in 2002, RAPD markers were used to analyze 28 accessions from various species and subspecies of *Fagopyrum*, revealing the proximity of *F. tataricum* to its wild ancestor, *F. tataricum* ssp. *potanini Batalin* (Sharma and Jana 2002). In the Western Himalayas, researchers employed inter-simple sequence repeats (ISSR) to assess the genetic diversity of 15 *F. tataricum* germplasms, finding a significant correlation between genetic diversity and altitude in this context (Kishore et al. 2013). Another study focused on rutin content variation among 195 Tartary buckwheat accessions using AFLP fingerprinting, and it identified two distinct groups, one with high rutin content and the other with low rutin content (Gupta et al. 2012). Simple repeat sequence (SSR) markers have gained popularity for high-throughput genotyping and map construction due to their abundance, random distribution across the genome, high polymorphism information content (PIC), and stable co-dominance (Zietkiewicz et al. 1994; Iwata et al. 2005). In 2020, a study utilized 15 SSRs to investigate polymorphism among 110 genotypes of buckwheat (*Fagopyrum* spp.), leading to the identification of two major groups through cluster analysis and STRUCTURE analysis. In summary, diverse molecular markers have been instrumental in studying the genetic diversity and relationships among various buckwheat accessions, shedding light on their origins, traits, and potential applications.

Table 6.2 List of *Fagopyrum* genus genetic variation studies, including population size, number, and type of markers

S.No.	Germplasm lines	Species/Type/Group	No. of markers	Type of marker	References
1	80	*F. esculentum Moench*	13	RAPD	Kump and Javornik (1996)
2	42	Cultivated Tartary buckwheat and wild Tartary buckwheat	10	AFLP	Tsuji and Ohnishi (2001)
3	40	Cultivated Tartary buckwheat and wild Tartary buckwheat	29	RAPD	Kump and Javornik (2002)
4	19	*F. esculentum*	5, 4	SSR, AFLP	Iwata et al. (2005)
5	51	*F. esculentum, F. tataricum, F. cymosum*	20	RAPD	Senthilkumaran et al. (2008)
6	19	*F. esculentum, F. tataricum*	7	RAPD	Deng et al. (2011)
7	195	*F. tataricum*	19	AFLP	Gupta et al. (2012)
8	15	*F. tataricum*	13	ISSR	Kishore et al. (2013)
9	64	*F. tataricum*	23	SSR	Hou et al. (2016)
10	31	*F. esculentum, F. tataricum*	26	ISSR	Shukla et al. (2018)
11	179	*F. esculentum*	10	SSR	Song et al. (2018)
12	52	*F. esculentum*	15	SSR	Bashir et al. (2021)
13	63	*F. esculentum, F. tataricum*	7, 7	SSR, ISSR	Nazir et al. (2021)
14	193	*F. tataricum*	62	SSR	Taoxiong et al. (2021)
15	14	*F. esculentum*	7	SSR	Grahić et al. (2022)
16	35	*F. esculentum, F. tataricum*	10	SCoT	Mikolášová et al. (2022)
17	112	*F. tataricum*	10	SSR	Song et al. (2022)

RAPD random amplified polymorphic DNA; *AFLP* amplified fragment length polymorphism; *SSR* simple sequence repeats; *ISSR* inter simple sequence repeats; *SCoT* start codon targeted polymorphism

6.4 Germplasm Conservation Repository

A number of efforts made to conserve the genetic material of buckwheat at the regional, national, and international levels. Food and Agriculture Organization of United Nations (FAO) started conservation activities for the first time in 1945 (Singh et al. 2020). With the objective of preserving genetic resources and ensuring their sustainability, they founded the International Board for Plant Genetic Resources (IBPGR) in 1974. There are almost 10,000 different accessions of buckwheat maintained globally (Zhou et al. 2018a, b). A list of institutes/organizations involved in

Table 6.3 Germplasm repositories containing *Fagopyrum* species located across various regions of the world

Institute/organization	Country	Majority of species	Germplasm collection	References
Gene Bank of the Crop Research Institute	Czech Republic	Mostly exotic accessions of *F. esculentum*	170	Cepková et al. (2009)
National Institute of Agrobiological Sciences (NIAS)	Japan	Landraces of *F. esculentum* and varieties of *F. tataricum*	226	Katsube-Tanaka (2016)
National Agriculture Genetic Resource Centre	Nepal	Indigenous landraces *F. esculentum* and *F. tataricum*	511	Paudel et al. (2016)
National Bureau of Plant Genetic Resources (NBPGR)	India	*F. esculentum, F. tataricum, F. sagittatum, F. cymosum, F. tataricum* ssp. *Himalianium* and *F. esculentum* ssp. *Emerginatum*	1050	Rana et al. (2016)
N I Vavilov All-Russian Institute of Plant Industry (VIR)	Russia	Landraces, cultivars, and wild forms of *F. esculentum* and *F. tataricum* subsp. Multholium	2230	Romanova et al. (2018)
Crop Genetic resources Institute of Chinese Academy of Agricultural Science	China	*F. esculentum, F. tataricum*	2804	Zhou et al. (2018a, b)
Podillya State Agricultural University	Ukraine	*F. esculentum* and *F. tataricum*	900	Zhou et al. (2018a, b)
University of Ljublijana	Slovenia	Landraces of *F. esculentum*, accessions of *F. tataricum*	378	Zhou et al. (2018a, b)
V Y Yuriev Institute of Plant Production	Ukraine	Landraces, cultivars, and wild forms of *F. esculentum* and *F. tataricum*	1600	Zhou et al. (2018a, b)

the conservation of buckwheat is provided in Table 6.3. The largest repository, with more than 2800 accessions, is located at the Crops Genetic Resources Institute of the Chinese Academy of Agricultural Sciences (CAAS) (Zhou et al. 2018a, b). Depending on the usage, buckwheat accessions are either preserved in long-term storage (LTS) at −20 °C with 5% relative humidity and mid-term storage (MTS) at 5 °C with 35% relative humidity (Singh et al. 2020). Buckwheat introduction, collection, and conservation initiatives have also been undertaken in India by NBPGR, which has been successful in acquiring 1050 accessions. They preserved their accessions through LTS method collections in New Delhi and at its regional station in Shimla. Recently on-farm conservation of buckwheat germplasm has been initiated globally (Rana et al. 2016).

6.5 Genetic Linkage Mapping of Buckwheat

Genotyping is a crucial step in identifying agronomically advantageous genes and studying population dynamics. Compared to the major food crops of corn, wheat, and rice, buckwheat has received more attention (Zhang and Chen 2004; Bindler et al. 2007; Grimmer et al. 2007). In common buckwheat, numerous types of genetic marker systems have been created (Table 6.4). Six allozyme and 15 morphological markers were included in the first linkage map to be available, which was based on

Table 6.4 List of markers used to investigate various objectives along with their results within the genus *Fagopyrum*

Marker type	Objective of the study	Results	References
Morphological and allozyme	To investigate linkage relationship between morphological and allozyme markers	1st linkage map of *F. esculentum* in the world was created using 15 morphological markers and six allozymes	Ohnishi and Ohta (1987)
Allozyme	To examine population differentiation and diversity	Using 160 worldwide populations (32,000 samples) and 19 loci, the results of allozyme analysis showed high variety and low population differentiation	Ohnishi (1993)
SCAR, RAPD	To investigate RAPD markers, associate with homostylar (Ho) gene	DNA markers for the Ho gene, a crucial self-compatible gene for agriculture, were created	Aii et al. (1998)
RAPD	To investigate evolutionary relationship between cultivated and wild Tartary buckwheat	The birth place of cultivated Tartary buckwheat is most likely is northwestern Yunnan, according to a phylogenetic tree that was created	Tsuji and Ohnishi (2001)
AFLP	Linkage studies of *F. esculentum* and *F. homotropicum*	Three morphological trait genes were mapped, and the first high density genetic map was created using genome-wide AFLP markers	Yasui et al. (2004)
RAPD	Studying the interspecific hybridization between common buckwheat and *Fagopyrum homotropicum*	RAPD markers were effective in identify F1 offsprings between *F. esculentum and F. homotropicum*	Wang et al. (2005)
SSR	To create SSR markers for *Fagopyrum esculentum*	For *Fagopyrum esculentum*, microsatellite markers (54 loci) were created	Konishi and Ohnishi (2006)
EST	For examine QTL using EST marker	Photoperiod sensitivity genes were revealed via QTL analysis utilizing 50 EST markers	Hara et al. (2011)

(continued)

Table 6.4 (continued)

Marker type	Objective of the study	Results	References
NGS based DNA array	To create a high-density linkage map in common buckwheat using a DNA micro array	High density linkage map (8884 markers) was created using DNA micro arrays	Yabe et al. (2014)
NGS	Construction of the buckwheat draft genome and its usage to find agriculturally valuable genes	The genome database (BGDB) was created, draft genome sequences were decoded, and several agronomically usable genes are found in the BGDB	Yasui et al. (2016)
NGS-based DNA array	An empirical study to raise the production of *F. esculentum* examines the potential of genomic selection in mass selection breeding of an allogamous crop	Genomic selection using 14,598–50,000 markers resulted in a 20.9% increase in the selection index compared to the initial population	Yabe et al. (2018)
GBS	Gene flow signature in the S-allele region of cultivated buckwheat	Using GBS, 46 cultivated common buckwheat plants provided 255,517 SNP sites, suggesting that there was likely gene transfer from wild to cultivated buckwheat	Mizuno and Yasui (2019)

SCAR sequenced characterized amplified region; *RAPD* random amplified polymorphic DNA; *AFLP* amplified fragment length polymorphism; *SSR* simple sequence repeats; *EST* expressed sequence tags; *NGS* next generation sequencing; *GBS* genotyping-by sequencing

allozymes variations and morphology (Ohnishi and Ohta 1987). Additionally, Ohnishi (1993) used 64 populations mostly collected from Asian nations such as Japan, China, Nepal, and Korea to clarify the global population structure of common buckwheat. 200 individuals per population from 19 allozyme loci were studied in this study. A cultured population's allozyme locus's average heterozygosity (which ranged from 0.110 to 0.138) was discovered to be fairly high (Ohnishi 1993). In the buckwheat species *F. homotropicum*, RAPD markers connected to the self-pollination gene were discovered and transformed into highly repeatable SCAR markers. DNA markers for the Ho gene, a crucial self-compatible gene for agriculture, are created (Aii et al. 1998). Different biparental mapping populations which segregates for different traits, developed in buckwheat as listed in Table 6.5. Sobano (Common buckwheat) × *F. esculentum var. homotropicum*'s F2 offspring were used to create linkage map, in which sobano is self-infertile and distylous and *F. esculentum var. homotropicum* is self-fertile and homostylous. This map had eight linkage groups with a total length 508.3 cM that includes 223 AFLP markers (Yasui et al. 2004). Pseudo-testcross strategy was used to conduct linkage mapping. C2002 (*F. esculentum* ssp. *esculentum)* (♀) × C2009 (♂) a wild species, (*F. esculentum* ssp. *Ancestrale*) F1 offspring were used to create linkage map. Male map had 12 linkage groups with a total length 909.0 cM including 34 AFLP and 37 microsatellite

Table 6.5 An overview of different biparental mapping populations (segregate for different traits) developed in buckwheat

Mapping population type	Parents information	Size	Trait	References
Mutant lines	Full-sib mating of common buckwheat	70	Morphological mutant lines	Ohnishi and Ohta (1987)
F2 population	*F. esculentum* (♀) × *F. homotropicum* (♂)	123	Heteromorphic and self-compatible	Nagano et al. (2001)
F2 population	*F. esculentum* (♀) × *F. homotropicum* (♂)	85	Non-abortion of mature seeds	Yasui and Wang (2001)
F1 population	*F. esculentum ssp. esculentum* (C2002) (♀) × *F. esculentum ssp. Ancestrale* (♂) a wild species, (C2009)	80	–	Konishi and Ohnishi (2006)
F2 population	*Fagopyrum esculentum Moench* (♀) × *F. esculentum var. homotropicum* (♂)	225	Homo/long style (H/s), shattering habit (Sht/sht), and acute/obtuse achene ridge (ac/ac)	Pan and Chen (2010)
F4 population	02AL113 (Kyukei SC2) LH. (KYU) (♀) × C0408-0RP (CAN) (♂)	100	Photoperiod sensitivity	Hara et al. (2011)
F4 population	Dianning 1(♀) × *F. tataricum* (♂)	119	–	Du et al. (2013)
F1 population	P1 × P2, which were selected from 92FE1-F4	178	Stem length	Yabe et al. (2014)
F1 population	Ukraine daliqiao (UD) (♀) × Youqiao 2 (YQ2) (♂)	217	Grain size	Fang et al. (2022)
F2 population	Yunqiao no. 1× Rice Tartary	335	Easily-shelled traits	Duan et al. (2022)

markers. The female map had 12 linkage groups with a total length of 911.3 cM includes 77 AFLP and 54 microsatellite markers (Konishi and Ohnishi 2006). Sobano×*F. esculentum var. homotropicum* F_2 offspring is used to create a linkage map. This map had ten linkage groups with a total length 655.2 cM including 12 STS, four seed protein subunit, three morphological alleles and 87 RAPD markers (Pan and Chen 2010). The hybrid between "Dianning 1" and a wild species of Tartary buckwheat produced F4 offspring. In order to create the molecular genetic linkage map, SSR markers were utilized. This map had 15 linkage group with an entire map length of 860.2 cM containing 89 SSR markers (Du et al. 2013). A high-density linkage map was created for *F. esculentum* through a novel marker system. The genetic map had eight linkage groups with 8884 markers. P1 linkage group's entire map length is 773.8 cM and P2 linkage map's entire map length is 800.4 cM (Yabe et al. 2014). A genetic map containing 12 linkage groups was constructed using 80 SSR molecular markers (Liang 2016). The high genetic variation of 46 cultivated *F. esculentum* plants was discovered using genome-wide genotyping data from GBS analysis, which detected 255,517 SNP locations (1 SNP per 28 base pair)

Table 6.6 Summarized the construction of linkage maps in *Fagopyrum* species including details about the number and types of molecular markers. Map length (in cM) and marker density (per cM) are also mentioned

Species type	Number of markers	Markers type	Map length (cM)	Marker density (per cM)	References
F. esculentum	71–131	AFLPs and SSRs	909–911.3	0.07–0.14	Konishi and Ohnishi (2006)
		RAPD, STS	692.4	0.017–0.12	Pan and Chen (2010)
	63	EST	311.6	0.2	Hara et al. (2011)
	8884	Novel-based markers	773.8–800.4	11.09–11.48	Yabe et al. (2014)
	39–93	SSR and InDel	1398.3	0.027–0.066	Fang et al. (2022)
F. esculentum and *F. homotropicum*	214–224	AFLPs	508.3–548.9	0.38–0.44	Yasui and Wang (2001)
	211–223	AFLPs	508.3–548.9	0.38–0.43	Yasui et al. (2004)
F. tataricum	89	SSR	860.2	0.1	Du et al. (2013)
	122,185	SNP_S	1444.2	84.6	Shi et al. (2021)

AFLP amplified fragment length polymorphism; *SSR* simple sequence repeats; *RAPD* random amplified polymorphic DNA; *STS* sequence-tagged sites; *EST* expressed sequence tag; *SNPs* single nucleotide polymorphism; *InDel* insertion/deletion

(Mizuno and Yasui 2019). A linkage map was created for common buckwheat. This genetic map had eight linkage group with 269 markers which have entire map length 752.5 cM. The distance among the closest markers was on average 2.8 cM (Hara et al. 2020). Using a RILs (Recombinant Inbred Lines) population based on RAD (restriction site associated DNA) sequencing; a high-density SNP map for Tartary buckwheat was created. The high-density map had eight linkage groups with 4151 bin markers, and the gap between adjacent bin markers was 0.35 cM on average (Shi et al. 2021). The Summary of the construction of linkage maps in *Fagopyrum* species that includes details about number and types of molecular markers. This has been detailed in Table 6.6.

6.6 QTL Based on Linkage Map

In buckwheat, QTLs were studied for numerous traits including grain size, stem length, photoperiod sensitivity (PS), preharvest sprouting, and flowering time and maturing time as detailed in Table 6.7. According to this study PS in *F. esculentum* is regulated by various genes, along with interactions among these genes. There are three PS candidate gene (*FeELF3*, *FeCOL3*, and *FeCCA1*) detected in *F. esculentum*. For PS, *FeELF3* gene was responsible. Fest_L0337_6 and Fest_L0606_4 were

Table 6.7 Summary of QTLs that have been identified for various traits in buckwheat, including details about the linked maker, chromosomal position and the proportion of phenotypic variation explained (PVE)

S. no.	Trait	Linked marker/QTL name	Chromosome, position (cM)	PVE (%)	References
1	Photoperiod sensitivity	Fest_L0337_6	7;0.0	14.2	Hara et al. (2011)
		Fest_L0606_4	3;34.0	20	
		FeCCA1	6;21.5	14.2	
		FECOL3	9;0.6	0.4	
		qFT12hl × E_1	4;24.7	28.6	Hara et al. (2020)
		qFT12hl × E_2	1;110.3	15.6	
		qFT12hL × E_2	8;61.2	4.7	
		qFT12hL × E_1	6;54.5	56.6	
		qFT12hL × E_3	3;100.5	3.8	
		qFT15hl × E_1	1;110	44.1	
		qFT15hl × E_2	5;42.6	19.2	
		qFT15hl × E_3	2;52.6	15.4	
		qFT15hL × E_1	5;43.6	24.5	
		qFT15hL × E_2	3;68.0	7.6	
		qFT12hL × E_3	8;49.9	6.9	
2	Stem length	P1–1	1;9.3	6.62	Yabe et al. (2014)
		P1–2	2;49.0	5.64	
		P1–5	5;9.0	8.51	
		P2–4	4;16.9	8.18	
3	Preharvest sprouting	qPHS1_KY29	1;71.2	11.7	Takeshima et al. (2021)
		qPHS6_KY29	6;59.5	15.6	
		qPHS4_KY29[a]	4;57.8	7.4	
		qPHS7_KY29[a]	7;49.6	7.5	
		qPHS3_KY28	3;3.5	10.6	
		qPHS1_KY28	1;45	9	
		qPHS8_KY28	8;7	15.7	
		qPHS4_KY28	4;5.2	22	
		qPHS5_KY28[a]	5;47.6	7.1	
		qPHS3_NF1	3;53.2	12	
		qPHS8_NF1	8;12.1	16	
		qPHS2_NF1[a]	2;24.1	10	

(continued)

Table 6.7 (continued)

S. no.	Trait	Linked marker/QTL name	Chromosome, position (cM)	PVE (%)	References
4	Thousand grain weight	qTGW-17-C1a	1;15.31	5	Shi et al. (2021)
		qTGW-18-C1a	1;15.31	3.4	
		qTGW-17-C1b	1;38.91	23.6	
		qTGW-18-C1b	1;38.91	47.5	
		qTGW-19-C1	1;38.91	24.4	
		qTGW-19-C4a	4;113.9	2.9	
		qTGW-19-C4b	4;122.1	6.6	
		qTGW-17-C4	4;122.5	10.9	
		qTGW-18-C4	4;126.9	3.1	
		qTGW1.1	1;15.3	4.96	Li et al. (2023)
		qTGW1.1	1;15.3	3.41	
		qTGW1.2	1; 38.5	7.58	
		qTGW1.2	1; 38.9	23.58	
		qTGW1.2	1; 38.9	47.51	
		qTGW1.2	1;38.9	24.45	
		qTGW4.1	4;113.6	5.74	
		qTGW4.1	4;117.0	5.81	
		qTGW4.2	4;122.1	9.33	
		qTGW4.3	4;127.3	7.6	
		qTGW4.3	4.131.3	9.2	
5	Maturity time	qMT6_KTW	6;96.6	21	Takeshima et al. (2022)
		qMT3_RCK	3;6.1	20.5	
6	Flowering time	qFT3_KTW	3;91.7	15.4	Takeshima et al. (2022)
		qFT5_KTW	5;40.8	11.1	
		qFT6_KTW	6;100.6	9.3	
		qFT7_RCK	7;143.0	13.6	
7	Grain length	qGL1	1;119.4	6.8	Fang et al. (2022)
		qGL2	2;89.6	6.3	
		qGL3	3;91.0	4.7	
		qGL5	5;46.7	6.2	
		qGL6	6;18.6	8	
		qGL7	7;82.1	6	
		qGL1.1	1;39.6	8.39	Li et al. (2023)
		qGL1.1	1;40.5	6.23	
		qGL1.1	1;40.8	15.97	
		qGL1.2	1;49.0	7.4	
		qGL1.3	1;77.8	9.81	
		qGL1.4	1;87.6	39.43	
		qGL1.4	1;87.6	34.17	
		qGL1.4	1;87.6	27.44	
		qGL1.5	1;93.1	18.4	
		qGL3.1	3;163.7	4.75	

Table 6.7 (continued)

S. no.	Trait	Linked marker/QTL name	Chromosome, position (cM)	PVE (%)	References
8	Grain width	qGW2	2;136.5	6.6	Fang et al. (2022)
		qGW4	4;115.5	5.7	
		qGW5	5;123.7	7	
		qGW1.1	1;1.4	6.32	Li et al. (2023)
		qGW1.2	1;6.9	4.16	
		qGW1.3	1;14.2	3.75	
		qGW1.4	1;20.9	4.18	
		qGW1.5	1;30.9	12.32	
		qGW1.6	1;36.6	6.58	
		qGW1.7	1;38.9	4.88	
		qGW1.7	1;38.9	29.7	
		qGW1.7	1;38.9	17.92	
		qGW1.8	1;69.2	8.06	
		qGW1.9	1;81.4	8.75	
		qGW1.10	1;87.6	16.18	
		qGW1.10	1;88.0	14.26	
		qGW1.10	1;88.0	14.24	
		qGW8.1	8;0.0	6.15	
		qGW8.1	8;0.3	5.39	
9	Hundred grain weight	qHGW1	1;217.8	7.4	Fang et al. (2022)
		qHGW2	2;136.5	10.7	
		qHGW4.1	4;67.5	6.3	
		qHGW4.2	4;115.5	4.6	
		qHGW7	7;125.7	11.9	
10	Grain length/width	qL/W1.1	1;69.2	7.5	Li et al. (2023)
		qL/W1.2	1;87.6	54.85	
		qL/W1.2	1;87.6	58.79	
		qL/W1.2	1;87.6	49.36	
		qL/W1.3	1;96.7	29.5	
		qL/W1.3	1;96.7	27.25	
11	Grain yield	qGY1.1	1;35.1	10.36	Li et al. (2023)
		qGY1.2	1;37.6	10.56	
		qGY1.2	1;38.0	9.2	
		qGY1.2	1;41.9	10.84	
		qGY1.2	1;42.9	7.51	
		qGY1.3	1;47.4	9.03	
		qGY1.4	1;163.8	5.7	
		qGY1.5	1.173.7	5.94	
		qGY7.1	7;65.1	5.16	
		qGY8.1	8;163.6	6.41	

two EST regions linked to PS that explained 14.2 and 20.0% of the PVE (phenotypic variance), respectively. Hara et al. (2020) studied PS in 15 accessions from various regions of Japan, and analyzed QTLs using F2 segregating population. They identified two QTLs in the KUZUU ZAIRAI (KUZ) × Buckwheat Norin PL1 (BNPL1) and three QTLs in the MIYAZAKI (MIZ) × Kyukei SC2 (KSC2) F2 segregating population with 12 h photoperiod. With the 15 h photoperiod three QTLs in each F2 segregating population are detected. Yabe et al. (2014) studied QTL mapping and identified four QTLs. The four QTLs explained 5.64–8.51% of the PVE (phenotypic variance) for main stem length. In the A 1 hybrid, they discovered two main and two minor QTLs. A study was conducted focused on mapping Quantitative Trait Loci (QTLs) associated with the tolerance of preharvest sprouting. In their study, they analyzed three different buckwheat crosses: A_1, B_1, and C. For the A_1 cross, they identified two minor and two major genetic regions (QTLs) that together explained 42.2% of the variation in the observed traits. In the B_1 cross, they found one minor and four major QTLs that explained 66.4% of the trait variance. In the C cross, one minor and two major QTLs were identified, explaining 38.0% of the trait variance (Takeshima et al. 2021). Similarly, another study focused on the thousand-grain weight (TGW) in buckwheat and discovered multiple QTLs located on Chromosomes 4 and 1. Specifically, they found nine TGW QTLs dispersed across these chromosomes. Within the region spanning from 38.2 to 39.8 centimorgans (cM) on Chromosome 1, a significant and reliable genetic locus was identified, accounting for a substantial portion of the phenotypic variance (ranging from 23.6% to 47.5%) (Shi et al. 2021). In a subsequent study by (Takeshima et al. 2022), two F2 populations of common buckwheat were created to investigate QTLs related to seed maturation time. They identified two major QTLs for seed maturation time and found QTLs at distinct loci affecting flowering time, suggesting separate genetic mechanisms controlling maturity time and flowering time. One significant QTL (qMT6_KTW) for seed maturation time was located on Linkage Group 6, explaining 21.0% of the phenotypic variance. Additionally, three QTLs associated with flowering time were identified in cross A, with the most influential one being qFT3_KTW, explaining 15.4% of the phenotypic variance. In cross B_1, one significant QTL for maturity time (qMT3_RCK) was found on Linkage Group 3, explaining 20.5% of the phenotypic variance, and one QTL for flowering time (qMT7_RCK) on Linkage Group 7, explaining 13.6% of the phenotypic variance. Later on, 14 QTL was identified in common buckwheat. These QTLs were distributed among various traits, including six associated with grain length (GL), three with grain width (GW), and five with hundred-grain weight (HGW). Interestingly, QTLs for both GW and HGW were found in close proximity to genetic markers known as SWU_Fe_InDel076 and Fe_InDel086 (Fang et al. 2022). Recently, 32 QTLs in Tartary buckwheat has been identified which were associated with various traits, including seven related to grain yield (GY), 5–1000 grain weight, 11 to grain width (GW), six to grain length, and three to the length-to-width ratio. Notably, on Chromosome Ft1, the researchers observed the presence of two distinct QTL clusters, named *qclu-1-5* and *qclu-1-3* (Li et al. 2023).

6.7 Whole Genome Sequencing

In 2016, the Buckwheat Genome DataBase (BGDB) has been developed (Yasui et al. 2016). Next-generation sequencing (NGS) was used to generate a preliminary assembly of the buckwheat genome, resulting in 387,594 scaffolds referred to as the draft genome sequence (FES r1.0). The N50 scaffolds had a size of 25,109 base pairs, and the total length of FES r1.0 amounted to 1,177,687,305 base pairs. This database has wide applications in molecular breeding programs, facilitating the rapid identification of gene homologues. It is particularly valuable for identifying genes involved in flavonoid biosynthesis pathways, genes encoding 2S albumin-type allergens, and granule-bound starch synthases (GBSSs). Some researchers have opted to create their own de novo transcriptome assemblies (Xu et al. 2017; Fang et al. 2019). In 2017, a study presented a high-quality, chromosome-scale Tartary buckwheat genome sequence totaling 489.3 megabases (Mb). They accomplished this by combining whole genome shotgun sequencing with single molecule real-time long reads, Illumina short reads, sequence tags from extensive Hi-C sequencing data, BioNano genome maps, and DNA insert library data (Zhang et al. 2017). Through expression data analysis, they confidently identified 33,366 protein-coding genes. More recently, in 2021, the common buckwheat genome has been assembled using long-read technology and a high-resolution expression atlas. In that study, 46 different organs and developmental stages were analyzed, revealing a substantial presence of transposable elements (TE) in the genome (Penin et al. 2021). The main reason for the threefold increase in the genome size of common buckwheat in comparison to its closely related species, *F. tataricum*, is primarily attributed to the difference in the content of TE. They characterized the genome of *F. esculentum*, measuring 1.5 gigabases (Gb) in size, and created a reference assembly that significantly enhanced the representation of protein-coding genes while achieving high contiguity.

6.8 Epigenetics and Epigenomics in Buckwheat

Epigenetics has recently emerged as a critical area of research due to the ongoing climate changes which affects the plant growth, development, and crop productivity. In response to these changes, plants have changed biological mechanism including transcription control, epigenetic regulation, physiological and metabolic reprogramming (Golldack et al. 2011; Sun et al. 2022). Epigenetic changes play a crucial role in the control of chromatin modelling, gene expression, genomic imprinting and finally phenotypes (Gahlaut et al. 2021). Many studies in various crops have indicated that epigenetic alterations, including as histone modification, miRNA and DNA methylation are quite significant and are contributing to a significant alteration in phenotype (Shi et al. 2015; Neto et al. 2017). In buckwheat, DNA methylation act as epigenetic regulator for *F. tataricum* in response to cold. To demonstrate the phenotypic and transcript abundance variations of HDACs under various low temperatures, the cold-tolerant variety Dingku 1 was used. OE-*AtHDA6* and OE-*FtHDA6-1* overexpression lines were created and found that *FtHDA6-1* positively regulated cold tolerance (Hou et al. 2022a, b). A DNA methylation inhibitor was applied, and this resulted in a shift in the free Lys level, indicating that methylation on DNA can influence the accumulation of metabolite for *F. tataricum* cold responses (Song et al. 2020).

Histone deacetylases (HDACs), which are extensively distributed in many types of eukaryotic cells, are essential for many biological processes, including plants' reactions to biotic and abiotic stress. In *F. tataricum*, 14 potential *FtHDAC* genes were discovered and annotated. It was discovered that the three main subgroups of Tartary buckwheat HDACs were HD2-like, RPD3/HDA1 and SRT which is related to Arabidopsis. Tartary buckwheat is resistant to cold and more drought resistance than the common buckwheat. Tartary buckwheat has drought endurance traits, whereas common buckwheat has drought avoidance qualities (Aubert et al. 2021). So, buckwheat species differ in their growth characteristics (Kasajima et al. 2019).

MicroRNAs are also involved in stress response, growth, ageing, death, and metabolism (Ma et al. 2020). A significant number of MicroRNAs, such as *miR156–SPL2* pair in cotton, are involved in responses to salt stress (Wang et al. 2013), *miR160* in *Populus euphratica* (Li et al. 2013a) but only a few studies of salt-responsive miRNAs in *F. tataricum* is done. Chuanqiao No. 2 is a salt sensitive variety of Tartary buckwheat. Various 5-azaC concentrations were employed to infect *F. tataricum* leaves, and when coupled with gene expression analyses, it was shown that 100 M was the right amount to use. The gene promoter region frequently experiences DNA methylation, which alters the transcription of genes and controls the physiological resistance to salt stress. It was discovered that DNA methylation strongly influenced the expression of the *FtNHX1* gene (Wang et al. 2022).

6.9 Breeding in Buckwheat

During the early twentieth century, worldwide efforts were initiated to improve the genetics of buckwheat, focusing on collecting, characterizing, and conserving buckwheat accessions. However, breeding programs for buckwheat face significant challenges. Common buckwheat, in particular, presents difficulties in developing purebred cultivars and stabilizing economically desirable traits. These challenges arise from factors such as the presence of various types of self-incompatibility, dimorphic characteristics, apomixis, low seed production, sterility, and genetic instability (Yasui et al. 2016). Buckwheat breeding programs have specific objectives, including reducing seed shattering, promoting determinate growth, achieving synchronized flowering, and reducing plant height for the purpose of automated harvesting (Fig. 6.4) (Campbell 1995). Traditional breeding methods face significant barriers to hybridizing *F. esculentum* and *F. tataricum* due to strong crossability barriers (Adachi et al. 1989; Samimy 1991; Asaduzzaman et al. 2009). In contrast to Tartary buckwheat, which is a self-pollinating species, common buckwheat (*F. esculentum*) is self-incompatible and requires cross-pollination (Woo et al. 2010). This makes it easier to generate diverse offspring and maintain the genetic

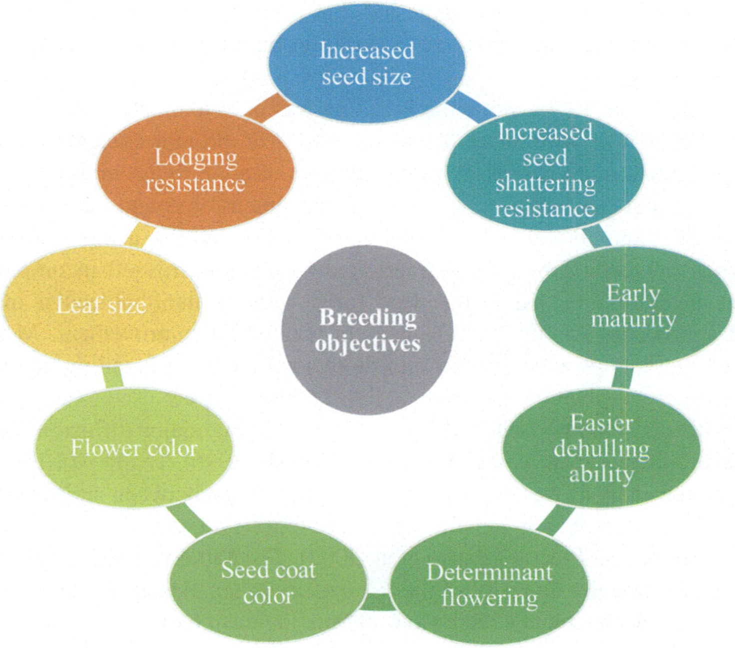

Fig. 6.4 Objectives related to the improvement of specific characteristics in buckwheat through breeding efforts

integrity of advanced breeding lines in Tartary buckwheat. Although there have been descriptions of common buckwheat lines that are homomorphic, self-compatible, and capable of self-fertilization (Marshall 1969), their use in breeding programs is often limited due to the occurrence of severe inbreeding depression. Many attempts were made to improve the buckwheat genus through interspecific hybridization were largely failed due to the presence of great incompatibility barriers. However, with the induction of embryo rescue techniques, ovule rescue techniques, pollen-anther related breeding programs and other molecular techniques for managing backcross generations increasing the feasibility of interspecific hybridization.

In previous study, it was reported that buckwheat anther cultivation method that may result in the induction of haploid or spontaneous diploid plants (Bohanec et al. 1993). Björkman (1995) investigates pollen competition in buckwheat, finding that it improves progeny performance. *Fagopyrum* interspecific crossings were done by in order to investigate pollen tube behavior in relation to self-incompatibility and found that species with longer styles have prominent pollen tube development and that dimorphic self-incompatibility affects interspecific crossovers (Hirose et al. 1995). Additionally, more attempts were made and succeeded while crossing via embryo rescue techniques between *F. esculentum* and *F. cymosum* tetraploid levels (Ujihara et al. 1990; Hirose et al. 1993; Suvorova et al. 1994; Wang and Campbell 1998; Woo et al. 2018). The first interspecific buckwheat hybridization at the diploid level was reported and suggested that progeny from this hybridization is viable, and subsequent backcrosses to common buckwheat have been conducted (Campbell 1995).

Beginning in the 1970s, mutation breeding was used to create genetic polymorphism in buckwheat with an emphasis on seed size, protein content, amino acid profile, and economic features. With the addition of a family group selection strategy, induced mutagenesis resulted in the advancement of potent cultivars such as Aelita, Lada, and Podolyanka (Alekseeva 1984). While using conventional approaches for breeding in buckwheat various difficulties such as self-incompatibility, allergens in grains etc. led to the need for development of molecular markers-assisted approaches for further advances in buckwheat improvement. Moreover, molecular markers provide quicker and more accurate plant breeding methods by identifying certain plant characteristics. RAPD markers were used to study the origin of cultivated common buckwheat and to assess the major diffusion routes of buckwheat cultivation in world (Murai and Ohnishi 1996). Using AFLP markers, bulk segregate analysis was conducted to identify relation between molecular markers and self-incompatibility gene (Nagano et al. 2001). Later, five AFLP markers associated to the *Sht 1* locus (Matsui et al. 2004). The 136 novel SSR markers were created to examine the diversity of the *F. esculentum ssp* (Ma et al. 2009). Rutin is a component of the plant's defensive system against biotic and abiotic stress (Treutter 2006). As a result, improving rutin content is a crucial breeding topic. SunRutin and Toyomusume are two high-rutin common buckwheat types that have so far been created through machine or hand selection (Minami 2001; Ito 2005). It is possible to create functional food products from Tartary and common buckwheat grains that have the right proportions of amino acids containing selenium. Buckwheat by-products are an important source of bioactive compounds which prevent colon cancer by limiting cell proliferation. A viable method to raise the amount of protein in Tartary and common buckwheat grain is to breed buckwheat with larger embryos because the embryo in grain contains a high number of proteins (Luthar et al. 2020).

6.10 Conclusion

In summary, the integration of genetics, genomics, epigenetics, and breeding techniques has advanced our understanding of buckwheat biology and paved the way for the development of improved buckwheat varieties that are better suited to meet the challenges of modern agriculture and the needs of consumers. This research not only benefits buckwheat production but also contributes to sustainable and nutritious food systems. All information, including molecular markers, linkage maps, QTLs, genes, mapping populations, natural germplasm and mutant lines, and genome/epigenome/transcriptome, is compiled. The advancement of specialized mapping population and molecular techniques such as map-based cloning, physical mapping, and genome editing technologies may open up new opportunities in buckwheat breeding programs.

References

Adachi T, Yamaguchi A, Miike Y, Hoffmann F (1989) Plant regeneration from protoplasts of common buckwheat (*Fagopyrum esculentum*). Plant Cell Rep 8:247–250

Ahmad M, Ahmad F, Dar EA, Bhat RA, Mushtaq T, Shah F (2018) Buckwheat (*Fagopyrum esculentum*)–a neglected crop of high altitude cold arid regions of ladakh: biology and nutritive value. Int J Pure Appl Bio 6(1):395–406

Ahmed A, Khalid N, Ahmad A, Abbasi NA, Latif MS, Randhawa MA (2014) Phytochemicals and biofunctional properties of buckwheat: a review. J Agric Sci 152(3):349–369

Aii J, Nagano M, Penner AG, Campbell GC, Adachi T (1998) Identification of RAPD markers linked to the homostylar (Ho) gene in buckwheat. Jap J Breed 48(1):59–62

Alekseeva ES (1984) Experimental mutagenesis as a method of breeding work in buckwheat. Fagopyrum 4:23–29

Asaduzzaman M, Minami M, Matsushima K, Nemoto K (2009) An in vitro ovule culture technique for producing interspecific hybrid between tartary buckwheat and common buckwheat. J Biol Sci 9(1):1–11

Aubert L, Decamps C, Jacquemin G, Quinet M (2021) Comparison of plant morphology, yield and nutritional quality of *Fagopyrum esculentum* and *Fagopyrum tataricum* grown under field conditions in Belgium. Plan Theory 10(2):258

Baljeet SY, Ritika BY, Roshan LY (2010) Studies on functional properties and incorporation of buckwheat flour for biscuit making. Int Food Res J 17(4)

Bashir EM, Ali AM, Ali AM, Mohamed ET, Melchinger AE, Parzies HK, Haussmann BI (2015) Genetic diversity of Sudanese pearl millet (*Pennisetum glaucum L.* R. Br.) landraces as revealed by SSR markers, and relationship between genetic and agro-morphological diversity. Genet Res Crop Evol 62:579–591

Bashir E, Mahajan R, Mir RA, Dar WA, Zargar SM (2021) Unravelling the genetic variability and population structure of buckwheat (*Fagopyrum* spp.) a collection of north western Himalayas. Nucleus 64:93–101

Bindler G, van der Hoeven R, Gunduz I, Plieske J, Ganal M, Rossi L, Gadani F, Donini P (2007) A microsatellite marker based linkage map of tobacco. Theor Appl Genet 114:341–349

Björkman T (1995) Gametophyte selection through pollen competition in buckwheat. Curr Adv Buckwheat Res:443–451

Bohanec B, Nešković M, Vujičić R (1993) Anther culture and androgenetic plant regeneration in buckwheat (*Fagopyrum esculentum Moench*). Plant Cell Tissue Organ Cult 35:259–266

Campbell CG (1995) Inter-specific hybridization in the genus *Fagopyrum*. Curr Adv Buckwheat Res I–III:255–263

Cepková PH, Janovská D, Stehno Z (2009) Assessment of genetic diversity of selected Tartary and common buckwheat accessions. Span J Agric Res 4:844–854

Chen QF (1999) A study of resources of *Fagopyrum* (Polygonaceae) native to China. Bot J Linn Soc 130(1):53–64

Chen QF, Hsam SL, Zeller FJ (2004) A study of cytology, isozyme, and interspecific hybridization on the big-achene Group of Buckwheat Species (*Fagopyrum*, Polygonaceae). Crop Sci 44(5):1511–1518

Dar FA, Pirzadah TB, Malik B, Tahir I, Rehman RU (2018) Molecular genetics of buckwheat and its role in crop improvement. In: Buckwheat Germplasm in the World AP, pp 271–286

Deng L, Zhang K, Huang K, Chen Q (2011) RAPD analysis for genetic diversity of nineteen common and Tartary buckwheat varieties. Agric Sci Technol-Hunan 12(1):65–69

Diels L (1901) Die flora von Central China. Bot Jahrb Syst Pflanzengesch Pflanzengeogr 29:169–659

Du X, Zhang Z, Wu B, Li Y, Yu AH (2013) Construction of genetic linkage map in tartary buckwheat (*Fagopyrum tataricum*) using SSR. Chin Agric Sci Bull 29(21):61–65

Duan Y, Yin G, He R, Yang X, Cai S, Wang Y, Lu W, Sun D, Wang L, Wang Y, Sun H (2022) Identification of candidate genes for easily-shelled traits in Tartary buckwheat based on BSA-Seq and RNA-Seq methods. Euphytica 218(7):91

Edwardson SE, Matano T, Ujihara A (1995) Using growing degree days to estimate optimum windrowing time in buckwheat. Curr Adv Buckwheat Res 70:509–514

Fabjan N, Rode J, Košir IJ, Wang Z, Zhang Z, Kreft I (2003) Tartary buckwheat (*Fagopyrum tataricum Gaertn.*) as a source of dietary rutin and quercitrin. J Agric Food Chem 51(22):6452–6455

Fang X, Zhang Y, Zhang Y, Huang K, Yang W, Li X, Zhang Z, Wu K, Xu X, Ruan R, Yuan X (2019) De novo transcriptome assembly and identification of genes related to seed size in common buckwheat (*Fagopyrum esculentum M.*). Breed Sci 69(3):487–497

Fang X, Zhang Y, Cui J, Yue L, Tao J, Wang Y, Zhang R, Liu J, Jiang A, Zhang J, Ruan R (2022) Genetic map construction, QTL mapping, and candidate genes screening of grain size traits in common buckwheat. (*Fagopyrum esculentum M.*). Agron 12(9):2062, vol 12

Gaertner J (1791) De fructibus et seminibus plantarum, vol 2. Typis Academiae Carolinae, Stuttgart, p 520

Gahlaut V, Kumari P, Jaiswal V, Kumar S (2021) Genetics, genomics and breeding in Rosa species. J Hortic Sci Biotechnol 96(5):545–559

Gao XD, Kim JH, Park CH, Hong SK (2010) Studies on genetic diversity of buckwheat germplasms. Korean J Plant Res 23(3):214–222

Giménez-Bastida JA, Piskula MK, Zielinski H (2015) Recent advances in processing and development of buckwheat derived bakery and non-bakery products—a review. Polish J Food Nutr Sci 65(1):9–20

Golldack D, Luking I, Yang O (2011) Plant tolerance to drought and salinity: stress regulating transcription factors and their functional significance in the cellular transcriptional network. Plant Cell Rep 30:1383–1391

Govindaraj M, Vetriventhan M, Srinivasan M (2015) Importance of genetic diversity assessment in crop plants and its recent advances: an overview of its analytical perspectives. Genet Res Int 2015:1–14

Grahić J, Okić A, Šimon S, Djikić M, Gadžo D, Pejić I, Gaši F (2022) Genetic relationships and diversity of common buckwheat accessions in Bosnia and Herzegovina. Agronomy 12(11):2676

Grimmer MK, Trybush S, Hanley S, Francis SA, Karp A, Asher MJ (2007) An anchored linkage map for sugar beet based on AFLP, SNP and RAPD markers and QTL mapping of a new source of resistance to beet necrotic yellow vein virus. Theor Appl Genet 114(7):1151–1160

Gross MH (1913) Remarques sur les Polygonées de l'Asie Orientale. Bull Géogr Bot 23:7–32

Gupta N, Sharma SK, Rana JC, Chauhan RS (2012) AFLP fingerprinting of tartary buckwheat accessions (*Fagopyrum tataricum*) displaying rutin content variation. Fitoterapia 83(6):1131–1137

Hara T, Iwata H, Okuno K, Matsui K, Ohsawa R (2011) QTL analysis of photoperiod sensitivity in common buckwheat by using markers for expressed sequence tags and photoperiod-sensitivity candidate genes. Breed Sci 61(4):394–404

Hara T, Shima T, Nagai H, Ohsawa R (2020) Genetic analysis of photoperiod sensitivity associated with difference in ecotype in common buckwheat. Breed Sci 70(1):101–111

Haraldson K (1978) Anatomy and taxonomy in Polygonaceae sub fam. PolygonoideaeMeissn emend Jaretzky Symbolae Botamicae Upsalienses 22:1–95

Hedberg O (1946) Pollen morphology in the genus Polygonum ls lat. and its taxonomic significance. Sven Bot Tidskr 40:371–404

Hirose T, Kitabayashi H, Ujihara A, Minami M (1993) Morphology and identification by isozyme analysis of interspecific hybrids in buckwheats. Buckwheat Newsletter 13:25–30

Hirose T, Ujihara A, Kitabayashi H, Minami M (1995) Pollen tube behavior related to self-incompatibility in interspecific crosses of *Fagopyrum*. Jap J Breed 45(1):65–70

Hou LL, Zhou ML, Zhang Q, Qi LP, Yang XB, Tang Y, Zhu XM, Shao JR (2015) *Fagopyrum luojishanense*, a new species of Polygonaceae from Sichuan, China. Novon 24(1):22–26

Hou S, Sun Z, Linghu B, Xu D, Wu B, Zhang B, Wang X, Han Y, Zhang L, Qiao Z, Li H (2016) Genetic diversity of buckwheat cultivars (*Fagopyrum tartaricum Gaertn.*) assessed with SSR markers developed from genome survey sequences. Plant Mol Biol Report 34:233–241

Hou Y, Lu Q, Su J, Jin X, Jia C, An L, Tian Y, Song Y (2022a) Genome-wide analysis of the HDAC gene family and its functional characterization at low temperatures in Tartary buckwheat (*Fagopyrum tataricum*). Int J Mol Sci 23(14):7622

Hou X, Chen M, Chen Y, Hou X, Jia Z, Yang X, Zhang J, Liu Y, Ye N (2022b) Identifying QTLs for grain size in a colossal grain Rice (*Oryza sativa L.*) line, and analysis of additive effects of QTLs. Int J Mol Sci 23(7):3526

Ikeda K (2002) Buckwheat composition, chemistry, and processing. Adv Food Nutri Res 44:395–434

Ikeda S, Yamashita Y (1994) Buckwheat as a dietary source of zinc, copper and manganese. Fagopyrum 14(29):34

Ito S (2005) A new common buckwheat variety "*Toyomusume*" with high yield and higher rutin content. Bull Natl Agric Res Cent 6:1–13

Iwata H, Imon K, Tsumura Y, Ohsawa R (2005) Genetic diversity among Japanese indigenous common buckwheat (*Fagopyrum esculentum*) cultivars as determined from amplified fragment length polymorphism and simple sequence repeat markers and quantitative agronomic traits. Genome 48(3):367–377

Jaiswal V, Gupta S, Gahlaut V, Muthamilarasan M, Bandyopadhyay T, Ramchiary N, Prasad M (2019) Genome-wide association study of major agronomic traits in foxtail millet (*Setariaitalica L.*) using ddRAD sequencing. Sci Rep 9:1–11

Kasajima S, Yoshimaru I, Itoh H (2019) Differentiation and growth of a growing point until the stage of flower bud appearance in leading common buckwheat and Tartary buckwheat varieties in northern Japan. Fagopyrum 36(2):37–41

Katsube-Tanaka T (2016) Buckwheat production, consumption, and genetic resources in Japan. In: Molecular breeding and nutritional aspects of buckwheat AP, pp 61–80

Kim SL, Kim SK, Park CH (2004) Introduction and nutritional evaluation of buckwheat sprouts as a new vegetable. Food Res Int 37(4):319–327

Kishore G, Pandey A, Dobhal R, Gupta S (2013) Population genetic study of *Fagopyrum*tataricum from Western Himalaya using ISSR markers. Biochem Genet 51:750–765

Konishi T, Ohnishi O (2006) A linkage map for common buckwheat based on microsatellite and AFLP markers. Fagopyrum 23(2):1–6

Kump B, Javornik B (1996) Evaluation of genetic variability among common buckwheat (*Fagopyrum esculentum Moench*) populations by RAPD markers. Plant Sci 114(2):149–158

Kump B, Javornik B (2002) Genetic diversity and relationships among cultivated and wild accessions of tartary buckwheat (*Fagopyrum tataricum Gaertn.*) as revealed by RAPD markers. Genet Resour Crop Evol 49(6):565–572

Li Q (1992) Preliminary investigation on buckwheat origin in Yunnan, China. In: Proc. 5th Intl. Symp. Buckwheat, Taiyuan, pp 44–48

Li SQ, Zhang QH (2001) Advances in the development of functional foods from buckwheat. Crit Rev Food Sci Nutr 41(6):451–464

Li B, Duan H, Li J, Deng XW, Yin W, Xia X (2013a) Global identification of miRNAs and targets in *Populus euphratica* under salt stress. Plant Mol Biol 81:525–539

Li FL, Zeller FJ, Huang KF, Shi TX, Chen QF (2013b) Improvement of fluorescent chromosome in situ PCR and its application in the phylogeny of the genus *FagopyrumMill.* using nuclear genes of chloroplast origin (cpDNA). Plant Syst Evol 299:1679–1691

Li R, Chen Z, Zheng R, Chen Q, Deng J, Li H, Huang J, Liang C, Shi T (2023) QTL mapping and candidate gene analysis for yield and grain weight/size in Tartary buckwheat. BMC Plant Biol 23(1):1–16

Liang LB (2016) The study of main agronomic traits and SSR molecular markers in genetic population of Tartary buckwheat. Guizhou Normal University, pp 1–88

Liu JL, Tang Y, Xia MZ, Shao JR, Cai GZ, Luo Q, Sun JX (2008) *Fagopyrum crispatifolium* J. L. Liu, a new species of Polygonaceae from Sichuan, China. J Syst Evol 46:929–932

Luthar Z, Zhou M, Golob A, Germ M (2020) Breeding buckwheat for increased levels and improved quality of protein. Plan Theory 10(1):14

Ma KH, Kim NS, Lee GA, Lee SY, Lee JK, Yi JY, Park YJ, Kim TS, Gwag JG, Kwon SJ (2009) Development of SSR markers for studies of diversity in the genus Fagopyrum. Theor Appl Genet 119(7):1247–1254

Ma F, Lu GA, Chen Q, Ruan Y, Li X, Lu X, Li C (2020) Dynamic global analysis of transcription reveals the role of miRNAs in synergistic stabilization of gene expression. Sci Bull 65:2130–2140

Malysheva-Otto L, Ganal MW, Law JR, Reeves JC, Röder MS (2007) Temporal trends of genetic diversity in European barley cultivars (*Hordeum vulgare L.*). Mol Breed 20:309–322

Marshall HG (1969) Isolation of self-fertile, homomorphic forms in buckwheat, *Fagopyrum sagittatum Gilib* 1. Crop Sci 9(5):651–653

Matsui K, Kiryu Y, Komatsuda T, Kurauchi N, Ohtani T, Tetsuka T (2004) Identification of AFLP makers linked to non-seed shattering locus (sht1) in buckwheat and conversion to STS markers for marker-assisted selection. Genome 47(3):469–474

Meisner CF (1857) Polygonaceae. In: de Candolle A (ed) Prodromus Systematis Naturalis Regni Vegetabilis, vol 14. Victoris Masson, Paris, pp 1–186

Mikolášová L, Hromadová Z, Vivodík M, Chňapek M, Gálová Z, Gregáňová RH, Balazova Z (2022) Genetic variability of *Fagopyrum SP.* genotypes determined by gene-targeted markers. J Microbiol Biotechnol Food Sci 12:e9308

Minami M (2001) Breeding of high rutin content common buckwheat "SunRutin." In: Advances in Buckwheat Research In: Proc 8th Intl Symp Buckwheat at Chunchon 2001

Mizuno N, Yasui Y (2019) Gene flow signature in the S-allele region of cultivated buckwheat. BMC Plant Biol 19(1):1–9

Moench C (1794) Methodus plantas horti botanici et agri Marburgensis: a staminum situ describendi. Marburg 2(790)

Murai M, Ohnishi O (1996) Population genetics of cultivated common buckwheat, *Fagopyrum esculentum Moench.* X. Diffusion routes revealed by RAPD markers. Genes Genet Syst 71(4):211–218

Nagano M, Aii J, Kuroda M, Campbell C, Adachi T (2001) Conversion of AFLP markers linked to the Sh allele at the S locus in buckwheat to a simple PCR based marker form. Plant Biotechnol 18(3):191–196

Nazir M, Mahajan R, Hashim MJ, Iqbal J, Alyemeni MN, Ganai BA, Zargar SM (2021) Deciphering allelic variability and population structure in buckwheat: an analogy between the efficiency of ISSR and SSR markers. Saudi J Biol Sci 28(11):6050–6056

Neto JRCF, Daniel da Silva M, Pandolfi V, Crovella S, Maria Benko-Iseppon A, Akio Kido E (2017) Epigenetic signals on plant adaptation: a biotic stress perspective. Curr Protein Pept Sci 18(4):352–367

Ohnishi O (1988) Search for the wild ancestor of buckwheat I. Description of new *Fagopyrum* (Polygonaceae) species and their distribution in China and the Himalayan hills. Fagopyrum 15:18–28

Ohnishi O (1993) Population genetics of cultivated common buckwheat, *Fagopyrum esculentum Moench.* IX. Concluding remarks on worldwide survey of allozyme variability. Jap J Genet 68(4):317–326

Ohnishi O, Matsuoka Y (1996) Search for the wild ancestor of buckwheat II. Taxonomy of *Fagopyrum* (Polygonaceae) species based on morphology, isozymes and cpDNA variability. Genes Genet Syst 71(6):383–390

Ohnishi O, Ohta T (1987) Construction of a linkage map in common buckwheat, *Fagopyrum esculentum Moench.* Jap J Genet 62(5):397–414

Ohnishi O, Yasui Y (1998) Search for wild buckwheat species in high mountain regions of Yunnan and Sichuan provinces of China. Fagopyrum 15:8–17

Ohsako T, Ohnishi O (1998) New *Fagopyrum* species revealed by morphological and molecular analyses. Genes Genet Syst 73:85–94

Ohsako T, Yamane K, Ohnishi O (2002) Two new *Fagopyrum* (Polygonaceae) species, *F. Gracilipedoides* and *F. Jinshaense* from Yunnan, China. Genes Genet Syst 77(6):399–408

Oshawa R, Tsitsumi T, Uehara H, Namai H, Ninomiya S (1998) Quantitative evaluation of common buckwheat (*Fagopyron esculentum Moench*) kernel shape by elliptic Fourier descriptors. Euphytica 101:175–183

Pan SJ, Chen QF (2010) Genetic mapping of common buckwheat using DNA, protein and morphological markers. Hereditas 147(1):27–33

Paudel MN, Joshi BK, Ghimire KH (2016) Management status of agricultural plant genetic resources in Nepal. Agron J Nepal 4:75–91

Penin AA, Kasianov AS, Klepikova AV, Kirov IV, Gerasimov ES, Fesenko AN, Logacheva MD (2021) High-resolution transcriptome atlas and improved genomeassembly of common buckwheat, *Fagopyrum esculentum*. Front Plant Sci 12:612382

Rana JC (2004) Buckwheat genetic resources management in India. In: Proc 9th Int Symp buckwheat, Prague, pp 271–282

Rana JC, Singh M, Chauhan RS, Chahota RK, Sharma TR, Yadav R, Archak S (2016) Genetic resources of buckwheat in India. In: Molecular breeding and nutritional aspects of buckwheat AP, pp 109–135

Romanova OI, Fesenko AN, Fesenko NN, Fesenko IN, Koshkin VA (2018) Buckwheat resources in the VIR (Russia) collection: the photoperiod response. In: Buckwheat germplasm in the world AP, pp 225–234

Samimy C (1991) Barrier to interspecific crossing of *Fagopyrum esculentum* with *Fagopyrum tataricum*: I. Site of pollen-tube arrest. II. Organogenesis from immature embryos of *F. Tataricum*. Euphytica 54:215–219

Senthilkumaran R, Bisht IS, Bhat KV, Rana JC (2008) Diversity in buckwheat (*Fagopyrum spp.*) landrace populations from north-western Indian Himalayas. Genet Resour Crop Evol 55:287–302

Shao JR, Zhou ML, Zhu XM, Wang DZ, Bai DQ (2011) *Fagopyrum wenchuanense* and *Fagopyrum qiangcai*, two new species of Polygonaceae from Sichuan, China. Novon 21(2):256–261

Sharma T, Jana S (2002) Species relationships in *Fagopyrum* revealed by PCR-based DNA fingerprinting. Theor Genet 105:306–312

Shi Z, Han C, Huang K (2011) Research on se content of different tartary buckwheat genotypes. Agric Sci Technol-Hunan 12(1):102–156

Shi J, Dong A, Shen WH (2015) Epigenetic regulation of rice flowering and reproduction. Front Plant Sci 5:803

Shi TX, Li RY, Zheng R, Chen QF, Li HY, Huang J, Zhu LW, Liang CG (2021) Mapping QTLs for 1000-grain weight and genes controlling hull type using SNP marker in Tartary buckwheat (*Fagopyrum tataricum*). BMC Genomics 22(1):1–12

Shukla A, Srivastava N, Suneja P, Yadav SK, Hussain Z, Rana JC, Yadav S (2018) Genetic diversity analysis in buckwheat germplasm for nutritional traits. IJEB 56(11):827–837

Singh M, Malhotra N, Sharma K (2020) Buckwheat (*Fagopyrum sp.*) genetic resources: what can they contribute towards nutritional security of changing world? Genet Resour Crop Evol 67(7):1639–1658

Skerritt JH (1986) Molecular comparison of alcohol-soluble wheat and buckwheat proteins. Cereal Chem 63(4):365–369

Song JY, Lee GA, Yoon MS, Ma KH, Choi YM, Lee JR, Jung YJ, Park HJ, Kim CK, Lee MC (2018) Analysis of genetic diversity and population structure of buckwheat (Fagopy*rum esculentum Moench.*) landraces of Korea using SSR markers. In: Buckwheat germplasm in the world AP, pp 315–331

Song Y, Jia Z, Hou Y, Ma X, Li L, Jin X, An L (2020) Roles of DNA methylation in cold priming in tartary buckwheat. Front Plant Sci 11:608540

Song Y, Cheng Z, Dong Y, Liu D, Bai K, Jarvis D, Feng J, Long C (2022) Diversity of Tartary buckwheat (*Fagopyrum tataricum*) landraces from Liangshan, Southwest China: evidence from morphology and SSR markers. Agronomy 12(5):1022

Sun M, Yang Z, Liu L, Duan L (2022) DNA methylation in plant responses and adaption to abiotic stresses. Int J Mol Sci 23(13):6910

Suvorova GN, Fesenko NN, Kostrubin MM (1994) Obtaining of interspecific buckwheat hybrid (*Fagopyrum esculentum Moench* × *Fagopyrum cymosumMeissn.*). Fagopyrum 14:13–16

Takeshima R, Ogiso-Tanaka E, Yasui Y, Matsui K (2021) Targeted amplicon sequencing+ next-generation sequencing–based bulked segregant analysis identified genetic loci associated with preharvest sprouting tolerance in common buckwheat (*Fagopyrum esculentum*). BMC Plant Biol 21(1):1–13

Takeshima R, Yabe S, Matsui K (2022) Genetic basis of maturity time is independent from that of flowering time and contributes to ecotype differentiation in common buckwheat (*Fagopyrum esculentum Moench*). BMC Plant Biol 22(1):353

Tang Z, Huang L, Gou J, Chen H, Han X (2014) Genetic relationships among buckwheat (*Fagopyrum*) species from Southwest China based on chloroplast and nuclear SSR markers. J Genet 93:849–853

Taoxiong SH, Ruiyuan LI, Fan PA, Juan HU, Liwei ZH, Yan WA, Chenggang LI (2021) Correlation between flavonoids content and SSR markers of Tartary buckwheat. Fujian J Agric Sci 36(8):884–891

Treutter D (2006) Significance of flavonoids in plant resistance: a review. Environ Chemi Lett 4(3):147–157

Tsuji K, Ohnishi O (2001) Phylogenetic relationships among wild and cultivated Tartary buckwheat (*Fagopyrum tataricum Gaert.*) populations revealed by AFLP analyses. Genes Genet Syst 76(1):47–52

Ujihara A, Nakamura Y, Minami M (1990) Interspecific hybridization in genus *Fagopyrum*-properties of hybrids (*F. Esculentum Moench*× *F. Cymosum Meissner*) through ovule culture. Gamma Field Symp 29(5):45–53

Wang YJ, Campbell CG (1998) Interspecific hybridization in buckwheat among *Fagopyrum esculentum, F. homotropicum* and *F. tataricum*. Adv Buckwheat Res 7:I1–I12

Wang Y, Scarth R, Campbell C (2005) Interspecific hybridization between diploid *Fagopyrum esculentum* and tetraploid *F. Homotropicum*. Can J Plant Sci 85(1):41–48

Wang M, Wang Q, Zhang B (2013) Response of miRNAs and their targets to salt and drought stresses in cotton (*Gossypium hirsutum L.*). Gene 530(1):26–32

Wang YQ, Song JN, Yang HB (2022) DNA methylation regulates the expression of salt tolerance gene FtNHX1 in Tartary buckwheat. Theor Exp Plant Physiol 34(2):185–195

Woo SH, Kamal AM, Tatsuro S, Campbell CG, Adachi T, Yun YH, Chung KY, Choi JS (2010) Buckwheat (*Fagopyrum esculentum Moench.*): concepts, prospects and potential. Eur J Plant Sci Biotechnol 4(1):1–16

Woo SH, Roy SK, Kwon SJ, Cho SW, Park JH (2018) Production of interspecific hybridization between *Fagopyrum tataricum* and *F. Cymosum* through embryo rescue. J Agr Sci 34(1)

Xu JM, Fan W, Jin JF, Lou HQ, Chen WW, Yang JL, Zheng SJ (2017) Transcriptome analysis of Al-induced genes in buckwheat (*Fagopyrum esculentum Moench)* root apex: new insight into Al toxicity and resistance mechanisms in an Al accumulating species. Front Plant Sci 8:1141

Yabe S, Hara T, Ueno M, Enoki H, Kimura T, Nishimura S, Yasui Y, Ohsawa R, Iwata H (2014) Rapid genotyping with DNA micro-arrays for high-density linkage mapping and QTL mapping in common buckwheat (*Fagopyrum esculentum Moench*). Breed Sci 64(4):291–299

Yabe S, Hara T, Ueno M, Enoki H, Kimura T, Nishimura S, Yasui Y, Ohsawa R, Iwata H (2018) Potential of genomic selection in mass selection breeding of an allogamous crop: an empirical study to increase yield of common buckwheat. Front Plant Sci 9:276

Yasui Y, Wang Y (2001) Construction of genetic maps of common buckwheat (*Fagopyrum esculentum Moench*) and its wild relative, *F. Homotropicum* Ohnishi based on amplified fragment length polymorphism (AFLP) markers. Proc. 8th Intl Symp Buckwheat 30:225–232

Yasui Y, Wang Y, Ohnishi O, Campbell CG (2004) Amplified fragment length polymorphism linkage analysis of common buckwheat (*Fagopyrum esculentum*) and its wild self-pollinated relative *Fagopyrum homotropicum*. Genome 47(2):345–351

Yasui Y, Hirakawa H, Ueno M, Matsui K, Katsube-Tanaka T, Yang SJ, Aii J, Sato S, Mori M (2016) Assembly of the draft genome of buckwheat and its applications in identifying agronomically useful genes. DNA Res 23(3):215–224

Yururi L, Zhongging W (1984) Production of buckwheat in the autonomous province of Inner Mongolia. Fagopyrum 4:7–13

Zhang YZ, Chen QF (2004) The actuality and expectation on buckwheat studies. Seeds 23:39–42

Zhang L, Li X, Ma B, Gao Q, Du H, Han Y, Li Y, Cao Y, Qi M, Zhu Y, Lu H (2017) The Tartary buckwheat genome provides insights into rutin biosynthesis and abiotic stress tolerance. Mol Plant 10(9):1224–1237

Zhang K, Fan Y, Weng W, Tang Y, Zhou M (2021) *Fagopyrum longistylum* (Polygonaceae), a new species from Sichuan, China. Phytotaxa 482(2):173–182

Zhou ML, Zhang Q, Zheng YD, Tang Y, Li FL, Zhu XM, Shao JR (2015) *Fagopyrum hailuogouense* (Polygonaceae), one new species from Sichuan, China. Novon 24(2):222–224

Zhou M, Kreft I, Suvorova G, Tang Y, Woo SH, editors (2018a) Buckwheat germplasm in the world. AP

Zhou M, Tang YU, Deng X, Ruan C, Kreft I, Tang Y, Wu Y (2018b) Overview of buckwheat resources in the world. In: Buckwheat germplasm in the world AP, pp 1–7

Zietkiewicz E, Rafalski A, Labuda D (1994) Genome fingerprinting by simple sequence repeat (SSR)-anchored polymerase chain reaction amplification. Genomics 20(2):176–183

Advancing Food Security with Genetic Resources of *Amaranthus*, Buckwheat, and *Chenopodium*

7

Kanishka R. C, Mithilesh Kumar, Gopal Katna, Kirti Rani, Krishan Prakash, Rakesh Kumar Bairwa, and Rahul Chandora

7.1 Introduction

Throughout history, human sustenance has relied on numerous plant species as vital sources of calories. However, a growing dependence has developed on just three major cereal crops: rice, wheat, and maize, which purportedly fulfil approximately half of the world's calorie requirements (Chang et al. 2019; Dawson et al. 2019). Intriguingly, the regions where these three crucial cereals originated are thought to have been the birthplaces of three of the oldest civilizations in human history. These major staple crops are sufficient to overcome hunger but fail to provide all the essential micronutrients, which are a must-have in our diets. Due to unbalanced diets with insufficient nutrition diversity, malnutrition and long-term health concerns are on an all-time rise. The economic and political forces as well as the COVID-19 pandemic

K. R. C · R. Chandora (✉)
ICAR-National Bureau of Plant Genetic Resources (NBPGR), Regional Station, Shimla, Himachal Pradesh, India
e-mail: rahul1@icar.gov.in

M. Kumar
Agricultural Research Station, Mandor of Agriculture University, Jodhpur, Rajasthan, India

G. Katna
Chaudhary Sarwan Kumar Himachal Pradesh Krishi Vishvavidyalaya (CSK HPKV), Palampur, Himachal Pradesh, India

K. Rani
ICAR-National Bureau of Plant Genetic Resources (NBPGR), Regional Station, Jodhpur, Rajasthan, India

K. Prakash
ICAR-Indian Agricultural Research Institute, Hazaribagh, Jharkhand, India

R. K. Bairwa
ICAR-Directorate of Mushroom Research, Solan, Himachal Pradesh, India

V. Gahlaut, V. Jaiswal (eds.), *Genetics and Genomics of High-Altitude Crops*, https://doi.org/10.1007/978-981-99-9175-4_7

have caused the food prices to soar to record highs in recent years, causing hardships across the globe.

UN's Sustainable Development Goals 2 (SDG2) to achieve zero hunger by 2030 can be targeted particularly by achieving food security and enhanced nutrition. Many countries in Asia, including India, have been classified as "serious" in the Global Hunger Index (GHI) because 20–34% of the populations in these nations are undernourished. In the current scenario, our agriculture and food systems face two major gaps viz., production and nutrition gap. To fill these two gaps, we need to follow the rules of diversification of our cropping as well as food systems. The narrowing of diversity within majorly cultivated crop species, in addition to the increasing consequences of changing climatic (drought, heat, floods, etc.) conditions, availability of limited natural resources as well as stagnancy in the productivity of major crops brought about by environmental and soil degradation given the inordinate use of chemical fertilizers, changing dietary preferences, a trend toward biofuels and an increasing population, have caused global food and nutritional insecurity.

The diversification of crop production plays a crucial role in achieving sustainable food systems as it can enhance both food security and nutritional balance (Khoury et al. 2014; Dawson et al. 2019). To address these challenges, orphan crops are a viable solution. These crops are underexplored and are often called "hidden treasures." Besides playing significant roles in the local agroecosystem, orphan crops also offer formidable opportunities for fighting hunger, malnutrition and inequalities. These crop varieties are often well-adapted to agricultural settings with minimal resource inputs, and in contrast to conventional industrial crop production, they offer a broader genetic diversity for potential future enhancements in crop quality (Mabhaudhi et al. 2019). Production of the few major staple crops usually requires high inputs, which has made farming more vulnerable to environmental shocks and thereby threatens human rights.

The Himalayan region harbors a high concentration of neglected and underutilized species. The major orphan species such as the grain amaranths (*Amaranthus* spp.), buckwheat (*Fagopyrum* spp.) and chenopods (*Chenopodium* spp.), that belong to three different families, viz., Amaranthaceae, Polygonaceae, and Chenopodiaceae, respectively, are considered as the third group of food grains. These are sometimes referred to as pseudocereals, being distinguished from true cereals which belong to the grass family viz., Gramineae. Pseudocereals are versatile non-cereal crops that find widespread cultivation in high-altitude regions and are commonly grown as catch crops in the foothills. These crops have served as essential life-support plants for traditional hill farmers in the Himalayan region for countless generations. Many of these pseudocereal species have their origins traced back to the Himalayan region, specifically the Hindustan gene center.

Agriculture provides the principal means of livelihood for the population living in the Western Himalayan Region. It has a great scope for enhancing farming incomes by utilizing the diverse Plant Genetic Resources available there, as a significant share of income in the area comes from agricultural livelihoods. Rural people are keepers of unique languages and beliefs and possess indispensable knowledge of practices invaluable for the sustainable management of these natural resources. The genetic resources of pseudocereals can address the present-day demanding situations such as food and nutritional security, environmental sustainability and protection of cultural identity. Naturally, the area harbors many wild relatives and landraces of these crop plants. Although there has been some decline in the cultivation of the pseudo-cereals, in recent times these crops continue to be grown in the hill region of both Old and New Worlds.

As a result, pseudo-cereals have proven to be well-suited for cultivation, particularly in elevated terrains with shorter growing seasons and less fertile soils, where it may be challenging to grow more essential cereals or other crops. The nature and magnitude of variability and distribution of species are remarkable in the hilly areas. They are highly nutritious, possess innate climate resilience and are even potentially valuable in global trade, making them highly suitable for a Zero Hunger future. Pseudocereal crop species are characterized by their elevated nutritional value, natural gluten-free nature, and potential to serve as a vital source of essential micronutrients, protein, dietary fiber, and beneficial plant compounds. These grains have been associated with a range of health advantages, such as better blood sugar level control and enhanced cardiovascular health when compared to typical staple grain foods. Furthermore, they are resistant to pests and diseases as well as are tolerant to various biotic and abiotic stresses.

For effective genetic improvement, genetic variability is a prerequisite. The existing collections of minor and pseudo-cereal cultivars are likely to have considerable genetic variability due to their very short or non-intensive breeding histories (Yabe and Iwata 2020). Compared to major crop species, the germplasm resources of minor and pseudo-cereals that are preserved in gene banks are both deficient and inadequate. Additionally, there is a lack of DNA polymorphism data for these genetic resources. Consequently, it is of utmost importance to collect valuable genetic resources before they are lost and to thoroughly characterize and assess the existing genetic reservoir, which may include DNA polymorphism data. Genomic-assisted breeding techniques, such as Genomic Selection and Marker Assisted Selection, prove valuable for enhancing the genetics of crops, addressing the increasing demand for food grains, and developing novel varieties capable of adapting to uncertain future climates (Varshney et al. 2018). Genomics-assisted breeding of food crops can also contribute significantly to global food security by concurrently enhancing crop diversity, yield, and quality. Therefore, substantial efforts should be dedicated to the ex-situ conservation of pseudocereals worldwide. This chapter offers a comprehensive perspective on the present status of worldwide efforts in the conservation, characterization, and utilization of genetic and genomic assets within the germplasm of pseudocereals. It also underscores the pivotal role played by these resources in endeavors aimed at enhancing genetic traits.

7.2 Amaranthus

Amaranth is a climate-resilient, fast-growing C_4 crop and is much needed for global nutrition and food security as well as attaining the sustainable developmental goal (SDG) 2: Zero Hunger. Amaranth is said to have been grown in Mexico by the Aztecs, well before the Spanish Conquest. Along with maize, it served as a significant food crop before the introduction of major cereals like wheat and rice from the Old World. Grain amaranth holds a unique value beyond its excellent baking and sensory qualities, primarily in its highly favorable amino acid composition. Notably, it boasts high lysine levels at 5.0% and contains sulfur-rich amino acids at 4.4%. Amaranth's protein profile complements that of other cereals owing to its lysine content, making it an excellent choice for protein supplementation when combined with other grains (Grubben and van Sloten 1981). The isolation of the AMA-1 gene in amaranth has even led to its introduction in crops like rice and potatoes, contributing to the production of high-quality protein. Additionally, amaranth is rich in folic acid, which aids in increasing blood hemoglobin levels. Its versatile applications range from being used in bread, biscuits, flakes, cakes, pastries, crackers, and even ice cream to serving as a key ingredient in lysine-rich baby foods. Furthermore, amaranth oil is valued for containing squalene, a cosmetic component and effective skin penetrant. Medicinally, it finds use in treating various conditions such as measles, snakebites, foot-and-mouth disease in animals, kidney stones, chest congestion, and piles (Joshi and Rana 1991).

Amaranth is not only adaptable to fragile environments and diverse soil types but also thrives in specialized agroecological niches. It exhibits resilience to various environmental stresses and contributes to stabilizing ecosystems. These underutilized, often ethnic, crops have the potential to play a crucial role in improving income, enhancing food security, and addressing nutritional needs. Amaranths are currently cultivated as pseudocereals in both the Old and New Worlds. Sauer (1950, 1967) has presented multiple lines of evidence, including archaeological, historical, and folkloric data, indicating the New World as the center of origin and domestication of grain amaranth. There are approximately 60 native species in the New World and about 15 in the Old World and Australia. Many of these species are pioneer annuals naturally found in open habitats like mountain and desert canyons, riverbanks, lakeshores, tidal marshes, and ocean beaches. Amaranth produces abundant seeds that are widely dispersed by water and birds. This characteristic allows it to thrive by continually colonizing sites with disturbed soil, ample sunlight, and minimal competition. Different regions have their dominant species: *A. hypochondriacus* (Prince's feather) in Mexico, *A. cruentus* (purple amaranth) in Guatemala, *A. caudatus* (Inca wheat) in the Andes, and *A. edulis* in Argentina. Among these, *A. hypochondriacus*, *A. cruentus*, and *A. caudatus* are the primary grain-producing species, while *A. dubius, A. bilitum, A. viridis,* and *A. tricolor* are mainly grown for vegetable purposes. *A. hybridus* serves as both a vegetable and fodder and *A. spinosus*, a wild species, is a valuable source of genes for various beneficial traits.

In recent years, the combination of high yield potential, rich protein content, and low cultivation costs has garnered significant interest from research organizations, scientists, and stakeholders worldwide. This attention has been focused on retrieving, researching, and disseminating information related to the production and extensive utilization of potential minor crops, such as amaranth. The goal is to transform fragmented and time-based farming practices into a more uniform and adaptable farming system that can better withstand both biotic and environmental stresses. Furthermore, global awareness of the importance of healthy diets and nutritious food has been steadily increasing. As a result, there is a growing demand for grain amaranth, making it a top choice for health-conscious consumers around the world. This surge in demand underscores the significance of exploring and promoting the cultivation and use of amaranth as a valuable and nutritious food source.

7.2.1 Genetic Resources and Its Utilization

Amaranth and its wild relatives are known for their impressive genetic diversity, leading to significant global efforts aimed at conserving their genetic resources. Presently, there exist 61 distinct collections of amaranth genetic resources in at least 11 countries, as documented by Das (2016). During the 1970s, the Rodale Research Center (RRC) in Pennsylvania, USA, took a pioneering role in amassing amaranth germplasm, accumulating a total of 1400 accessions from various species, which are now part of the United States Department of Agriculture's (USDA) National Plant Germplasm System, as reported by Kauffman and Weber (1990). The USDA collection stands as the largest in the world today, housing 3300 accessions gathered from 40 nations and spanning 42 species within the amaranth genus, according to Brenner et al. (2000). In India, the national gene bank at ICAR-NBPGR has diligently collected and conserved nearly 3000 accessions of grain amaranth. Additional collections have been made by GB Pant Agricultural University in Pant Nagar, VYPKAS in Almora, and the Northeastern complex of ICAR in Shillong. However, there are notable gaps in these collections, particularly in relation to wild species and landraces from regions where amaranth is not a prominent crop. To address these gaps and ensure the preservation of landraces before they vanish, it is imperative to conduct further research and expand these germplasm collections.

Taxonomic keys and descriptors have been developed to facilitate the classification of amaranth collections for the identification and characterization of genetic resources, as noted by Grubben and Van Sloten (1981). Phenotypic assessments of germplasm collections in India, Peru, the United States, and Mexico have unveiled significant variability within the amaranth genus, as reported by Espitia Rangel (1986), Joshi (1986), and Das (2016). Germplasm analysis of both indigenous and foreign amaranth accessions has revealed a wide range of diversity in crucial morpho-agronomic traits. In diversity-rich regions like Mexico and Guatemala, *A. cruentus* and *A. hypochondriacus* landraces showcase remarkable diversity, as highlighted by Espitia (1992). In grain amaranth, the average plant height measures 248.18 cm, ranging from 100.55 to 330.50 cm. The mean inflorescence length is 75.58 cm, with a range from 29.35 to 197.40 cm. The mean 1000-seed weight is 0.69 g, varying from 0.30 to 0.95 g, and the average seed yield per plant stands at

Fig. 7.1 Inflorescence types in grain amaranth (**a**) drooping type in *A. caudatus* and (**b**) erect type in *A. hypochrondriacus*

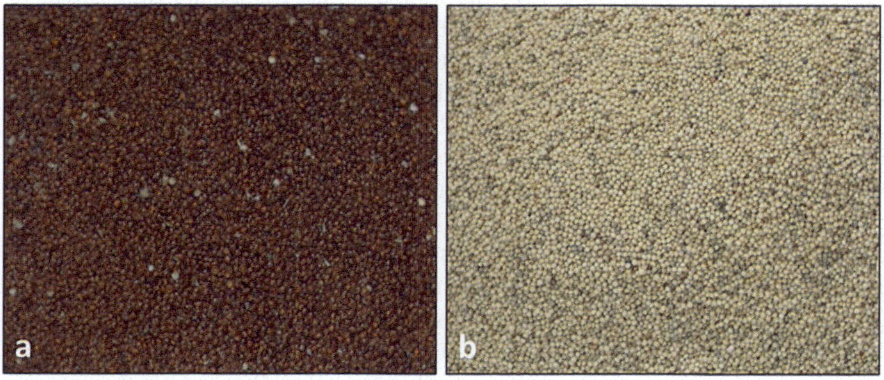

Fig. 7.2 Seeds of (**a**) *A. cruentus* and (**b**) *A. hypochrondriacus*

49.24 g, ranging from 10.00 to 186.39 g. The genetic resources gathered from a diverse array of amaranth species hold substantial genetic diversity, as observed in research by Kietlinski et al. (2014), Stetter et al. (2017), and Wu and Blair (2017). This genetic diversity holds significant promise for the adaptation and enhancement of grain amaranth, as illustrated in Figs. 7.1 and 7.2.

7.2.2 Genomic Resources

One of the biggest benefits of Genomic Selection, QTL Mapping and Association Studies (GWAS) is that they do not necessarily rely on a reference genome. However, having access to reference genomes can significantly enhance the effectiveness of genomics-assisted breeding. For instance, GWAS depends on marker location information and an annotated genome to identify candidate genes responsible for the relationships between markers and phenotypes. Therefore, the recent development and validation of draft genomes for grain amaranth (*A. hypochondriacus* L.; Clouse et al. 2016) focused on drought resistance have proved invaluable. Furthermore, the establishment of genomic resources and databases for various minor and pseudocereals (Yasui et al. 2016a, b) has further expanded our capabilities in genomics-assisted breeding. The continuous advancements in sequencing technologies have made whole-genome sequence analysis more cost-efficient and high-throughput, contributing to the rapid progress in this field. These developments highlight the synergy between genomics and breeding efforts, ultimately leading to more effective and targeted crop improvement strategies.

7.2.2.1 Omics-Driven Approaches for Nutraceutical Development

The term "omics" underscores the idea of "all" or "whole" and is often used as a suffix when referring to cellular molecules like genes, proteins, transcripts, ions, and metabolites. These include genome (all genes), proteome (all proteins), transcriptome (all transcripts), ionome (all ions), and metabolome (all metabolites). Analyzing individual biomolecules alone cannot provide a comprehensive understanding of cellular processes, but an omics-based approach encompasses the entire molecular landscape, offering a more holistic perspective. A "multiomics approach" involves using bioinformatics to simultaneously analyze all these omics layers, providing a more integrated and comprehensive view of complex biological properties. While such integrated approaches have not been widely applied to amaranth research, recent breakthroughs in Next-Generation Sequencing (NGS) technology, phenotyping platforms, and computational biology skills are drawing researchers' attention to their potential collaborative applications.

Grain amaranth has gained increasing attention in recent years due to its nutritional characteristics, especially the nutritional value of its seed protein (Bressani et al. 1992; Tucker 1986). Amaranth seeds contain higher fiber content (8%) compared to other cereal grains (Breene 1991; Pedersen et al. 1987). On a dry-matter basis, the crude protein content ranges from 12.5 to 22.5%, which is over 50% higher than grains like wheat, rice, and maize (Gupta and Gudu 1991). Additionally, amaranth seeds have relatively high levels of lysine, a commonly limiting amino acid in cereal crops, ranging from 0.73 to 0.84% (Bressani et al. 1987; Písaříková et al. 2005). The significant amounts of nutrients like iron (Fe), magnesium (Mg), and calcium (Ca) in amaranth flour make it an attractive candidate for wheat flour fortification (Alvarez-Jubete et al. 2009). The high iron content also holds promise for addressing anemia in underdeveloped regions (Caselato-Sousa and Amaya-Farfán 2012). Moreover, amaranth seed protein is gluten-free, making it a valuable protein source for individuals with celiac disease (Alvarez-Jubete et al. 2009; Kupper 2005; Pagano 2006). Amaranth seeds also contain approximately 5% oil with relatively high squalene concentrations, which has been associated with potential anti-cancer properties and the reduction of cholesterol levels in epidemiological studies (Smith 2000; Berger et al. 2003; Martirosyan et al. 2007).

Various research approaches and resources have been developed for studying grain amaranth, including genetic markers (e.g., random amplified polymorphic DNA, simple-sequence repeats, and single nucleotide polymorphisms), genetic linkage maps, bacterial artificial chromosomes, and draft genome sequences (Sunil et al. 2014; Maughan et al. 2008, 2011; Transue et al. 1994; Chan and Sun 1997; Mallory et al. 2008). Although progress has been made, there is still a need for further improvement in generating a high-quality reference genome for amaranth. Clouse et al. (2016) reported the first physical map assembly of the species, which significantly improved the assembly's N50. Physical maps are valuable for identifying structural rearrangements at the chromosome level and enhancing genome assemblies. The study of grain amaranth's evolution and speciation provides a unique opportunity to investigate the early stages of genome differentiation associated with post-zygotic separation, as hybridization between grain amaranths and wild relatives has shown varying degrees of hybrid breakdown (Gupta and Gudu 1991). Comparing physical maps across species could reveal early patterns of structural genome changes linked to evolution and speciation at an unprecedented resolution. To improve the existing genome assembly, two promising technologies are (1) Pacific Biosciences long-read sequencing, which can consolidate smaller scaffolds and reduce gaps within larger ones, and (2) genome conformation capture (Hi-C) technology, as described by Putnam et al. (2016). Hi-C technology leverages proximity ligation of DNA based on in vitro reconstituted chromatin to generate read pairs. This approach can bridge gaps in assemblies that conventional sequencing technologies struggle with, such as variable number tandem repeats and centromeres, while also providing valuable haplotype phasing information.

7.2.2.2 Genetic Transformation and Genome Editing

Only a few successful instances of genetic manipulation in amaranth have been reported, as detailed in various studies. For instance, Jofre-Garfias et al. (1997) achieved the transformation of *A. hypochondriacus* through the inoculation of mature embryos with *Agrobacterium tumefaciens*, leading to plant regeneration. They established an in vitro *A. tumefaciens*-mediated culture system capable of regenerating transformed *A. hypochondriacus* plants. Swain et al. (2010) outlined a technique for genetic transformation of *A. tricolour* using *Agrobacterium rhizogenes*. They emphasized the key parameters affecting the genetic transformation of leafy amaranth roots. Notably, internode explants were employed to maximize hairy root emergence, followed by a 5-day co-cultivation period post-infection with *A. rhizogenes*. Positive tests for agropine and mannopine confirmed successful transformation. Pal et al. (2013a) introduced a genetic transformation strategy for *A. tricolour* utilizing epicotyl explants and mediated by *A. tumefaciens*. This method involved a binary plasmid for constitutive gene expression, carrying the neomycin phosphotransferase II (nptII) and glucuronidase (uidA) genes. It helped standardize essential parameters for genetic transformation. *A. spinosus*, a nutritional amaranth variety, was also transformed by Pal et al. (2013b) using a combination of *A. rhizogenes* strains and various explants. Despite mixed outcomes in different experiments, they established a practical experimental methodology for creating transgenic plants capable of accumulating opiates. Munusamy et al. (2013) introduced a floral dip transformation method similar to one routinely used for *Arabidopsis thaliana*. This technology successfully transformed amaranth, with the changes efficiently inherited by the next generation. Kuluev et al. (2017) employed *A. tumefaciens*-mediated transformation to create transgenic *A. retroflexus* plants expressing the *A. thaliana* ARGOS-LIKE (ARL) gene. This led to enhanced vegetative growth and improved forage crop performance. They reported a 1.4% transformation efficiency, with genetically modified amaranth plants showing significant increases in stem and leaf length as well as fresh weight. Genome editing, a cutting-edge technology, is anticipated to be highly valuable for crop improvement in the near future. When combined with QTL mapping, GWAS, and GS, genome editing can accelerate genetic enhancement in amaranth. It holds promise for addressing agronomic concerns, such as shattering behavior, which still pose challenges for minor cereals and pseudocereals. Varshney et al. (2018) highlighted that genome editing can precisely modify one or more genes in the genome to achieve desired traits. Various genome editing tools, including CRISPR/Cas9, ZFNs, TALENs, and Meganucleases, are available. CRISPR/Cas9, in particular, has shown potential for creating transgene-free plants and enhancing yield-related and stress-responsive features in major crops. While genome editing has not yet been systematically explored in amaranth, existing genetic transformation techniques (Munusamy et al. 2013) suggest its applicability in this crop. Potential applications of genome editing in amaranth include improving herbicide resistance, understanding the genetic mechanisms behind it, and enhancing nutraceutical characteristics. For instance, precise nucleotide alterations in herbicide-resistance genes could shed light on the evolution of herbicide resistance in species like *A. palmeri*. Additionally, targeted molecular stacking of herbicide resistance genes using CRISPR/Cas9 in grain amaranth could reduce weed management costs and increase yield with minimal input expenses. Genome editing techniques can also be employed to enhance genes responsible for accumulating higher levels of nutraceutical compounds in plant tissues.

7.2.2.3 Transcriptomics

Transcriptomics, the comprehensive study of all cellular transcripts, has gained increasing prominence due to its direct detection of expressed genes. While genomics tools can furnish genome-level sequence data, transcriptome research holds critical significance because gene transcription is conditional, with not all genes being active throughout an organism's entire growth and development. Therefore, the transcriptome serves as a valuable resource for investigating genes associated with various traits in a specific organism during a particular growth stage. Additionally, the alignment of RNA-seq reads, whether done de novo or using a reference, yields valuable information about the expression levels of relevant genes, quantified as RPKM (Reads Per Kilobase Million), FPKM (Fragments Per Kilobase Million), or TPM (Transcripts Per Kilobase Million) values.

In the past, cDNA-AFLP was employed to explore transcriptome dynamics across different species, samples, and time points; however, this method had limited resolution. Microarrays represented a more reliable alternative for studying the transcriptome and conducting comparative transcriptomics research. Nevertheless, microarrays suffered from limited coverage and a tendency to produce false positive results. They remained a consistent platform for well-studied crops like *A. thaliana*, mainly because sequencing information was necessary for producing the hybridization chips. Another technique for transcriptome analysis was serial analysis of gene expression (SAGE), but the advent of Next-Generation Sequencing (NGS) revolutionized transcriptomics research by enabling large-scale sequencing of coding and non-coding RNA samples without the need for prior genomic knowledge.

Transcriptome sequencing and analysis of various crops, including foxtail millet, finger millet, sweet potato, soybean, and cowpea, have unveiled genes displaying differential expression patterns during growth, development, and responses to stress. These identified genes can subsequently undergo further characterization, including functional evaluation (Zhang et al. 2012; Qi et al. 2013; Ji et al. 2017; Khan et al. 2017; Chen et al. 2017). Notably, within the Amaranthaceae family, the genera *Amaranthus, Beta, Spinacia,* and *Chenopodium* produce betalain pigments derived from tyrosine (Brockington et al. 2011). Transcriptome technology, in conjunction with other molecular biology techniques, has been instrumental in identifying three pivotal genes involved in betalain biosynthesis: (1) a cytochrome P450 enzyme with tyrosine hydroxylase activity (Polturak et al. 2016); (2) l-DOPA 4,5-dioxygenase (DODA) (Christinet et al. 2004), with all known DODA enzymes to date belonging to the DODAα orthologous group (Sheehan et al. 2020); and (3) l-DOPA oxidase, catalyzed by the cytochrome P450 enzyme CYP76AD1, with the α clade being specific to betalain synthesis (Brockington et al. 2015). In B. vulgaris, the betalain-specific isoforms of DODAα1 and CYP76AD1 were found to be co-located on chromosome 2, with a single gene positioned between them (Sheehan et al. 2020). Orthologs of these two genes were similarly co-located on chromosome 16 in *A. hypochondriacus* (Sheehan et al. 2020). The analysis of the *A. cruentus* genome revealed co-localization of DODAα1 and CYP76AD1 orthologs on chromosome 16, and a cluster of these biosynthetic genes was also observed in the *S. oleracea* genome (Ma et al. 2021).

7.2.2.4 Proteomics and Metabolomics

Not all transcripts are translated into proteins, highlighting the crucial role of post-transcriptional modifications in the synthesis of specific proteins. Some transcripts also lack direct connections to the cell's proteins. While SDS-PAGE can be employed for the quantitative measurement of protein content, two-dimensional gel electrophoresis (2DE) becomes essential for protein identification (Eldakak et al. 2013). In the first dimension of 2DE, protein samples are separated across a pH gradient, causing proteins to migrate based on their isoelectric pH values. Subsequently, SDS-PAGE is performed perpendicular to the initial direction. This process results in the separation of proteins by both pH and molecular weight. The protein spots on the gel are then extracted, subjected to trypsin digestion, and identified using mass spectrometry combined with database searches (Eldakak et al. 2013). When estimating the molecular weight of proteins, MALDI-TOF offers a higher level of resolution (Baggerman et al. 2005).

Proteomics and metabolomics analyses provide valuable insights into synthesized proteins and offer quantitative data on protein abundance. Additionally, these studies can elucidate the post-translational modifications experienced by proteins, such as phosphorylation, glycosylation, sulfation, prenylation, acetylation, and ubiquitination. Furthermore, proteomics investigations can unveil organelle-specific proteins and illuminate protein-protein interactions occurring within a cell. Consequently, comparative proteomics studies can identify proteins uniquely expressed in stress-tolerant genotypes. These proteins can then undergo further exploration using reverse genetics techniques to ascertain their precise roles in the molecular stress response machinery.

7.3 Buckwheat

Buckwheat is an underutilized, ancient pseudocereal well known for its nutraceutical properties. Its cultivation gained momentum in recent years due to its balanced amino acid profile, gluten-free protein and bioactive compounds. The increased awareness among consumers and rising demand for healthier foods also contributed to the growth in the production of buckwheat. It belongs to the dicotyledonous family Polygonaceae and genus *Fagopyrum* (Joshi et al. 2021). The genus *Fagopyrum* includes both annual and perennial species. All the species under *Fagopyrum* have diploid genome (2n = 16) except *F. cymosum* and *F. gracilipes*, which are tetraploid (2n = 32).

The genus exhibits both autogamy and allogamy, with the presence of self-incompatibility caused by dimorphic heterostyly in certain species (Chrungoo et al. 2012; Farooq et al. 2016). The term "Buckwheat" commonly refers to two cultivated species: common buckwheat (*Fagopyrum esculentum* Moench) and Tartary buckwheat (*Fagopyrum tataricum* Moench) (Figs. 7.3 and 7.4) (Zhou et al. 2018).

Although it includes the word "wheat" in its name, buckwheat is taxonomically unrelated to traditionally cultivated wheat and is classified among pseudo-cereals (Joshi and Rana 1995). Incorporating pseudo-cereals like buckwheat into cropping systems is essential for addressing hidden hunger (Boukid et al. 2018). Historically, it has been cultivated primarily for subsistence farming as a reliable crop. Buckwheat possesses significant potential due to its short life cycle, adaptability to higher altitudes, suitability for tribal and marginal ecosystems, and its valuable nutraceutical properties. It serves various purposes, including food, feed, medicine, and even as a natural fertilizer (Li and Zhang 2001). Moreover, it contributes to improved soil structure and increased phosphorus availability. Its rapid, vigorous growth makes it an effective cover crop for weed suppression. While buckwheat may not achieve the status of a major crop due to its relatively lower yields compared to other cereal crops, its short growth cycle enables its use as a fallback crop during periods of food scarcity (Ohsawa 2020).

Fig. 7.3 Buckwheat in fields (**a**) *Fagopyrum esculentum* and (**b**) *Fagopyrum tataricum*

7.3.1 Genetic Resources

7.3.1.1 Status of Buckwheat Conserved in Gene Banks

Buckwheat genetic resources consist of landraces, wild species, improved and traditional varieties, advanced breeding lines and special genetic stocks. In modern agriculture, these genetic resources are prone to genetic erosion due to replacement by high-yielding varieties. Therefore, the conservation of buckwheat genetic

Fig. 7.4 Seeds of (**a**) *Fagopyrum tataricum* and (**b**) *Fagopyrum esculentum*

resources (cultivars and wild species) is critical for the food and nutritional security of the global population. Significant efforts have been made in this direction. The systemic collection of buckwheat germplasm started in the 1980s by botanists, with the aid of IBPGR, Italy, currently known as Bioversity International. More than 10,000 buckwheat germplasm accessions are conserved in long and medium-term storage conditions in different gene banks all over the world (Table 7.1), with about 50% of the germplasm accessions coming from South and East Asia (Zhou et al. 2018). The largest collection of 2804 germplasm accessions is maintained by the Institute of Crop Sciences of the Chinese Academy of Agricultural Sciences (CAAS), Beijing.

Table 7.1 Ex-situ collections of pseudocereals conserved in major gene banks globally

Crop	Country	No. of accessions	Name of the institution/organization
Amaranth	USA	3300	North Central Regional Plant Introduction Station (NCRPIS)
	India	3081	ICAR-National Bureau of Plant Genetic Resources, New Delhi
	Peru	740	Univ. Nacional San Antonio Abad del Cusco (UNSAAC/CICA)
	Mexico	495	Instituto Nacional de Investigaciones Forestales y Agropecuarias (INIFAP)
	Peru	440	Universidad Nacional del Altiplano Puno
	China	438	Institute of Crop Germplasm Resources, CAAS, Beijing
	Taiwan	532	World Vegetable Center
Buckwheat	China	2804	Institute of Crop Sciences, Chinese Academy of Agricultural Sciences (CAAS), Beijing
	Russia	2230	N I Vavilov All-Russian Institute of Plant Industry
	Ukraine	1600	V Y Yuriev Institute of Plant Production
	Ukraine	900	Podillya State Agricultural University
	India	1134	ICAR-National Bureau of Plant Genetic Resources, New Delhi
	Nepal	511	National Agriculture Genetic Resource Center
	Slovenia	378	University of Ljubljana
	Japan	746	National Institute of Agrobiological Sciences
Quinoa	Bolivia	3178	Instituto Nacional de Innovación Agropecuaria y Forestal - INIAF
	Peru	2089	Universidad Nacional Agraria La Molina
	Peru	1873	Universidad Nacional del Altiplano
	Ecuador	673	Departamento Nacional de Recursos Fitogenéticos y Biotecnologí
	Germany	984	Leibniz Institute of Plant Genetics and Crop Plant Research

Source: Joshi et al. (2021); Romanova et al. (2018); Chauhan et al. (2010)

7.3.1.2 Buckwheat Cultivars and Promising Accessions for Various Traits

Considerable diversity has been documented among different accessions within both cultivated species, *F. esculentum* and *F. tataricum*, of buckwheat (Keli 1992; Baniya 1995). Numerous studies have been conducted with different accessions of buckwheat for diversity and variability studies, agronomic performance, agro-morphological traits, nutritional and medicinal values and other characteristics. (Lu et al. 1996; Tang et al. 2011; Rauf et al. 2020). Various pharmacological properties and the bioactive compounds responsible for medicinal and nutritive values in buckwheat is present in higher concentration in *F. tataricum* accessions compared to *F. esculentum* accessions (Jiang et al. 2007), which can be exploited further to develop elite chemotypes of buckwheat. Wild buckwheat species also possess a significant reservoir of genetic diversity, encompassing numerous agronomic and nutritional attributes. Among these wild species, *F. cymosum* stands out for its superior nutritive qualities compared to other buckwheat species (Zhang et al. 1999; Tang et al. 2011). Being an underutilized crop, the genetic improvement work in buckwheat was ignored until the twenty-first century. However, in recent years, concerns about nutritious and functional food options led the breeders to carry out focused research towards buckwheat genetic improvement for various traits. In recent years a rapid advancement has been noticed globally in the development and registration of buckwheat genotypes. "Bogatyr" was the first buckwheat variety developed in Russia by mass selection of landrace population (Suvorova and Zhou 2018). In the last decade, hundreds of varieties of buckwheat have been registered with the respective national repositories of buckwheat breeding countries. India also developed some improved varieties for cultivation within the country.

7.3.2 Genomic Resources of Buckwheat

In the early twenty-first century, modern biotechnological tools were applied to enhance buckwheat through various improvement programs. Molecular marker techniques played a crucial role in advancing molecular genetics and breeding research in buckwheat. These techniques facilitated tasks such as quantifying polymorphism, studying genetic diversity, constructing linkage maps, assisting in breeding programs, and unravelling the evolutionary history of buckwheat (Hou et al. 2009; Yabe et al. 2014; Song et al. 2022). Table 7.2 describes a list of trait-specific genes identified in buckwheat (as well as other pseudocereals) and Table 7.3 provides a summary of the development and use of molecular marker systems in buckwheat and its wild relatives.

The initial linkage map for buckwheat was constructed by Ohnishi and Ohta (1987) using six allozymes and 15 morphological markers. Back in 2002, Sharma and Jana utilized RAPD markers to study species relationships and conduct diversity analysis in buckwheat. Yasui et al. (2004) created an interspecific linkage map of *F. esculentum* and *F. homotropicum* using amplified fragment length polymorphism (AFLP) markers, while Du et al. (2013) developed an SSR-based linkage

Table 7.2 Identified trait-specific genes form pseudocereals

Crop	Genes	Roles	Model/ major crop	References
Amaranth (*Amaranthus hypochondriacus*)	Seed albumin gene *AmA1*	Improve growth, production and protein content	Potatoes	Chakraborty et al. (2000)
	Antimicrobial peptide gene *Ah-AMP*	Pathogen/disease resistance	Tobacco	Chen et al. (2003)
	Nuclear factor-Y NF-YC subunits gene *AhNFY-C*	Drought tolerance	Arabidopsis	Palmeros-Suárez et al. (2015)
	Group VII ethylene response factor transcription factor *AhERF*	Water-deficit tolerance	Arabidopsis	Massange-Sanchez et al. (2016)
Buckwheat (*Fagopyrum esculentum*)	Dehydration-responsive element (DREB) transcription factors *FeDREB1*	Enhanced freezing and drought tolerance	Arabidopsis	Fang et al. (2015)
	Metallothionein type 3 FeMT3	Drought and oxidative stress defense gene		Samardžić et al. (2010)
	Basic helix-loop-helix *FtbHLH3*	Drought/ oxidative stress		Yao et al. (2017)
	R2R3-MYB transcription factor gene *FtMYB9*	Drought and salt stresses		Gao et al. (2017)
	R2R3-MYB transcription factors gene *FtMYB13*	Drought/salt tolerance	Arabidopsis	Huang et al. (2018)
Quinoa (*Chenopodium quinoa*)	Salt Overly Sensitive 1 (SOS1) genes *cqSOS1A* and *cqSOS1B*	Salt tolerance		Maughan et al. (2009b)
	Sodium transporter genes *CqSOS1* and *CqNHX*	Salt tolerance		Ruiz-Carrasco et al. (2011)
	Gamma-glutamylcysteine synthetase genes *BrECS1* and *BrECS2*	Tolerance to abiotic stress and enhance growth and development	Rice	Bae et al. (2013)

Modified from: Kamenya et al. (2021)

map for Tartary buckwheat. Microsatellite markers were employed in buckwheat research, revealing a higher level of polymorphism and heterozygosity (Li et al. 2007; Ma et al. 2009). Yasui et al. (2008) constructed a BAC (Bacterial Artificial Chromosome) library to rapidly enrich buckwheat genetic resources. Additionally, Yabe et al. (2014) created a high-density linkage map for common buckwheat (*F. esculentum*) consisting of 751 loci and 8884 markers, with an average distance

Table 7.3 Genetic diversity studies in buckwheat

Sr. No.	Marker	Salient Findings	References
1.	RAPD	RAPD markers were employed to study the genetic basis of self-compatibility. Three RAPD markers OPB141250, OPP81000, and OPQ7800 were identified to be linked with the Homostylar (Ho) gene	Aii et al. (1998)
2.	AFLP	F_2 progenies of *F. esculentum* and *F. homotropicum* cross were screened to identify linkage of homostylar locus with marker	Nagano et al. (2001)
3.	RAPD	Genetic variation and inter-relationship between cultivated and wild species of Tartary buckwheat	Kump and Javornik (2002)
4.	RAPD	RAPD markers were employed for analysis of species relationship in 28 accessions of *Fagopyrum* belonging to 14 species and two subspecies	Sharma and Jana (2002)
5.	AFLP and SSR	19 common buckwheat indigenous cultivars were assessed for genetic diversity by 1 AFLP and 5 SSR markers	Iwata et al. (2005)
6.	AFLP	312 primer combinations were screened and a linkage map was constructed around the *sht1* locus by using 102 F2 plants. Five AFLP markers linked to the *sht1* locus (genes linked to brittle pedicle in buckwheat) were identified	Matsui et al. (2004)
7.	SSR	Microsatellite markers were developed for diversity analysis of Tartary buckwheat	Li et al. (2007)
8.	SSR	136 new SSR markers developed in *F. esculentum* ssp. *esculentum*. Among these 10 polymorphic SSR markers were utilized for genetic diversity analysis of a common buckwheat population consisting of 41 accessions of diverse origin	Ma et al. (2009)
9.	SSR	Microsatellite markers (54 loci) for common buckwheat were developed	Konishi et al. (2006)
10.	RAPD	51 buckwheat landraces comprised *of F. esculentum* and *F. tataricum* were characterized using Random Amplified Polymorphic DNA markers	Senthilkumaran et al. (2008)
11.	SRAP	26 sequence-related amplified polymorphism markers were employed to analyze the genetic diversity of 10 accessions of Tartary buckwheat from China	Li et al. (2009)
12.	AFLP	20 AFLP markers were employed for deciphering the genetic diversity of 165 accessions of Tartary buckwheat from 14 different geographical regions of China	Hou et al. (2009)
13.	RAPD	The genetic variation of 19 common and Tartary buckwheat varieties from China was assessed by 7 RAPD markers	Deng et al. (2011)
14.	SSR	179 common buckwheat landrace accessions from Korea were assessed by 10 SSR markers	Song et al. (2018)
15.	EST	QTL analysis using 50 EST markers uncovered photoperiod-sensitivity genes.	Hara et al. (2011)

(continued)

Table 7.3 (continued)

Sr. No.	Marker	Salient Findings	References
16.	AFLP	DNA fingerprinting was performed for 18 Tartary buckwheat accessions with high and low rutin content by 19 AFLP markers	Gupta et al. (2012)
17.	SSR	50 Tartary buckwheat accessions from China were studied for genetic diversity using 19 SSR markers	Gao et al. (2012)
18.	SSR	23 polymorphic SSR markers were employed for the assessment of genetic diversity in 64 accessions of Tartary buckwheat	Hou et al. (2016)
19.	DNA array	Genomic selection using 14,598–50,000 markers resulted in a 20.9% increase in the selection index compared to the initial population	Yabe et al. (2018)
20.	SSR	A total of 26,549 SSR markers were designed from the flanking sequences of 37,572 microsatellites in *F. tataricum*	Fang et al. (2019)
21.	GBS	GBS detected 255,517 SNP sites from 46 cultivated common buckwheat plants, pointing to the likelihood of gene flow from wild to cultivated buckwheat	Mizuno and Yasui (2019)
22.	SSR	112 Tartary buckwheat accessions from China were studied for assessment of genetic diversity by 10 SSR markers	Song et al. (2022)
23.	SSR	14 common buckwheat accessions from the Gene bank of the Faculty of Agriculture and Food Sciences in Sarajevo were studied for genetic variation by 10 SSR markers	Grahić et al. (2022)

of 2.13 cM between adjacent loci. Fang et al. (2019) developed a high-density physical map for buckwheat with an average marker density of 58.82 markers per mega base. Joshi et al. (2006) generated Expressed Sequence Tags (ESTs), which are more likely to be conserved across populations and species. They reported that the transferability of EST markers decreased as genetic distances between buckwheat species increased. ESTs were also used to identify Quantitative Trait Loci (QTL) governing photoperiod sensitivity in buckwheat (Hara et al. 2011), and DNA microarray techniques were applied for QTL mapping of stem length in *Fagopyrum* (Yabe et al. 2014).

In recent years, the genomes of two cultivated buckwheat species, *F. esculentum* (Yasui et al. 2016b) and *F. tataricum* (Zhang et al. 2017a), have been sequenced. The common buckwheat genome size is approximately 1.2 gigabase pairs (Gbp), while Tartary buckwheat has a smaller genome of around 0.48 Gbp (Table 7.4). These genome sizes are relatively smaller compared to major cereal crops like maize and wheat. The genome of golden buckwheat (*F. cymosum* or *F. dibotrys*), a wild relative of cultivated species, was also sequenced, revealing a genome size of 1.08 Gbp (He et al. 2022). The complete chloroplast (cp) genome sequence of *F. esculentum* ssp. ancestrale, the wild ancestor of common buckwheat, was reported by Logacheva et al. (2008). This cp genome consists of 159,599 nucleotide bases

Table 7.4 Whole genome sequences of pseudocereals

Family	Species and ploidy	Assembled Genome size	N50	Reference
Amaranthaceae	*Amaranthus hypochondriacus* (2×)	403.9 Mb	1.25 Mb	Lightfoot et al. (2017)
Polygonaceae	*Fagopyrum tataricum* (2×)	489.3 Mb	550.7 Kb	Zhang et al. (2017a)
Polygonaceae	*Fagopyrum esculentum* (2×)	1.2 Gb	25,109 bp	Yasui et al. (2016b)
Amaranthaceae	*Chenopodium quinoa* (4×)	1.39 Gb	1.6 Mb	Jarvis et al. (2017)

and is approximately 9 kilobases larger than the cp genome of its closest relative, *Spinacia oleracea.* Cho et al. (2015) reported the chloroplast genome of Tartary buckwheat, which was 327 base pairs shorter than the common buckwheat cp genome, with a total length of 159,272 base pairs. Wang et al. (2017) sequenced the chloroplast genomes of two wild buckwheat species, *F. luojishanense* and *F. dibotrys*, which had lengths of 159,265 base pairs and 159,320 base pairs, respectively. Additionally, Fan et al. (2021) sequenced the cp genomes of three *Fagopyrum* species, namely *F. longistylum*, *F. leptopodum*, and *F. urophyllum*, to gain insights into genome evolution and the phylogenetic relationships between *Fagopyrum* species. Furthermore, recent transcriptome studies in buckwheat have characterized potential genes associated with various traits. Logacheva et al. (2011), Xu et al. (2017), Wu et al. (2017), and Huang et al. (2017) conducted comprehensive transcriptome analyses for floral structure, aluminium toxicity, salt tolerance, and seed developmental stages, respectively. Unlike cereal crops where rapid genetic advancements have been possible and RILs (Recombinant Inbred Lines), NILs (Near Isogenic Lines), NAMs (Nested Association Mapping), MAGIC (Multiple Advanced Generation Intercross) populations have been developed, however, where buckwheat is concerned, limited efforts have been made in this area. Recently, in many cereal crops, speed breeding techniques have also been employed for the rapid advancement of generations, to grow more crop cycles per year (Watson et al. 2018), and this technique can also be employed in the case of buckwheat to take up up to six crop cycles per year. Moreover, the use of CRISPR/Cas9-mediated genome editing technique can also enhance some oligogenic trait expression besides nutraceutical value and yield in buckwheat.

7.4 Chenopodium

Chenopodium album, commonly known as Lamb's quarters, has a long history of cultivation in the Himalayan region and was once considered a common weed in the plains (Hooker 1885, 1952; Partap 1982). *Chenopodium murale* L. is a plant species primarily located in lowland areas and has been grown for its edible leaves, which are commonly used as a culinary green. It is often utilized as a pot herb in cooking.

Chenopodium berlandieri subsp. *nuttaliae*, originally believed to be a wild species in the United States of America, has found utility in Mexico and is named after Mrs. Zelia Nuttall of Mexico. It has been studied for its potential to cross with cultivated *C. quinoa* to develop tolerance to high temperatures. Two closely related species, canihua or caniqua (*Chenopodium pallidicaule*) and huauzontle (*Chenopodium nuttaliae*), also contribute to human nutrition and share their ecological niche with *Chenopodium quinoa*. Quinoa exhibits adaptability to various challenging environments due to its resistance to salt and drought tolerance (Wilson and Heiser Jr 1979). While *Chenopodium pallidicaule* (2n = 18) and *Chenopodium berlandieri* subsp. *nuttalliae* (2n = 36) have been used for both grains and vegetables, *Chenopodium quinoa* (2n = 36) is primarily grown as a grain crop. *Chenopodium album* (2n = 18, 36, 54) is mainly used as a forage crop and a source of green vegetables. In some Himalayan regions, both *C. album* and *C. quinoa* are cultivated for their leaves and grains, respectively (Figs. 7.5 and 7.6).

The first form of wild quinoa was the result of hybridization between two diploid ancestors. The tetraploid ancestor in the New World emerged through a natural hybridization process, originating from the union of a female relative, *Chenopodium standleyanum* from temperate America, with a male relative, *Chenopodium album* from Eurasia (refer to Fig. 7.7).

The domestication of the precursor to modern quinoa and the subsequent phases of its evolution were facilitated by tetraploid varieties like *C. berlandieri* and *C. hircinum*, which descended from this tetraploid ancestor, as detailed in the work of Jellen et al. (2013).

Fig. 7.5 Inflorescences of (**a**) *Chenopodium quinoa* and (**b**) *Chenopodium album*

Fig. 7.6 Seeds of (**a**) *Chenopodium quinoa* and (**b**) *Chenopodium album*

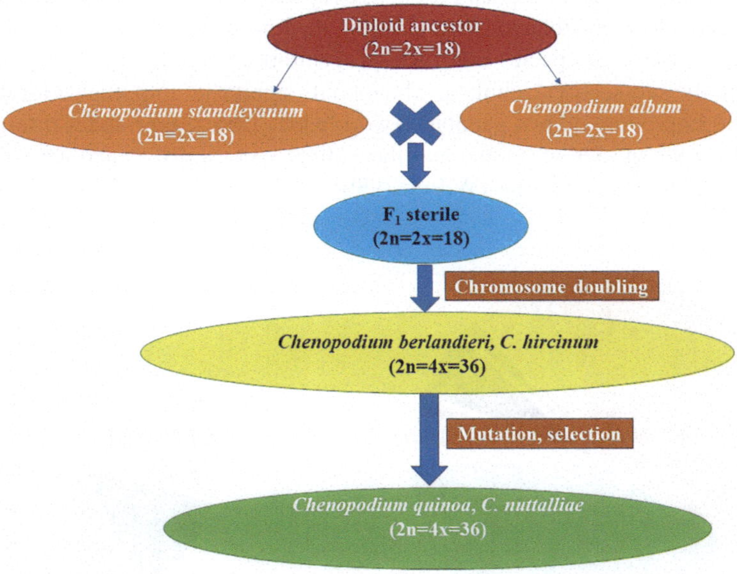

Fig. 7.7 Schematic representation showing the origin of quinoa

Recent data on ex-situ collections of quinoa and its wild relatives, in collaboration with FAO, Bioversity International, and experts specializing in quinoa collections, indicate the presence of approximately 16,422 accessions from a range of *Chenopodium* species, including *Chenopodium quinoa, C. album, C. berlandieri, C. hircinum, C. petiolare, C. murale*, and others, which are conserved globally. These accessions are conserved in 59 gene banks across 30 countries. The Americas host 10 of these genebanks (Argentina, Bolivia, Brazil, Canada, Colombia, Chile, Ecuador, Peru, and the United States of America), Europe has 11 (Germany, Austria, Slovakia, Spain, Hungary, the Czech Republic, Portugal, the United Kingdom,

Sweden, Turkey, and Romania), Africa has 5 (Ethiopia, Kenya, Lesotho, Zambia, and South Africa), and Asia has 3 (India, Japan, and Jordan), along with one in Australia. Among these, Bolivia and Peru, both Andean nations, exhibit the highest diversity, followed by Ecuador, Argentina, and Chile. Germany possesses 987 accessions of quinoa and its wild relatives, while India has 294 accessions, the United States has 229, and Japan has 191 accessions (Rojas et al. 2015) (Fig. 7.8).

Quinoa's wild relatives, *C. berlandieri* and *C. hircinum*, encompass a primary gene pool that harbors valuable alleles crucial for conferring tolerance to biotic stresses and heat in lowland subtropical and temperate regions. These alleles are sought after by breeders aiming to adapt quinoa for global cultivation. Wild and cultivated quinoa germplasm accessions harbor essential genes related to agro-morphological traits, agronomy, quality attributes, and tolerance to various biotic and abiotic stresses, as summarized in Table 7.5. In light of climate change, it would be prudent for quinoa growers and stakeholders to reconsider their adherence to stringent international germplasm exchange regulations, as these regulations could hinder access to stress-tolerant germplasm in the future.

Many of the wild species and their progenitors in South American countries belong to the sub-section Cellulata, which holds significant importance for diversifying and improving quinoa genotypes worldwide. Table 7.6 provides information about the wild species of quinoa and their allied species distributed across North and South American countries (Wilson 1980).

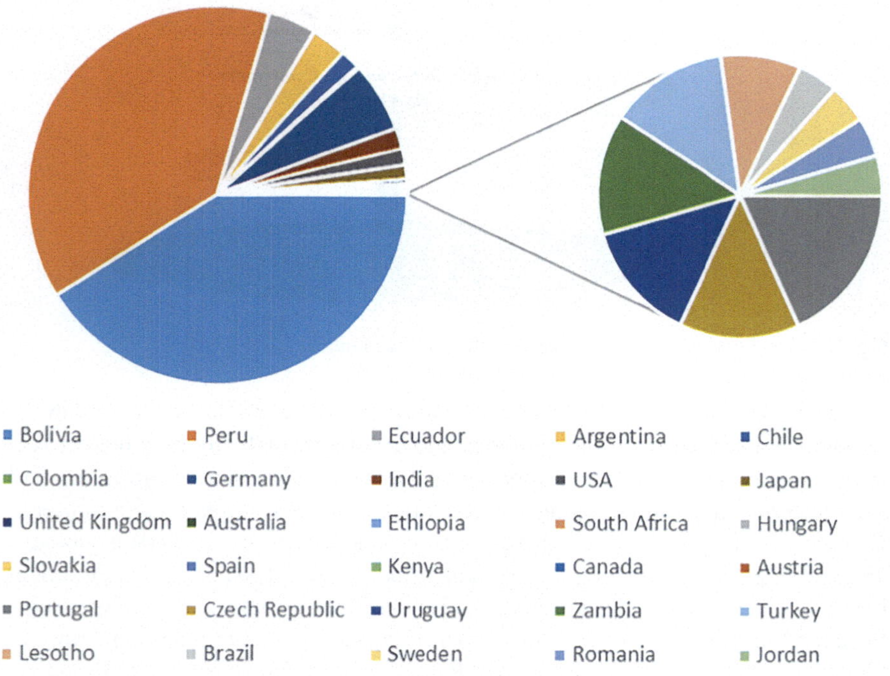

Fig. 7.8 Quinoa accessions conserved throughout the world

Table 7.5 Diverse quinoa wild and cultivated germplasm at the center of diversity for various traits

Wild/cultivated *Chenopodium* spp.	Attributes	Reference
Chenopodium hircinum, *C. pallidicaule*, *C. papulosum*, *C. zobelli*	Early	Jacobsen (2001)
C. quinoa var. *melanospermum*, *C. album*, *C. murale*, *C. hircinum*, *C. ambrosioides*, *C. pallidicaule*, *C. petiolare*, *C. carnosolum*	Water stress and soil salinity stress tolerance, large grains, high saponin content	Gandarillas (1967)
Chenopodium philippianum, *C. carnosolum*, *C. petiolare*	Small yellow seeds, translucent perisperm, spatulate leaves with dentated margins	Wilson (1990)
Chenopodium ambrosiodes, *C. petiolare*	Higher in biomass productivity	Pulgar Vidal (1954)
Chenopodium melanospermum, *C. album*, *C. ambrosioides*, *C. quinoa* var. *melanospermum*, *C. album*, *C. murale*, *C. hircinum*, *C. ambrosioides*, *C. pallidicaule*, *C. petiolare*, *C. carnosolum*	Heavily branched plants, medium seed size, white and late in maturity	Jacobsen (2001)
Chenopodium hircinum, *C. nuttalliae*, *C. petiolare*, *C. album* and *C. ambrosioides*	Downy mildew resistance	Jellen et al. (2015)
Chenopodium hircinum	Heat- and biotic stress-tolerant	Jellen et al. (2015)
Chenopodium standleyanum	Free-threshing (utriculate) pericarp character	Jellen et al. (2015)
Chenopodium nevadense	Extreme sodium tolerance	Jellen et al. (2015)
Chenopodium desiccatum, *C. hians*, *C. incanum*, *C. leptophyllum*, *C. petiolare and C. pratericola*	Extreme drought tolerance	Jellen et al. (2015)
Hybrids swarm between the tetraploid *Chenopodium berlandieri* and diploids like *C. incanum* and *C. pratericola*	Tolerant to air temperature of >40 °C	Jellen et al. (2015)
Chenopodium ficifolium	Resistance to biotic stresses	Jellen et al. (2015)

Table 7.6 Wild species of quinoa distributed across the North and South American countries

Subsection	Wild/cultivated *Chenopodium* spp.	Distribution
South America	*Chenopodium hircinum*	Peru, Bolivia, Ecuador, Argentina
Cellulata	*C. quinoa* var. *melanospermum*	Peru, Bolivia, Ecuador, Colombia
(2n = 4x = 36)	*C. papulosum*	Argentina
Liosperma	*C. zobelli*	Argentina
(2n = 2x = 18)	*C. petiolare*	Peru, Bolivia, Ecuador, Colombia, Chile
	C. philippianum	
North America	*Chenopodium* ssp. *sinuatum*	NW America
Cellulata	*C. berlandieri* ssp. *zschackei*	NW America
(2n = 4x = 36)	*C. bushianum*	NE America
	C. macrocalycium	NE America

Susceptibility to downy mildew has been reported in *Chenopodium murale* L., a wild amaranthaceous species in India. Meanwhile, sources of downy mildew resistance might be present in wild species such as *C. hircinum, C. nuttalliae, C. petiolare, C. album*, and *C. ambrosioides*, which coexist with cultivated quinoa (Bonifacio 1995; Verma et al. 1964).

7.4.1 Quinoa Germplasm Evaluation and Characterization

A comprehensive characterization has been carried out to differentiate and classify various quinoa germplasm accessions, evaluate their potential utility, establish core collections, and pinpoint any duplicates present within the collection. This characterization encompasses the agro-morphological, nutritional, and agro-industrial attributes of quinoa. Throughout the growth cycle of the crop, an extensive analysis of quinoa germplasm has been undertaken, examining various characteristics, including, Growth Characteristics: These encompass growth habits, such as simple/unbranched, branched to the bottom third (suitable for mechanical harvesting), branched to the second third (typically used for forage), and branched with an undefined main panicle. Plant Color: Evaluation is performed at different stages, including between panicle emergence and the start of flowering, and at the time of grain formation and physiological maturity. Varieties may exhibit colors such as green, purple, mixtures, red, and others. Panicle Attributes: This covers panicle shape, categorized as amarantiform (elongated glomerules), glomerulate (globose glomerules), or intermediate (a mix of elongated and globose shapes), as well as panicle density, classified as lax/loose or compact (characterized by short secondary axes and pedicels). Grain Properties: These include grain color options such as cream, white, orange, yellow, red, pink, light coffee, purple, greenish coffee, dark coffee, and black. Grain shape is described as lenticular, cylindrical, ellipsoid, or conical, while grain diameter is categorized as extra-large or Quinoa real (>2.2 mm), large (1.75–2.2 mm), medium (1.35–1.75 mm), and small (<1.35 mm).

Table 7.7 provides an overview of quinoa's morphological, nutritional, and industrial characteristics. Integrating the notion of "genetic diversity" into the manufacturing of processed products maximizes the efficient harnessing of quinoa's genetic capabilities. Optimal results in producing uniform popped or puffed quinoa grains can be achieved by selecting varieties with smaller starch granule diameters (≤ 3). Furthermore, certain quinoa varieties exhibit consistent levels of amylase and amylopectin, making them well-suited for crafting custard desserts, instant sauces, noodles, and jellies. For products requiring higher protein content ($\geq 18\%$), suitable varieties are available to enhance their appeal. The preservation and utilization of genetic diversity align seamlessly with the multitude of ways quinoa germplasm can be consumed and harnessed for various applications.

Table 7.7 Agro-morphological, nutritional and industrial value of quinoa germplasm accessions

Attributes	Mean	Minimum	Maximum
50% flowering (days)	93.5	60	145
Days to physiological maturity (days)	176.89	119	209
Harvest index	0.4	0.16	0.87
Stem diameter (mm)	17.12	10.16	26.26
Panicle length (cm)	37.41	15.4	62.8
Panicle diameter (cm)	6.85	2.86	19.42
Plant height (cm)	110.84	54	174.2
Grain diameter (mm)	1.96	1.36	2.66
100-g weight (g)	0.27	0.12	0.6
Saponin content (cc)	3.16	0	10.88
Protein (%)	14.33	10.21	18.39
Fat (%)	6.46	2.05	10.88
Fibre (%)	7.01	3.46	9.68
Ash (%)	3.63	2.12	5.21
Carbohydrates (%)	58.96	52.31	72.98
Energy (kcal/100 g)	353.36	312.92	401.27
Starch grain (μ)	4.47	1	28
Invert sugar (%)	16.89	10	35
Fused water (%)	28.92	16	66

Source: Rojas (2003); Rojas et al. (2007); Rojas and Pinto (2008); Rojas et al. (2009); Rojas and Pinto (2013); Bioversity International et al. (2013)

7.4.2 Genomic Resources

Recent advances in next-generation sequencing technologies have brought about a significant transformation in the field of crop genomics. These advancements have led to a substantial increase in the volume of sequencing data generated from various species, ushering in a new era in crop genomics research. Notably, these technologies have enabled the cost-effective production of molecular markers even in species with limited or no existing genetic information. In the case of quinoa, molecular studies commenced with the use of allozyme markers to explore molecular diversity among wild and cultivated species, shedding light on the co-evolution of these southern highland wild species and their cultivated counterparts (Wilson 1988a, b). Subsequently, DNA markers made their debut in quinoa research, initially employing random amplified polymorphic DNA (RAPD) markers. These markers were pivotal in distinguishing true hybrids resulting from intergeneric crossings and uncovering genetic variations among various quinoa and related chenopod species (Fairbanks et al. 1993; Ruas et al. 1999; Maughan et al. 2004; Del Castillo et al. 2007; Jarvis et al. 2008). The appeal of simple sequence repeat (SSR) markers in quinoa research soared due to their ability to detect high levels of polymorphism among closely related quinoa species, along with their reproducibility and co-dominant nature (Mason et al. 2005) (refer to Table 7.8 for details). Furthermore, quinoa research expanded into the realm of single nucleotide

Table 7.8 Genomic information of quinoa using protein, DNA, RNA, and sequence data

Detailed information	Marker(s)/technique(s) used	Reference(s)
21 isozyme loci-based analysis was used to characterize 99 populations of quinoa and its relatives (SA-LC-HLIAV), including the glutamate oxaloacetate transaminase (GOT), isocitrate dehydrogenase (IDH), leucine aminopeptidase (LAP), malic dehydrogenase (MDH), phosphoglucoisomerase (PGI), and phosphoglucomutase (PGM)	Isozyme markers	Wilson (1988a, b)
Characterization of quinoa seed proteins reveals the presence of three polymorphic polypeptides from the globulin fraction of approximately 34.3, 35.6, and 36.2 kDa across a wide genetic base	Protein analysis	Fairbanks et al. (1990)
In 16 randomly chosen quinoa accessions, the characterization of 30 RAPD markers revealed 26 polymorphic markers, indicating the occurrence of numerous polymorphic markers	RAPD markers	Fairbanks et al. (1993)
The 19 accessions of the six species of *Chenopodium* germplasm collections were divided into five separate clusters using a total of 33, 10-mer primers, which produced 399 alleles with an average of 12 polymorphic bands per RAPD primer	RAPD markers	Ruas et al. (1999)
Employing the fluorescent in-situ hybridization (FISH) technique, retrotransposon homologous sequences (15-5D, 21-5D, and 22-19A) and transposase genes (20-20I) were identified	FISH technique	Kolano (2004)
AFLP, SSR, and RAPD markers were used to create the first genetic map of quinoa. 80 F_2 populations from the genotypes, KU-2 (L/C) and 90,654 (HL) were included in the map population, which produced 35 genetic linkage groups with a total length of 1020 cM (an average density of 4.0 cM per marker)	Linkage map construction using AFLP, SSR, and RAPD markers	Maughan et al. (2004)
Assessment of 31 quinoa accessions using 208 SSR markers produced from Colombia, Ecuador, Bolivia, Chile, Argentina, Peru, and Chile (SA-L/C-HLIAV)	SSR markers	Mason et al. (2005)
Evaluation of the seed protein profiles of 40 *Chenopodium* species (SA-L/C-HL-IAV), both domesticated and wild. *C. quinoa* accessions were shown to be genetically related to the closely related *C. bushianum* and *C. berlandieri* subsp. *nuttalliae*	Protein profiling	Bhargava et al. (2005)
424 EST sequences and 51 SNP markers were created using immature seed and floral EST libraries, respectively. With an average of one SNP per 462 base pairs (bp) and one Indel per 1812 bp, SNP markers contained 38 single-base substitutions and 13 insertions-deletions	SNP/EST-SSR markers	Coles et al. (2005)

Table 7.8 (continued)

Detailed information	Marker(s)/technique(s) used	Reference(s)
Using the fluorescent in-situ hybridization (FISH) technique, the genomic organisation and distribution of the 45S (NOR, nucleolus organizer region) and 5S ribosomal RNA (rRNA) genes were characterized	FISH technique	Maughan et al. (2006)
From the "real" quinoa type, two libraries (*Bam*HI and *Eco*RI) produced 26,880 and 48,000 clones, respectively, with an average insert size of around 123 kb	BAC libraries	Stevens et al. (2006)
Using 36 SSR markers which produced 420 alleles, 151 quinoa accessions from Ecuador, Peru, Bolivia, Chile, and Argentina were evaluated	SSR markers	Christensen et al. (2007)
Use of 10 RAPD markers to characterize 87 Bolivian quinoa accessions which yielded 38 alleles	RAPD markers	Del Castillo et al. (2007)
Twenty-three quinoa accessions from Ecuador, Colombia, Bolivia, Argentina, Peru and Chile (SA-L/C-HL-IAV) were characterized with newly developed 216 SSR markers as well as 6 SSR markers developed from bacterial artificial chromosome-end sequences (BES-SSRs). Using SSR, AFLP, loci for the 11S seed storage protein, a nuclear organizing region (NOR), and a betalain color locus, a second quinoa genetic map was created. A recombinant inbred lines (RILs) population made up of 82 F_5 individuals from the KU-2 (L/C) and 90,654 (HL) genotypes produced 38 genetic linkage groups with a total map length of 913 cM	Genetic linkage map using SSR, AFLP, loci for the 11S seed storage protein, a NOR, and a betalain color locus	Jarvis et al. (2008)
Fluorescent in-situ hybridization (FISH) technique was used to characterize the repetitive sequence (pTaq10) obtained from the TaqI digest of quinoa genomic DNA	FISH technique	Kolano et al. (2008)
20 SSR markers are used to characterize 59 Chilean quinoa accessions (SA-L/C) that produced 150 alleles	SSR markers	Fuentes et al. (2009a, b)
Salt Overly Sensitive 1 (*SOS1*) gene (Na^+/H^+ antiporter) cloning and characterization, resulting in two homeologous SOS1 loci (cqSOS1A and cqSOS1B) from *C. quinoa*	Molecular characterization of two homologous loci	Maughan et al. (2009a)
Microarray assay is used to annotate a sizable EST collection from mature quinoa seed tissues that express saponins to clarify the genetic elements of its synthesis. 39,366 unigenes, comprised of 16,728 contigs and 22,638 singletons, were described	EST	Reynolds (2009)
Use of three AFLP markers to characterize 18 Chilean quinoa accessions (L/C-SA) that produce 130 alleles	AFLP markers	Leonardo and Max (2009)

(continued)

Table 7.8 (continued)

Detailed information	Marker(s)/technique(s) used	Reference(s)
Employing 12 RAPD and 4 mini-satellite or variable number of tandem repeat (VNTR) markers, characterize 55 accessions from 14 Chenopod species (350 polymorphic bands produced)	RAPD and directed amplification of minisatellite DNA (DAMD) markers	Rana et al. (2010)
Using the fluorescent in-situ hybridization (FISH) technique, a repetitive DNA sequence (18–24 J) and 12-13P sequence have been characterized	FISH technique	Kolano et al. (2011)
Analysis of the sodium transporter genes *CqSOS1* and *CqNHX* and their expression in the root and shoot tissues of genotypes (L/C-HL) in response to salt were performed to characterize the *NHX1* gene (vacuolar Na^+/H^+ antiporter) through quantitative real-time PCR	QRT-PCR method	Ruiz-Carrasco et al. (2011)
Use of 22 SSR markers to characterise 35 Argentinian quinoa accessions (SA) which reported 354 alleles) and heterozygosity ranged from 0.58 to 0.93	SSR markers	Costa Tártara et al. (2012)
Utilizing 20 SSR markers, 34 quinoa accessions from Ecuador, Colombia, Peru, Bolivia, Chile, and Argentina (SA-L/C-HL-IAV) were characterized (118 alleles reported)	SSR markers	Fuentes et al. (2012)
In 113 quinoa accessions from Ecuador, Peru, Bolivia, Chile, and Argentina (SA-L/CHL-IAV), the generation of 427 SNP markers (854 alleles) was evaluated and using these markers a third genetic map of quinoa was constructed. A map population consisting of 128 individuals from two advanced $F_{2:8}$ RIL populations having a common male parent that produced 29 genetic linkage groups spanning 1404 cM map length (3.1 cM per marker)	Genetic linkage map using genomic reduction protocol and SNP markers	Maughan et al. (2012)
Nutritional characterization of 21 Argentinean quinoa germplasm and their diversity and interrelationships among nutritional traits	Protein and fatty acids composition analysis	Vidueiros et al. (2015)
An optimized mix of high-throughput next-generation sequencing on the Illumina Hiseq 2500 and PacBio RS II sequencers was used to create the draft genome sequence of the inbred line, Kd. According to these results, the de novo genome assembly had 25 k scaffolds that each comprised 1 Gbp and had an N50 length of 86 kbp and hence, free-access Quinoa Genome Data Base (QGDB) was developed	NGS technique	Yasui et al. (2016a)
QTL analysis was done with 1076 SNPs to identify two major QTLs, one for days to floral bud appearance on chromosome Cq6B and the other for seed characters on chromosome Cq2B	QTL analysis using SNP markers	Cervantes and van Loo (2017)

Table 7.8 (continued)

Detailed information	Marker(s)/technique(s) used	Reference(s)
Using the coastal Chilean quinoa accession PI 614886 (QQ74) and its wild diploid progenitors, as well as resequencing of 15 other quinoa accessions, the first high-quality reference genome was established using single-molecule real-time (SMRT) sequencing in combination with optical, chromosome-contact, and genetic maps. According to the current study, the quinoa genome is estimated to be between 1.45 and 1.50 Gb	NGS, optical and in-vitro chromosome-contact maps using BioNano Genomics and Dovetail Genomics technologies, respectively	Jarvis et al. (2017)
In response to salinity tolerance, 219 quinoa-specific genes with several putative transmembrane domains showed differential expression	Using physiological data, RNAseq, and SNP analyses	Schmöckel et al. (2017)
Identification of SSR, SNP, and insertion/deletion (INDEL) markers in 11 quinoa accessions using a whole-genome resequencing approach	Whole-genome re-sequencing approach	Zhang et al. (2017b)
The genome size of quinoa was estimated to be approximately 1.45 Gbp using a hybrid technique to genome assembly that combined 284 Gb of paired-end reads from HiSeq2500 (Illumina) and 50 Gb of single-molecule long reads from RS II (PacBio)	Hybrid techniques like HiSeq2500 (Illumina) and RS II (PacBio)	Zou et al. (2017)
The study presents the complete mitochondrial and chloroplast genome sequences of quinoa accession PI 614886, as well as the identification of sequence variants in additional quinoa and related species accessions	Purely read-mapping approaches	Maughan et al. (2019)
In this study, 15 species-specific SSR markers were used to characterize 84 accessions representing the species' cultivated range in the Ecuadorian Andes. The extent of allelic richness (196 alleles) and genetic heterozygosity ($H_E = 0.71$) detected for these accessions demonstrate that Ecuadorian quinoa is highly diverse	SSR markers	Salazar et al. (2019)
Molecular diversity and population structure of world quinoa germplasm originating from 8 countries were carried out through the 11 most polymorphic iPBS-retrotransposon marker system which recorded a total of 235 bands and mean polymorphism information content (PIC) of 0.410	iPBS-retrotransposon marker	Barut et al. (2020)
Under Saudi Arabian conditions, 32 introduced quinoa genotypes were evaluated using 17 qualitative and 11 quantitative traits along with 21 primer pair combinations for sequence-related amplified polymorphism (SRAP) markers	SRAP markers	EL-Harty et al. (2021)

(continued)

Table 7.8 (continued)

Detailed information	Marker(s)/technique(s) used	Reference(s)
Following whole-genome resequencing of 310 accessions, 2.9 million polymorphic high-confidence single nucleotide polymorphism (SNP) loci were discovered. Highland and lowland quinoa were separated into two major groups, with FST divergence of 0.36 and LD decay of 6.5 and 49.8 kb, respectively. A multi-year phenotyping trial-based genome-wide association study discovered 600 SNPs that were stably associated with 17 traits	SNP markers	Patiranage et al. (2022)
A recent QTLs study using Skim-seq to genotype the F_2 population constructed a high-density genetic map with 133,923 single nucleotide polymorphism (SNPs) markers and found 15 QTLs in quinoa	QTLs analysis using the Skim-seq method to genotype and construct a high-density genetic map with SNP markers	Maldonado-Taipe et al. (2022)

polymorphism (SNP) markers, which were generated from specific tissues to create expressed sequence tag (EST) libraries and draw parallels with protein-encoding genes from other plant species (Coles et al. 2005). The development of the first genetic linkage map of quinoa was a significant milestone, involving various molecular resources such as SSR, amplified fragment length polymorphism, and RAPD markers, as well as loci related to seed storage proteins, chromosomal regions (NOR), and the morphological betalain color locus (Maughan et al. 2004; Jarvis et al. 2008). As quinoa research continued to evolve, the genome was extensively explored using single nucleotide polymorphism (SNP) markers, revealing millions of polymorphic loci through whole-genome resequencing (Patiranage et al. 2022). Additionally, sequence-related amplified polymorphism (SRAP) markers were employed to characterize introduced quinoa genotypes under specific environmental conditions (EL-Harty et al. 2021). The creation of bacterial artificial chromosome (BAC) libraries was yet another strategy aimed at enhancing the utility of the quinoa genome (Stevens et al. 2006). Several high-quality reference genome sequences have been established, providing valuable insights into the genetic makeup of quinoa (Yasui et al. 2016a; Zou et al. 2017; Maughan et al. 2012). These studies have not only uncovered the size and structure of the quinoa genome but have also elucidated its evolutionary history. In-depth investigations into quinoa's genetic diversity, population structure, and gene expression profiles have been carried out, utilizing various molecular techniques, including RNA sequencing (RNAseq) and SNP analyses (Jarvis et al. 2017; Zhang et al. 2017b; Schmöckel et al. 2017). These studies have identified specific genes associated with traits such as salt tolerance and have paved the way for the development of valuable markers. Recent research efforts have also delved into the mitochondrial and chloroplast genome sequences of quinoa accessions, further enhancing our understanding of quinoa's evolutionary journey (Maughan et al. 2019). A comprehensive summary of additional molecular research conducted in the field of quinoa is provided in Table 7.8.

References

Aii J, Nagano M, Penner AG, Campbell GC, Adachi T (1998) Identification of RAPD markers linked to the homostylar (Ho) gene in buckwheat. Japn J Breed 48(1):59–62

Alvarez-Jubete L, Holse M, Hansen Å, Arendt E, Gallagher E (2009) Impact of baking on vitamin E content of pseudocereals amaranth, quinoa, and buckwheat. Cereal Chem 86(5):511–515

Bae M-J, Kim Y-S, Kim I-S, Choe Y-H, Lee E-J, Kim Y-H, Park H-M, Yoon H-S (2013) Transgenic rice overexpressing the Brassica juncea gamma-glutamylcysteine synthetase gene enhances tolerance to abiotic stress and improves grain yield under paddy field conditions. Mol Breed 31:931–945

Baggerman G, Vierstraete E, De Loof A, Schoofs L (2005) Gel-based versus gel-free proteomics: a review. Comb Chem High Throughput Screen 8(8):669–677

Baniya BK (1995) Present status of buckwheat genetic resources in Nepal. Curr Adv Buckwheat Res:47–53

Barut M, Nadeem MA, Karaköy T, Baloch FS (2020) DNA fingerprinting and genetic diversity analysis of world quinoa germplasm usingiPBS-retrotransposon marker system. Turk J Agric For 44(5):479–491

Berger A, Monnard I, Dionisi F, Gumy D, Hayes K, Lambelet P (2003) Cholesterol-lowering properties of amaranth flakes, crude and refined oils in hamsters. Food Chem 81(1):119–124

Bhargava A, Rana TS, Shukla S, Ohri D (2005) Seed protein electrophoresis of some cultivated and wild species of Chenopodium. Biol Plant 49:505–511

Bioversity International, FAO, PROINPA, INIAF, IFAD (2013) Descriptores para quinua (*Chenopodium quinoa* Willd.) y sus parientes silvestres. Rome, Bioversity International & FAO; La Paz, Fundación PROINPA; La Paz, Instituto Nacional de Innovación Agropecuaria y Forestal, Rome, IFAD

Bonifacio A (1995) Interspecific and intergeneric hybridization in chenopod species. Department of Botany and Range Sciences, Brigham Young University

Boukid F, Folloni S, Sforza S, Vittadini E, Prandi B (2018) Current trends in ancient grains-based foodstuffs: insights into nutritional aspects and technological applications. Compr Rev Food Sci Food Saf 17(1):123–136. https://doi.org/10.1111/1541-4337.12315

Breene WM (1991) Food uses of grain Amaranth. Cereal Foods World 36(5):426–430

Brenner D, Baltensperger D, Kulakow P, Lehmann J, Myers R, Slabbert M, Sleugh B (2000) Genetic resources and breeding of Amaranthus. Plant Breed Rev 19:227–285

Bressani R, Gonzales J, Zuniga J, Breuner M, Elias L (1987) Yield, selected chemical composition and nutritive value of 14 selections of amaranth grain representing four species. J Sci Food Agric 38(4):347–356

Bressani R, Sánchez-Marroquín A, Morales E (1992) Chemical composition of grain amaranth cultivars and effects of processing on their nutritional quality. Food Rev Intl 8(1):23–49

Brockington SF, Walker RH, Glover BJ, Soltis PS, Soltis DE (2011) Complex pigment evolution in the Caryophyllales. New Phytol 190(4):854–864

Brockington SF, Yang Y, Gandia-Herrero F, Covshoff S, Hibberd JM, Sage RF, Wong GK, Moore MJ, Smith SA (2015) Lineage-specific gene radiations underlie the evolution of novel betalain pigmentation in Caryophyllales. New Phytol 207(4):1170–1180

Caselato-Sousa VM, Amaya-Farfán J (2012) State of knowledge on amaranth grain: a comprehensive review. J Food Sci 77(4):R93–R104

Cervantes DP, van Loo E (2017) QTL mapping for agromorphological traits in Quinoa (*Chenopodium quinoa* Willd.). Wageningen University

Chakraborty S, Chakraborty N, Datta A (2000) Increased nutritive value of transgenic potato by expressing a nonallergenic seed albumin gene from Amaranthus hypochondriacus. Proc Natl Acad Sci 97(7):3724–3729

Chan K, Sun M (1997) Genetic diversity and relationships detected by isozyme and RAPD analysis of crop and wild species of Amaranthus. Theor Appl Genet 95:865–873

Chang Y, Liu H, Liu M, Liao X, Sahu SK, Fu Y, Song B, Cheng S, Kariba R, Muthemba S (2019) The draft genomes of five agriculturally important African orphan crops. GigaScience 8(3):giy152

Chauhan RS, Gupta N, Sharma SK, Rana JC, Sharma TR, Jana S (2010) Genetic and genome resources in buckwheat–present status and future perspectives. Buckwheat 2:33–44

Chen Y, Guo H, Jiang Y, Li C, Zhao W, Tian Y (2003) Disease-tolerance of transgenic tobacco plants expressing Ah-AMP gene of Amaranthus hypochondriacus. Prog Nat Sci 13(5):362–366

Chen H, Wang L, Liu X, Hu L, Wang S, Cheng X (2017) De novo transcriptomic analysis of cowpea (*Vigna unguiculata* L. Walp.) for genic SSR marker development. BMC Genet 18:1–12

Cho K-S, Yun B-K, Yoon Y-H, Hong S-Y, Mekapogu M, Kim K-H, Yang T-J (2015) Complete chloroplast genome sequence of tartary buckwheat (Fagopyrum tataricum) and comparative analysis with common buckwheat (F. esculentum). PLoS One 10(5):e0125332

Christensen S, Pratt DB, Pratt C, Nelson P, Stevens M, Jellen EN, Coleman CE, Fairbanks DJ, Bonifacio A, Maughan PJ (2007) Assessment of genetic diversity in the USDA and CIP-FAO international nursery collections of quinoa (Chenopodium quinoa Willd.) using microsatellite markers. Plant Genet Resour 5(2):82–95

Christinet L, Burdet FX, Zaiko MA, Hinz U, Zrÿd J-P (2004) Characterization and functional identification of a novel plant 4, 5-extradiol dioxygenase involved in betalain pigment biosynthesis in Portulaca grandiflora. Plant Physiol 134(1):265–274

Chrungoo N, Kreft I, Sangma S, Devadasan N, Dohtdong L, Chetri U (2012) Genetic diversity in Himalayan buckwheats: a perspective for use in crop improvement programmes. In: Proceedings of the 12th International Symposium on Buckwheat, Pernica, pp 198–211

Clouse J, Adhikary D, Page J, Ramaraj T, Deyholos M, Udall J, Fairbanks D, Jellen E, Maughan P (2016) The amaranth genome: genome, transcriptome, and physical map assembly. Plant Genome 9(1):plantgenome2015.2007.0062

Coles N, Coleman C, Christensen S, Jellen E, Stevens M, Bonifacio A, Rojas-Beltran J, Fairbanks D, Maughan P (2005) Development and use of an expressed sequenced tag library in quinoa (*Chenopodium quinoa* Willd.) for the discovery of single nucleotide polymorphisms. Plant Sci 168(2):439–447

Costa Tártara SM, Manifesto MM, Bramardi SJ, Bertero HD (2012) Genetic structure in cultivated quinoa (Chenopodium quinoa Willd.), a reflection of landscape structure in Northwest Argentina. Conserv Genet 13:1027–1038

Das S (2016) Amaranthus: a promising crop of future. Springer

Dawson IK, Powell W, Hendre P, Bančič J, Hickey JM, Kindt R, Hoad S, Hale I, Jamnadass R (2019) The role of genetics in mainstreaming the production of new and orphan crops to diversify food systems and support human nutrition. New Phytol 224(1):37–54

del Castillo C, Winkel T, Mahy G, Bizoux J-P (2007) Genetic structure of quinoa (*Chenopodium quinoa* Willd.) from the Bolivian altiplano as revealed by RAPD markers. Genet Resour Crop Evol 54:897–905

Deng L, Zhang K, Huang K, Chen Q (2011) RAPD analysis for genetic diversity of nineteen common and tartary buckwheat varieties. Agric Sci Technol Hunan 12(1):65–69

Du X, Zhang Z, Wu B, Li Y, Yu A (2013) Construction of genetic linkage map in tartary buckwheat (*Fagopyrum tataricum*) using SSR. Chinese Agric Sci Bull

Eldakak M, Milad SI, Nawar AI, Rohila JS (2013) Proteomics: a biotechnology tool for crop improvement. Front Plant Sci 4:35

El-Harty EH, Ghazy A, Alateeq TK, Al-Faifi SA, Khan MA, Afzal M, Alghamdi SS, Migdadi HM (2021) Morphological and molecular characterization of quinoa genotypes. Agriculture 11(4):286

Espitia E (1992) Amaranth germplasm development and agronomic studies in Mexico. Food Rev Intl 8(1):71–86

Espitia Rangel E (1986) Caracterizacion y evaluacion preliminar de germoplasma de (Amaranthus spp.)

Fairbanks D, Burgener K, Robison L, Andersen W, Ballon E (1990) Electrophoretic characterization of quinoa seed proteins. Plant Breed 104(3):190–195

Fairbanks D, Waldrigues A, Ruas C, Ruas P, Maughan P, Robison L, Andersen W, Riede C, Pauley C, Caetano L (1993) Efficient characterization of biological diversity using field DNA extraction and random amplified polymorphic DNA markers. Rev Braz Genét 16:11–11

Fan Y, Yn J, Ding M, Tang Y, Cheng J, Zhang K, Zhou M (2021) The complete chloroplast genome sequences of eight Fagopyrum species: insights into genome evolution and phylogenetic relationships. Front Plant Sci 12:799904

Fang Z, Xu X, Gao J, Wang P, Liu Z, Feng B (2015) Characterization of FeDREB1 promoter involved in cold-and drought-inducible expression from common buckwheat (Fagopyrum esculentum). Genet Mol Res 14(3):7990–8000

Fang X, Huang K, Nie J, Zhang Y, Zhang Y, Li Y, Wang W, Xu X, Ruan R, Yuan X (2019) Genome-wide mining, characterization, and development of microsatellite markers in Tartary buckwheat (*Fagopyrum tataricum* Garetn.). Euphytica 215:1–10

Farooq S, Rehman RU, Pirzadah T, Malik B, Dar FA, Tahir I (2016) Cultivation, agronomic practices, and growth performance of buckwheat. In: Molecular breeding and nutritional aspects of buckwheat. Elsevier, pp 299–319

Fuentes F, Maughan P, Jellen E (2009a) Diversidad genética y recursos genéticos para el mejoramiento de la quinoa (Chenopodium quinoa Willd.). Revista geográfica de Valparaíso 42:20–33

Fuentes FF, Martinez EA, Hinrichsen P, Jellen E, Maughan PJ (2009b) Assessment of genetic diversity patterns in Chilean quinoa (Chenopodium quinoa Willd.) germplasm using multiplex fluorescent microsatellite markers. Conserv Genet 10:369–377

Fuentes F, Bazile D, Bhargava A, Martinez EA (2012) Implications of farmers' seed exchanges for on-farm conservation of quinoa, as revealed by its genetic diversity in Chile. J Agric Sci 150(6):702–716

Gandarillas H (1967) Observaciones sobre la biología reproductiva de la quinua. Rev Sayaña (Bolivia) 5:26–29

Gao F, Zhang Z, Wu B (2012) Construction and application of SSR molecular markers system for genetic diversity analysis of Chinese tartary buckwheat germplasm resources. Sci Agric Sin 45(6):1042–1053

Gao F, Zhou J, Deng R-Y, Zhao H-X, Li C-L, Chen H, Suzuki T, Park S-U, Wu Q (2017) Overexpression of a tartary buckwheat R2R3-MYB transcription factor gene, FtMYB9, enhances tolerance to drought and salt stresses in transgenic Arabidopsis. J Plant Physiol 214:81–90

Grahić J, Okić A, Šimon S, Djikić M, Gadžo D, Pejić I, Gaši F (2022) Genetic Relationships and Diversity of Common Buckwheat Accessions in Bosnia and Herzegovina. Agronomy 12(11):2676

Grubben G, Van Sloten D (1981) Genetic resources of amaranths. Genetic resources of amaranths (AGP: IBPGR/80/2)

Gupta V, Gudu S (1991) Interspecific hybrids and possible phylogenetic relations in grain amaranths. Euphytica 52(1):33–38

Gupta N, Sharma SK, Rana JC, Chauhan RS (2012) AFLP fingerprinting of tartary buckwheat accessions (Fagopyrum tataricum) displaying rutin content variation. Fitoterapia 83(6):1131–1137

Hara T, Iwata H, Okuno K, Matsui K, Ohsawa R (2011) QTL analysis of photoperiod sensitivity in common buckwheat by using markers for expressed sequence tags and photoperiod-sensitivity candidate genes. Breed Sci 61(4):394–404

He M, He Y, Zhang K, Lu X, Zhang X, Gao B, Fan Y, Zhao H, Jha R, Huda MN (2022) Comparison of buckwheat genomes reveals the genetic basis of metabolomic divergence and ecotype differentiation. New Phytol 235(5):1927–1943

Hooker JD (1885) The flora of british India, vol 5. L. Reeve

Hooker JD (1952) Chenopods. Himalayan J 1:8

Hou Y, Zhang Z, Wu B, Li Y (2009) Genetic diversity in tartary buckwheat revealed by AFLP analysis. Sci Agric Sin 42(12):4166–4174

Hou S, Sun Z, Linghu B, Xu D, Wu B, Zhang B, Wang X, Han Y, Zhang L, Qiao Z (2016) Genetic diversity of buckwheat cultivars (Fagopyrum tartaricum Gaertn.) assessed with SSR markers developed from genome survey sequences. Plant Mol Biol Report 34:233–241

Huang J, Deng J, Shi T, Chen Q, Liang C, Meng Z, Zhu L, Wang Y, Zhao F, Yu S (2017) Global transcriptome analysis and identification of genes involved in nutrients accumulation during seed development of rice tartary buckwheat (Fagopyrum Tararicum). Sci Rep 7(1):11792

Huang Y, Zhao H, Gao F, Yao P, Deng R, Li C, Chen H, Wu Q (2018) A R2R3-MYB transcription factor gene, FtMYB13, from Tartary buckwheat improves salt/drought tolerance in Arabidopsis. Plant Physiol Biochem 132:238–248

Iwata H, Imon K, Tsumura Y, Ohsawa R (2005) Genetic diversity among Japanese indigenous common buckwheat (Fagopyrum esculentum) cultivars as determined from amplified fragment length polymorphism and simple sequence repeat markers and quantitative agronomic traits. Genome 48(3):367–377

Jacobsen S-E (2001) Genetic resources and breeding of the Andean grain crop quinoa (Chenopodium quinoa Willd.). Submitted Plant Genetic Resources Newsletter

Jarvis D, Kopp O, Jellen E, Mallory M, Pattee J, Bonifacio A, Coleman C, Stevens M, Fairbanks D, Maughan P (2008) Simple sequence repeat marker development and genetic mapping in quinoa (Chenopodium quinoa Willd.). J Genet 87:39–51

Jarvis DE, Ho YS, Lightfoot DJ, Schmöckel SM, Li B, Borm TJ, Ohyanagi H, Mineta K, Michell CT, Saber N (2017) The genome of Chenopodium quinoa. Nature 542(7641):307–312

Jellen EN, Maughan PJ, Bertero D, Munir H (2013) Prospects for quinoa (Chenopodium Quinoa Willd.) improvement through biotechnology. In: Biotechnology of Neglected and Underutilized Crops. Springer, pp 173–201

Jellen EN, Maughan PJ, Fuentes F, Kolano BA (2015) Botany, phylogeny and evolution. In: Bazile D, Bertero D, Nieto C (eds) State of the art report on quinoa around the world in 2013. FAO & CIRAD, Rome, pp 12–23

Ji CY, Chung W-H, Kim HS, Jung WY, Kang L, Jeong JC, Kwak S-S (2017) Transcriptome profiling of sweetpotato tuberous roots during low temperature storage. Plant Physiol Biochem 112:97–108

Jiang P, Burczynski F, Campbell C, Pierce G, Austria J, Briggs C (2007) Rutin and flavonoid contents in three buckwheat species Fagopyrum esculentum, F. tataricum, and F. homotropicum and their protective effects against lipid peroxidation. Food Res Int 40(3):356–364

Jofre-Garfias A, Villegas-Sepúlveda N, Cabrera-Ponce J, Adame-Alvarez R, Herrera-Estrella L, Simpson J (1997) Agrobacterium-mediated transformation of Amaranthus hypochondriacus: light-and tissue-specific expression of a pea chlorophyll a/b-binding protein promoter. Plant Cell Rep 16(12):847–852

Joshi B (1986) Genetic variability in grain amaranth. Indian J Agric Sci 56(8):574–576

Joshi BD, Rana R (1991) Grain amaranths: the future food crop. Grain amaranths: the future food crop

Joshi BD, Rana RS (1995) Buckwheat (Fagopyrum esculentum). Cereals and pseudocereal, vol 85. Chapman Hall, London, p 127

Joshi BK, Okuno K, Ohsawa R, Hara T (2006) Common buckwheat-based EST primers in the genome of other species of Fagopyrum. Nep Agric Res J 7:27–36

Joshi D, Meena R, Chandora R (2021) Genetic resources: collection, characterization, conservation, and documentation. In: Millets and pseudo cereals. Elsevier, pp 19–31

Kamenya SN, Mikwa EO, Song B, Odeny DA (2021) Genetics and breeding for climate change in Orphan crops. Theor Appl Genet 134:1787–1815

Kauffman CS, Weber LE (1990) Grain amaranth. Advances in new crops. Timber Press, Portland, OR, pp 127–139

Keli Y (1992) Research on cultivated buckwheat germplasm resources in China. In: Proceedings of 5th international symposium on buckwheat, pp 20–26

Khan F, Chai HH, Ajmera I, Hodgman C, Mayes S, Lu C (2017) A transcriptomic comparison of two Bambara groundnut landraces under dehydration stress. Genes 8(4):121

Khoury CK, Bjorkman AD, Dempewolf H, Ramirez-Villegas J, Guarino L, Jarvis A, Rieseberg LH, Struik PC (2014) Increasing homogeneity in global food supplies and the implications for food security. Proc Natl Acad Sci 111(11):4001–4006

Kietlinski KD, Jimenez F, Jellen EN, Maughan PJ, Smith SM, Pratt DB (2014) Relationships between the weedy Amaranthus hybridus (Amaranthaceae) and the grain amaranths. Crop Sci 54(1):220–228

Kolano B (2004) Genome analysis of a few Chenopodium species. Ph. D. thesis, University of Silesia, Katowice, Poland

Kolano B, Plucienniczak A, Kwasniewski M, Maluszynska J (2008) Chromosomal localization of a novel repetitive sequence in the Chenopodium quinoa genome. J Appl Genet 49:313–320

Kolano B, Gardunia B, Michalska M, Bonifacio A, Fairbanks D, Maughan P, Coleman C, Stevens M, Jellen E, Maluszynska J (2011) Chromosomal localization of two novel repetitive sequences isolated from the Chenopodium quinoa Willd. genome. Genome 54(9):710–717

Konishi T, Iwata H, Yashiro K, Tsumura Y, Ohsawa R, Yasui Y, Ohnishi O (2006) Development and characterization of microsatellite markers for common buckwheat. Breed Sci 56(3):277–285

Kuluev B, Mikhaylova E, Taipova R, Chemeris A (2017) Changes in phenotype of transgenic amaranth Amaranthus retroflexus L., overexpressing ARGOS-LIKE gene. Russ J Genet 53:67–75

Kump B, Javornik B (2002) Genetic diversity and relationships among cultivated and wild accessions of tartary buckwheat (Fagopyrum tataricum Gaertn.) as revealed by RAPD markers. Genet Resour Crop Evol 49(6):565–572

Kupper C (2005) Dietary guidelines and implementation for celiac disease. Gastroenterology 128(4):S121–S127

Leonardo AR, Max TI (2009) Comparative analysis of genetic and morphologic diversity among quinoa accessions (Chenopodium quinoa Willd.) of the South of Chile and highland accessions. J Plant Breed Crop Sci 1(5):210–216

Li S-q, Zhang QH (2001) Advances in the development of functional foods from buckwheat. Crit Rev Food Sci Nutr 41(6):451–464

Li Y-Q, Shi T-L, Zhang Z-W (2007) Development of microsatellite markers from tartary buckwheat. Biotechnol Lett 29:823–827

Li Y, Fan X, Shi T, Zhang Q, Zhang Z (2009) SRAP marker reveals genetic diversity in tartary buckwheat in China. Front Agric China 3:383–387

Lightfoot D, Jarvis DE, Ramaraj T, Lee R, Jellen E, Maughan P (2017) Single-molecule sequencing and Hi-C-based proximity-guided assembly of amaranth (Amaranthus hypochondriacus) chromosomes provide insights into genome evolution. BMC Biol 15:1–15

Logacheva MD, Samigullin TH, Dhingra A, Penin AA (2008) Comparative chloroplast genomics and phylogenetics of Fagopyrum esculentum ssp. ancestrale–a wild ancestor of cultivated buckwheat. BMC Plant Biol 8:1–15

Logacheva MD, Kasianov AS, Vinogradov DV, Samigullin TH, Gelfand MS, Makeev VJ, Penin AA (2011) De novo sequencing and characterization of floral transcriptome in two species of buckwheat (Fagopyrum). BMC Genomics 12(1):1–17

Lu G, Zhang Y, Li Y, Lu G, He R, Bai Y (1996) Study on the nutritional components of Fagopyrum cymosum. Chin J Vet Drug 30:19–21

Ma K-H, Kim N-S, Lee G-A, Lee S-Y, Lee JK, Yi JY, Park Y-J, Kim T-S, Gwag J-G, Kwon S-J (2009) Development of SSR markers for studies of diversity in the genus Fagopyrum. Theor Appl Genet 119:1247–1254

Ma X, Vaistij FE, Li Y, Jansen van Rensburg WS, Harvey S, Bairu MW, Venter SL, Mavengahama S, Ning Z, Graham IA (2021) A chromosome-level Amaranthus cruentus genome assembly highlights gene family evolution and biosynthetic gene clusters that may underpin the nutritional value of this traditional crop. Plant J 107(2):613–628

Mabhaudhi T, Chimonyo VGP, Hlahla S, Massawe F, Mayes S, Nhamo L, Modi AT (2019) Prospects of orphan crops in climate change. Planta 250:695–708

Maldonado-Taipe N, Barbier F, Schmid K, Jung C, Emrani N (2022) High-density mapping of quantitative trait loci controlling agronomically important traits in quinoa (Chenopodium quinoa willd.). Front Plant Sci 13:916067

Mallory MA, Hall RV, McNabb AR, Pratt DB, Jellen EN, Maughan PJ (2008) Development and characterization of microsatellite markers for the grain amaranths. Crop Sci 48(3):1098–1106

Martirosyan DM, Miroshnichenko LA, Kulakova SN, Pogojeva AV, Zoloedov VI (2007) Amaranth oil application for coronary heart disease and hypertension. Lipids Health Dis 6:1–12

Mason S, Stevens M, Jellen E, Bonifacio A, Fairbanks D, Coleman C, McCarty R, Rasmussen A, Maughan P (2005) Development and use of microsatellite markers for germplasm characterization in quinoa (Chenopodium quinoa Willd.). Crop Sci 45(4):1618–1630

Massange-Sanchez JA, Palmeros-Suarez PA, Espitia-Rangel E, Rodriguez-Arevalo I, Sanchez-Segura L, Martinez-Gallardo NA, Alatorre-Cobos F, Tiessen A, Delano-Frier JP (2016) Overexpression of grain amaranth (Amaranthus hypochondriacus) AhERF or AhDOF transcription factors in Arabidopsis thaliana increases water deficit-and salt-stress tolerance, respectively, via contrasting stress-amelioration mechanisms. PLoS One 11(10):e0164280

Matsui K, Kiryu Y, Komatsuda T, Kurauchi N, Ohtani T, Tetsuka T (2004) Identification of AFLP makers linked to non-seed shattering locus (sht1) in buckwheat and conversion to STS markers for marker-assisted selection. Genome 47(3):469–474

Maughan PJ, Bonifacio A, Jellen E, Stevens M, Coleman C, Ricks M, Mason S, Jarvis D, Gardunia B, Fairbanks D (2004) A genetic linkage map of quinoa (Chenopodium quinoa) based on AFLP, RAPD, and SSR markers. Theor Appl Genet 109:1188–1195

Maughan P, Kolano B, Maluszynska J, Coles N, Bonifacio A, Rojas J, Coleman C, Stevens M, Fairbanks D, Parkinson S (2006) Molecular and cytological characterization of ribosomal RNA genes in Chenopodium quinoa and Chenopodium berlandieri. Genome 49(7):825–839

Maughan PJ, Sisneros N, Luo M, Kudrna D, Ammiraju JS, Wing RA (2008) Construction of an Amaranthus hypochondriacus bacterial artificial chromosome library and genomic sequencing of herbicide target genes. Crop Sci 48:S-85–S-94

Maughan P, Turner T, Coleman C, Elzinga D, Jellen E, Morales J, Udall J, Fairbanks D, Bonifacio A (2009a) Characterization of Salt Overly Sensitive 1 (SOS1) gene homoeologs in quinoa (Chenopodium quinoa Willd.). Genome 52(7):647–657

Maughan PJ, Yourstone SM, Jellen EN, Udall JA (2009b) SNP discovery via genomic reduction, barcoding, and 454-pyrosequencing in amaranth. Plant Genome 2(3)

Maughan P, Smith S, Fairbanks D, Jellen E (2011) Development, characterization, and linkage mapping of single nucleotide polymorphisms in the grain amaranths (Amaranthus sp.). Plant Genome 4(1)

Maughan P, Smith S, Rojas-Beltrán J, Elzinga D, Raney J, Jellen E, Bonifacio A, Udall J, Fairbanks D (2012) Single nucleotide polymorphism identification, characterization, and linkage mapping in quinoa. Plant Genome 5(3)

Maughan PJ, Chaney L, Lightfoot DJ, Cox BJ, Tester M, Jellen EN, Jarvis DE (2019) Mitochondrial and chloroplast genomes provide insights into the evolutionary origins of quinoa (Chenopodium quinoa Willd.). Sci Rep 9(1):185. https://doi.org/10.1038/s41598-018-36693-6

Mizuno N, Yasui Y (2019) Gene flow signature in the S-allele region of cultivated buckwheat. BMC Plant Biol 19(1):1–9

Munusamy U, Abdullah SNA, Aziz MA, Khazaai H (2013) Female reproductive system of Amaranthus as the target for Agrobacterium-mediated transformation. Adv Biosci Biotechnol 4(02):188–192

Nagano M, Aii J, Kuroda M, Campbell C, Adachi T (2001) Conversion of AFLP markers linked to the Sh allele at the S locus in buckwheat to a simple PCR based marker form. Plant Biotechnol 18(3):191–196

Ohnishi O, Ohta T (1987) Construction of a linkage map in common buckwheat, Fagopyrum esculentum Moench. Japanese J Genet 62(5):397–414

Ohsawa R (2020) Current status and prospects of common buckwheat breeding in Japan. Breed Sci 70(1):3–12

Pagano AE (2006) Whole grains and the gluten-free diet. Pract Gastroenterol 30(10):66

Pal A, Swain SS, Das AB, Mukherjee AK, Chand PK (2013a) Stable germ line transformation of a leafy vegetable crop amaranth (Amaranthus tricolor L.) mediated by Agrobacterium tumefaciens. In Vitro Cell Develop Biol Plant 49:114–128

Pal A, Swain SS, Mukherjee AK, Chand PK (2013b) Agrobacterium pRi T^ sub L^-DNA rolB and T^ sub R^-DNA Opine Genes Transferred to the Spiny Amaranth (Amaranthus spinosus L.), A nutraceutical crop. Food Technol Biotechnol 51(1):26

Palmeros-Suárez PA, Massange-Sánchez JA, Martínez-Gallardo NA, Montero-Vargas JM, Gómez-Leyva JF, Délano-Frier JP (2015) The overexpression of an Amaranthus hypochondriacus NF-YC gene modifies growth and confers water deficit stress resistance in Arabidopsis. Plant Sci 240:25–40

Partap T (1982) Cultivated grain chenopods of Himachal Pradesh _ distribution variations and ethnobotany

Patiranage DS, Rey E, Emrani N, Wellman G, Schmid K, Schmöckel SM, Tester M, Jung C (2022) Genome-wide association study in quinoa reveals selection pattern typical for crops with a short breeding history. elife 11:e66873

Pedersen B, Kalinowski LS, Eggum BO (1987) The nutritive value of amaranth grain (Amaranthus caudatus). Plant Foods Hum Nutr 36(4):309–324. https://doi.org/10.1007/BF01892352

Písaříková B, Kráčmar S, Herzig I (2005) Amino acid contents and biological value of protein in various amaranth species. Czeh J Anim Sci 50(4):169–174

Polturak G, Breitel D, Grossman N, Sarrion-Perdigones A, Weithorn E, Pliner M, Orzaez D, Granell A, Rogachev I, Aharoni A (2016) Elucidation of the first committed step in betalain biosynthesis enables the heterologous engineering of betalain pigments in plants. New Phytol 210(1):269–283

Pulgar Vidal J (1954) La quinua o suba, alimento básico de los Chibchas. Economía Colombiana, Bogotá, Colombia

Putnam NH, O'Connell BL, Stites JC, Rice BJ, Blanchette M, Calef R, Troll CJ, Fields A, Hartley PD, Sugnet CW (2016) Chromosome-scale shotgun assembly using an in vitro method for long-range linkage. Genome Res 26(3):342–350

Qi X, Xie S, Liu Y, Yi F, Yu J (2013) Genome-wide annotation of genes and noncoding RNAs of foxtail millet in response to simulated drought stress by deep sequencing. Plant Mol Biol 83:459–473

Rana T, Narzary D, Ohri D (2010) Genetic diversity and relationships among some wild and cultivated species of Chenopodium L.(Amaranthaceae) using RAPD and DAMD methods. Curr Sci:840–846

Rauf M, Yoon H, Lee S, Hyun DY, Lee M-C, Oh S, Choi Y-M (2020) Evaluation of Fagopyrum esculentum Moench germplasm based on agro-morphological traits and the rutin and quercetin content of seeds under spring cultivation. Genet Resour Crop Evol 67:1385–1403

Reynolds DJ (2009) Genetic dissection of triterpenoid saponin production in Chenopodium quinoa using microarray analysis. Brigham Young University

Rojas W (2003) Multivariate analysis of genetic diversity of Bolivian quinoa germplasm. Food Rev Intl 19(1–2):9–23

Rojas W, Pinto M (2008) Evaluación del valor nutritivo de accesiones de quinua y cañahua silvestre. In: Pinto M (ed) Proyecto Implementation of the UNEP-GEF project, "In situ conservation of crop wild relatives through enhanced information management and field application." Informe de Fase 2005–2008. Fundación PROINPA. La Paz, Bolivia, pp 54–60

Rojas W, Pinto M (2013) La diversidad genética de quinua de Bolivia. In: Congreso Científico de la Quinua (Memorias). La Paz, Bolivia, pp 77–92

Rojas W, Pinto M, Alcócer E (2007) Diversidad genética del valor nutritivo y agroindustrial del germoplasma de quinua. Universidad Mayor de San Simón, Cochabamba (Bolivia). Facultad de Ciencias ...

Rojas W, Pinto M, Mamani E (2009) Logro e impactos del subsistema granos altoandinos, periodo 2003-2008. Encuentro Nacional de Innovación Tecnológica, Agropecuaria y Forestal INIAF Cochabamba 29:58–65

Rojas W, Pinto M, Alanoca C, Gomez Pando L, Leon-Lobos P, Alercia A, Diulgheroff S, Padulosi S, Bazile D (2015) Quinoa genetic resources and ex situ conservation

Romanova OI, Fesenko AN, Fesenko NN, Fesenko IN, Koshkin VA (2018) Buckwheat resources in the VIR (Russia) collection: the photoperiod response. In: Zhou M, Kreft I, Suvorova G, Tang Y, Woo SH (eds) Buckwheat germplasm in the world. Academic Press, pp 225–234

Ruas PM, Ruas PM, Bonifacio A, Bonifacio A, Ruas CF, Ruas CF, Fairbanks DJ, Andersen WR (1999) Genetic relationship among 19 accessions of six species of Chenopodium L., by Random Amplified Polymorphic DNA fragments (RAPD). Euphytica 105:25–32

Ruiz-Carrasco K, Antognoni F, Coulibaly AK, Lizardi S, Covarrubias A, Martínez EA, Molina-Montenegro MA, Biondi S, Zurita-Silva A (2011) Variation in salinity tolerance of four lowland genotypes of quinoa (Chenopodium quinoa Willd.) as assessed by growth, physiological traits, and sodium transporter gene expression. Plant Physiol Biochem 49(11):1333–1341

Salazar J, de Lourdes TM, Gutierrez B, Torres AF (2019) Molecular characterization of Ecuadorian quinoa (Chenopodium quinoa Willd.) diversity: implications for conservation and breeding. Euphytica 215:1–11

Samardžić JT, Nikolić DB, Timotijević GS, Jovanović ŽS, Milisavljević MĐ, Maksimović VR (2010) Tissue expression analysis of FeMT3, a drought and oxidative stress related metallothionein gene from buckwheat (Fagopyrum esculentum). J Plant Physiol 167(16):1407–1411

Sauer JD (1950) The grain amaranths: a survey of their history and classification. Ann Mo Bot Gard 37(4):561–632

Sauer JD (1967) The grain amaranths and their relatives: a revised taxonomic and geographic survey. Ann Mo Bot Gard 54(2):103–137

Schmöckel SM, Lightfoot DJ, Razali R, Tester M, Jarvis DE (2017) Identification of putative transmembrane proteins involved in salinity tolerance in Chenopodium quinoa by integrating physiological data, RNAseq, and SNP analyses. Front Plant Sci 8:1023

Senthilkumaran R, Bisht I, Bhat K, Rana J (2008) Diversity in buckwheat (Fagopyrum spp.) landrace populations from north-western Indian Himalayas. Genet Resour Crop Evol 55:287–302

Sharma T, Jana S (2002) Species relationships in Fagopyrum revealed by PCR-based DNA fingerprinting. Theor Appl Genet 105:306–312

Sheehan H, Feng T, Walker-Hale N, Lopez-Nieves S, Pucker B, Guo R, Yim WC, Badgami R, Timoneda A, Zhao L (2020) Evolution of l-DOPA 4, 5-dioxygenase activity allows for recurrent specialisation to betalain pigmentation in Caryophyllales. New Phytol 227(3):914–929

Smith TJ (2000) Squalene: potential chemopreventive agent. Expert Opin Investig Drugs 9(8):1841–1848

Song JY, Lee G-A, Yoon M-S, Ma K-H, Choi Y-M, Lee J-R, Jung Y-J, Park H-J, Kim C-K, Lee M-C (2018) Analysis of genetic diversity and population structure of buckwheat (Fagopyrum esculentum Moench.) landraces of Korea using SSR markers. In: Buckwheat germplasm in the world. Elsevier, pp 315–331

Song Y, Cheng Z, Dong Y, Liu D, Bai K, Jarvis D, Feng J, Long C (2022) Diversity of Tartary buckwheat (Fagopyrum tataricum) landraces from Liangshan, Southwest China: evidence from morphology and SSR markers. Agronomy 12(5):1022

Stetter MG, Müller T, Schmid KJ (2017) Genomic and phenotypic evidence for an incomplete domestication of South American grain amaranth (*Amaranthus caudatus*). Mol Ecol 26(3):871–886

Stevens M, Coleman C, Parkinson S, Maughan P, Zhang H-B, Balzotti M, Kooyman D, Arumuganathan K, Bonifacio A, Fairbanks D (2006) Construction of a quinoa (*Chenopodium quinoa* Willd.) BAC library and its use in identifying genes encoding seed storage proteins. Theor Appl Genet 112:1593–1600

Sunil M, Hariharan AK, Nayak S, Gupta S, Nambisan SR, Gupta RP, Panda B, Choudhary B, Srinivasan S (2014) The draft genome and transcriptome of Amaranthus hypochondriacus: a C4 dicot producing high-lysine edible pseudo-cereal. DNA Res 21(6):585–602

Suvorova G, Zhou M (2018) Distribution of cultivated buckwheat resources in the world. In: Buckwheat germplasm in the world. Elsevier, pp 21–35

Swain SS, Sahu L, Barik DP, Chand PK (2010) Agrobacterium× plant factors influencing transformation of 'Joseph's coat'(*Amaranthus tricolor* L.). Sci Hortic 125(3):461–468

Tang Y, Sun J, Peng D, Liu J, Shao J (2011) Study on the nutrients and medical elements in wild buckwheat. J Sichuan Higher Inst Cuisine 6(28):1

Transue D, Fairbanks D, Robison L, Andersen W (1994) Species identification by RAPD analysis of grain amaranth genetic resources. Crop Sci 34(5):1385–1389

Tucker JB (1986) Amaranth: the once and future crop. Bioscience 36(1):9–13

Varshney RK, Singh VK, Kumar A, Powell W, Sorrells ME (2018) Can genomics deliver climate-change ready crops? Curr Opin Plant Biol 45:205–211

Verma S, Chauhan L, Mathur R (1964) Peronospora farinosa (Fr.) Fr. on Chenopodium murale L.-A new record for India. Curr Sci 33(23):720–721

Vidueiros S, Curti RN, Dyner LM, Binaghi MJ, Peterson G, Bertero HD, Pallaro A (2015) Diversity and interrelationships in nutritional traits in cultivated quinoa (Chenopodium quinoa Willd.) from Northwest Argentina. J Cereal Sci 62:87–93

Wang C-L, Ding M-Q, Zou C-Y, Zhu X-M, Tang Y, Zhou M-L, Shao J-R (2017) Comparative analysis of four buckwheat species based on morphology and complete chloroplast genome sequences. Sci Rep 7(1):6514

Watson A, Ghosh S, Williams MJ, Cuddy WS, Simmonds J, Rey M-D, Asyraf MD, Hatta M, Hinchliffe A, Steed A, Reynolds D (2018) Speed breeding is a powerful tool to accelerate crop research and breeding. Nat Plants 4(1):23–29

Wilson HD (1980) Artificial hybridization among species of Chenopodium sect. Chenopodium. Syst Botany 5:253–263

Wilson HD (1988a) Quinua biosystematics I: domesticated populations. Econ Bot 42(4):461–477

Wilson HD (1988b) Quinua biosystematics II: free-living populations. Econ Bot 42:478–494

Wilson HD (1990) Quinua and relatives (Chenopodium sect. Chenopodium subsect. Celluloid). Econ Bot 44(Suppl 3):92–110

Wilson HD, Heiser CB Jr (1979) The origin and evolutionary relationships of 'huauzontle'(Chenopodium nuttalliae Safford), domesticated chenopod of Mexico. Am J Bot 66(2):198–206

Wu X, Blair MW (2017) Diversity in grain amaranths and relatives distinguished by genotyping by sequencing (GBS). Front Plant Sci 8:1960

Wu Q, Bai X, Zhao W, Xiang D, Wan Y, Yan J, Zou L, Zhao G (2017) De novo assembly and analysis of tartary buckwheat (Fagopyrum tataricum Garetn.) transcriptome discloses key regulators involved in salt-stress response. Genes 8(10):255

Xu JM, Fan W, Jin JF, Lou HQ, Chen WW, Yang JL, Zheng SJ (2017) Transcriptome analysis of Al-induced genes in buckwheat (Fagopyrum esculentum Moench) root apex: new insight into Al toxicity and resistance mechanisms in an Al accumulating species. Front Plant Sci 8:1141

Yabe S, Iwata H (2020) Genomics-assisted breeding in minor and pseudo-cereals. Breed Sci 70(1):19–31

Yabe S, Hara T, Ueno M, Enoki H, Kimura T, Nishimura S, Yasui Y, Ohsawa R, Iwata H (2014) Rapid genotyping with DNA micro-arrays for high-density linkage mapping and QTL mapping in common buckwheat (Fagopyrum esculentum Moench). Breed Sci 64(4):291–299

Yabe S, Hara T, Ueno M, Enoki H, Kimura T, Nishimura S, Yasui Y, Ohsawa R, Iwata H (2018) Potential of genomic selection in mass selection breeding of an allogamous crop: an empirical study to increase yield of common buckwheat. Front Plant Sci 9:276

Yao PF, Li CL, Zhao XR, Li MF, Zhao HX, Guo JY, Cai Y, Chen H, Wu Q (2017) Overexpression of a Tartary Buckwheat Gene, FtbHLH3, Enhances Drought/Oxidative Stress Tolerance in Transgenic Arabidopsis. Front Plant Sci 8:625. https://doi.org/10.3389/fpls.2017.00625

Yasui Y, Wang Y, Ohnishi O, Campbell CG (2004) Amplified fragment length polymorphism linkage analysis of common buckwheat (Fagopyrum esculentum) and its wild self-pollinated relative Fagopyrum homotropicum. Genome 47(2):345–351

Yasui Y, Mori M, Matsumoto D, Ohnishi O, Campbell CG, Ota T (2008) Construction of a BAC library for buckwheat genome research—an application to positional cloning of agriculturally valuable traits. Genes Genet Syst 83(5):393–401

Yasui Y, Hirakawa H, Oikawa T, Toyoshima M, Matsuzaki C, Ueno M, Mizuno N, Nagatoshi Y, Imamura T, Miyago M (2016a) Draft genome sequence of an inbred line of Chenopodium quinoa, an allotetraploid crop with great environmental adaptability and outstanding nutritional properties. DNA Res 23(6):535–546

Yasui Y, Hirakawa H, Ueno M, Matsui K, Katsube-Tanaka T, Yang SJ, Aii J, Sato S, Mori M (2016b) Assembly of the draft genome of buckwheat and its applications in identifying agronomically useful genes. DNA Res 23(3):215–224

Zhang Z, Wang Z, Lin R, Wang R (1999) Analysis of nutritional components in the seeds of *Fagopyrum cymosum* (Trevir.) Meissn. Acta Nutr Sin 21(480):482

Zhang G, Liu X, Quan Z, Cheng S, Xu X, Pan S, Xie M, Zeng P, Yue Z, Wang W (2012) Genome sequence of foxtail millet (Setaria italica) provides insights into grass evolution and biofuel potential. Nat Biotechnol 30(6):549–554

Zhang L, Li X, Ma B, Gao Q, Du H, Han Y, Li Y, Cao Y, Qi M, Zhu Y (2017a) The tartary buckwheat genome provides insights into rutin biosynthesis and abiotic stress tolerance. Mol Plant 10(9):1224–1237

Zhang T, Gu M, Liu Y, Lv Y, Zhou L, Lu H, Liang S, Bao H, Zhao H (2017b) Development of novel InDel markers and genetic diversity in Chenopodium quinoa through whole-genome resequencing. BMC Genomics 18:1–15

Zhou M, Tang Y, Deng X, Ruan C, Tang Y, Wu Y (2018) Classification and nomenclature of buckwheat plants. In: Buckwheat germplasm in the world. Elsevier, pp 9–20

Zou C, Chen A, Xiao L, Muller HM, Ache P, Haberer G, Zhang M, Jia W, Deng P, Huang R (2017) A high-quality genome assembly of quinoa provides insights into the molecular basis of salt bladder-based salinity tolerance and the exceptional nutritional value. Cell Res 27(11):1327–1340

Genetic Resources and Breeding Approaches for Improvement of *Ferula assa-foetida* (*Heeng*) 8

Niketa Yadav, Satbeer Singh, Ramesh Chauhan,
Sanatsujat Singh, and Ashok Kumar

8.1 Introduction

Ferula assa-foetida L. (*Heeng*) is a monoecious and monocarpic perennial herb (Sahranavard et al. 2014) belonging to Apiaceae family (Hassani et al. 2008; Otroshy and Roozbeh 2015). This family consists of about 170 species worldwide (Mozaffarian 1996; Bahrami et al. 2013), out of them 30 species were found in Iran (Mozaffarian 1996; Bahrami et al. 2013), 15 species were identified in Pakistan (Nasir 1972; Rahman et al. 2008), and other species are native to Mediterranean region and central Asia (Rahman et al. 2008; Bahrami et al. 2013). Whereas, some other studies showed that this family consists of 140 species worldwide (Nasir 1972; Rahman et al. 2008). Basically, this species is native to Iran and Afghanistan (Abd El-Razek et al. 2001; Sahranavard et al. 2014; Tajbakht et al. 2018; Karimian et al. 2020) and this is one of the important species of the genus *Ferula* which produces oleo gum-resin (called as asafoetida or Heeng). It is used in kitchens as a condiment and in pharmaceutical industries as a medicine (Rahman et al. 2008; Bahrami et al. 2013). It is mainly used to treat digestive and other gastro-intestinal disorders. The oleo-gum resin is pungent, carminative, anthelmintic, antiperiodic, expectorant, antispasmodic, deobstruent, and stomachic (Khajeh et al. 2005). The oleo-gum resin has a position as a traditional medicine in Nepal. It also increases the sexual appetite (Abd El-Razek et al. 2001). It is a main household condiment and its components are used for many prescriptions in traditional therapies. This plant has glucuronic acid, galactose, arabinose, and rhamnose, which have been isolated from the oleo-gum resin (Abd El-Razek et al. 2001).

N. Yadav · S. Singh · R. Chauhan · S. Singh · A. Kumar (✉)
Division of Agrotechnology, Council of Scientific and Industrial Research - Institute of Himalayan Bioresource Technology, Palampur, Himachal Pradesh, India
e-mail: ashok@ihbt.res.in

V. Gahlaut, V. Jaiswal (eds.), *Genetics and Genomics of High-Altitude Crops*,
https://doi.org/10.1007/978-981-99-9175-4_8

As a dryland plant, this species is very valuable due to its chemical components extracted from the roots (Iranshahy and Iranshahi 2011). It is used for many treatments such as for hysteria and stomach pain (Iranshahy and Iranshahi 2011; Tajbakht et al. 2018). Some researchers reported that this plant has antidiabetic and antispasmodic activities (Fatehi et al. 2004; Abu-Zaiton 2010; Tajbakht et al. 2018). For the golden age, Heeng has been well used to rehabilitate various diseases, including urinary, respiratory diseases, gastrointestinal, asthma, epilepsy, stomachache, flatulence, intestinal parasite, poor metabolism, and influenza, as well as cure the intestinal worm infection, an aphrodisiac, an emmenagogue and to treat snake and insect bites.

It grows up to 2 m height (Abd El-Razek et al. 2001; Karizi 2008) and flowers only once in life cycle (Roozbeh et al. 2012). Each flowering stem of *Ferula assa-foetida* plant produces a lot of seed in umbel inflorescence. Seeds are light in weight (disperse easily by wind) with side wings. Seed length and width including side wings are 9–12 mm and 5–7.5 mm, respectively. 1000-seed weight and hectoliter weight are 12.5 g and 196.4 g/cm^3, respectively (Zare et al. 2011). After selecting the plants of proper age and size (12.5–15 cm diameter roots (Mahendra and Bisht 2012)) for extraction of oleo gum resin, after 4–5 years of sowing, first task is removing of leaves then aerial part is cut down just before the flowering stage. A pit is made around the taproot. A sharp cut is made horizontally (traditional method) on the top of the taproot and then cover it with grass or stone to protect it from direct sunlight. A milky juice (oleo-gum resin) ooze out from the incision made in the fleshy tap root followed by several cuts on root to ooze out the total oleo-gum resin. The ooze out resin is called asafoetida and collected in a container for processing after drying on root itself, after then it prepared for export. Oleo-gum resin arise in two major forms, i.e., mass and tears. The popular form is mass type, which is mostly sale out in the market, whereas tears type is not a common commercial form (Evans 2009). Asafoetida consists of mainly gum, resin, and essential oil in concentration of 25%, 40–64% and 10–17%, respectively (Upadhyay et al. 2017). The oleo-gum resin can be collected from the roots during June to October (before onset of winters) (Haghi et al. 2017).

At present, India is the main importer of the oleo-gum resin, followed by United States, United Kingdom and countries of the Persian Gulf from Iran, Afghanistan and Uzbekistan. The use of the oleo-gum resin is very limited in Iran, and it is mainly exported to the other countries as raw/unprocessed material (Haghi et al. 2017). Oleo-gum resin is locally called by different names in Iran such as Anghouzeh, Anguzakoma, and Khorakoma (Iranshahy and Iranshahi 2011). Oleo-gum resin have commercial value, so *F. assa-foetida* cultivation and oleo-gum resin collection is a significant source of income for local people in natural/wild habitat (Haghi et al. 2017). Most of the demand of asafoetida is fulfilled by collection of oleo gum-resin of *Ferula assa-foetida* grown in wild in Iran and Afghanistan. Continuous collection of oleo gum-resin/asafoetida by destructive method from wild, can extinct this species, as the plants are harvested before seed set to get oleo-gum resin. There is no literature on its vegetative propagation, because plants of *Ferula assa-foetida* traditionally propagated through seeds, however, germination is very poor due to seed dormancy (Karizi 2008). It is an endangered species [Red Data Book of Iran (Karizi 2008)] due to over harvesting and unsustainable harvesting methods (Gholami and Faravani 2014). The conservation and sustainable utilization of this species is strongly required to protect its diversity in natural habitats (Karizi 2008). Therefore, genetic resources and breeding approaches for their sustainable utilization, cultivation, and genetic enhancement are discussed in this chapter.

8.2 Genetic Resources

Most of the demand is being fulfilled by a ruthless collection of oleo-gum resin from its roots in wild habitats which leads to depletion of genetic resources of this species. Conservation, characterization and sustainable utilization of the genetic resources of this species are very important task of the time. The accessibility and evaluation of genetic resources and their variation are essential for improving the species genetically. A list of genetic resources of this species characterized for various chemical constituents is given in Table 8.1. In India, this species was introduced from Iran and Afghanistan in 2018 and its cultivation was initiated in cold desert regions of Himachal Pradesh, Jammu and Kashmir and Uttarakhand. A diverse germplasm is being maintained and characterized at CSIR-Institute of Himalayan Bioresource Technology, Palampur, India (Fig. 8.1).

This herb is widely distributed in Iran and which had various numbers of available germplasm resources. Iran had wide range of ecosystems, although its genetic diversity to be extinct continuously. This diversity should be used in selection and further utilization in future breeding and conservation programs (Poczai et al. 2013). Bitter and sweet types of asafoetida have been recognized in Iran. The sweet and bitter type *Ferula* species wildly distributed in the highlands and lowlands, respectively (Amalraj and Gopi 2017). The central Iranian plateau and desert regions are the main wildly grown areas of its occurrence. It is also found in the Zagros Mountains and southern Alborz, in Fars, Kerman, Khorasan, Yazd, Semnan, Hormozgan, Sistan and Baluchestan, Isfahan, Lorestan, Kohkiluyeh Boyer Ahmad and Bushehr provinces (Zare et al. 2011). Genetic similarity of *Ferula assa-foetida* population within the group was higher (Jafari et al. 2019). The Feragheh and Karaj populations exhibited were richest in flavonoids and phenolic acids contents (Tajbakht et al. 2018) which can be utilized future trait specific breeding programs (Tajbakht et al. 2018). In Kerman population, it proved that a significant high genetic diversity present within and between wild relatives and it could be used for future breeding programme (Sadeghi et al. 2021a).

Ferula assa-foetida leaf tissues and seeds were collected by Haghi et al. (2017) from a total of six populations (Kerman, Zarand, Tang-Sorkh, Boyer-Ahmad, Maymeh, Bastak) from Iran. Tang-Sorkh and Maymeh were grouped together, whereas Bastak and Geno populations were in another group through cluster analysis using neighbor-joining method (NJ) and UPGMA methods. While, the Bastak population was separated from Geno population using phylogenetic methods and these all results proved genetic distances were low/neutrality in across all the Iranian

Table 8.1 Characterized genetic resources of *Ferula assa-foetida*

Sr. no.	Population	Origin	Mean tem. (°C)	Average altitude (m)	Constituents	Concentration	Reference
1.	Feragheh population 1	Yazd province, Iran	–	–	P-coumaric acid	0.47 mg/100 g	Tajbakht et al. (2018)
2.	Feragheh population 2	Yazd province, Iran	–	–	1,5-Caffeoylquinic acid	13.67 mg/g	Tajbakht et al. (2018)
3.	Feragheh population 3	Yazd province, Iran	–	–	Gallic acid	0.0002 mg/100 g	Tajbakht et al. (2018)
4.	Garmab population	Zanjan province, Iran	–	–	Apigenin	4.23 mg/g	Tajbakht et al. (2018)
5.	Siyah bisheh population 1	Mazandaran province, Iran	–	–	1,5-Caffeoylquinic acid	1.40 mg/g	Tajbakht et al. (2018)
6.	Siyah bisheh population 2	Mazandaran province, Iran	–	–	Gallic acid	2.912 mg/100 g	Tajbakht et al. (2018)
7.	Abarkooh population	Yazd province, Iran	–	–	P-coumaric acid	14.76 mg/100 g	Tajbakht et al. (2018)
8.	Karaj population	Alborz province, Iran	–	–	Apigenin	7.8 mg/g	Tajbakht et al. (2018)
9.	Kerman population	Toghrol Jerd region, Iran	–	–	Sesquiterpenoid coumarin (galbanic acid)	5.0×10^{-5} mol L^{-1}	Ahmadi et al. (2017)
10.	Kerman population	Iran	–	–	E-1-Propenyl sec-butyl disulfide	40.0%	Haghi et al. (2017); Bamoniri and Mazoochi (2009)
					Germacrene B	7.8%	
11.	Plants (Parvand protected area)	Khorasan Razavi province, Iran	–	–	Rosmarinic acid	64.11 mg g^{-1}	Parvandi et al. (2021)

12.	Wild population 1	Koohpaye, Iran	10.5 ± 2.1	2600	(E)-propenyl sec-butyl disulfide	40.41%	Moghaddam and Farhadi (2015)
					(Z)-propenyl sec-butyl disulfide	23.10%	
					β-Pinene	9.66%	
					(E)-β-Ocimene	5.50%	
13.	Wild population 2	Jangale Ghaem, Iran	15.4 ± 1.5	1900	(E)-propenyl sec-butyl disulfide	40.31%	Moghaddam and Farhadi (2015)
					(Z)-propenyl sec-butyl disulfide	22.10%	
					β-Pinene	10.69%	
					α-Pinene	4.96%	
14.	Wild population 3	Joopar, Iran	11.5 ± 1.8	2450	(E)-propenyl sec-butyl disulfide	44.39%	Moghaddam and Farhadi (2015)
					(Z)-propenyl sec-butyl disulfide	22.80%	
					β-Pinene	9.58%	
					(E)-β-Ocimene	6.28%	
15.	Wild population 4	Khomroot, Iran	10.5 ± 1.7	2600	(E)-propenyl sec-butyl disulfide	50.03%	Moghaddam and Farhadi (2015)
					β-Pinene	14.87%	
					(Z)-propenyl sec-butyl disulfide	13.50%	
					α-Pinene	5.06%	

(continued)

Table 8.1 (continued)

Sr. no.	Population	Origin	Mean tem. (°C)	Average altitude (m)	Constituents	Concentration	Reference
16.	Wild population 5	Pabdana, Iran	11.9 ± 1.6	2400	(E)-propenyl sec-butyl disulfide	49.09%	Moghaddam and Farhadi (2015)
					(Z)-propenyl sec-butyl disulfide	12.10%	
					β-Pinene	11.97%	
					α-Pinene	6.17%	
17.	Wild population 6	Rayen, Iran	7.0 ± 2.5	3100	(E)-propenyl sec-butyl disulfide	37.30%	Moghaddam and Farhadi (2015)
					(Z)-propenyl sec-butyl disulfide	17.80%	
					β-Pinene	11.80%	
					α-Pinene	6.70%	
18.	Wild population 7	Sardoo, Iran	10.8 ± 1.7	2550	(E)-propenyl sec-butyl disulfide	42.62%	Moghaddam and Farhadi (2015)
					(Z)-propenyl sec-butyl disulfide	17.20%	
					β-Pinene	14.43%	
					α-Pinene	5.13%	
19.	Wild population 8	Sirjan, Iran	13.3 ± 1.6	2200	(E)-propenyl sec-butyl disulfide	52.24%	Moghaddam and Farhadi (2015)
					(Z)-propenyl sec-butyl disulfide	13.20%	
					β-Pinene	9.52%	
					α-Pinene	4.17%	

20.	Wild population 9	Shahr babak, Iran	15.7 ± 1.6	1850	(E)-propenyl sec-butyl disulfide	53.99%	Moghaddam and Farhadi (2015)
					(Z)-propenyl sec-butyl disulfide	12.70%	
					β-Pinene	7.96%	
					α-Pinene	5.65%	
21.	Wild population	Tangsorkh rangelands (Yasouj province), Iran	–	–	α-Pinene	18.2–48.6%	Karimian et al. (2020)
22.	Yasuj population	Tange sorkh, Kohgiluyeh and Boyer-Ahmad province, Iran	15.18 ± 0.52	2199	α-Pinene	0.04–27.79%	Karimi et al. (2020)
23.	Azeran population	Azeran-Kamoo road, Isfahan province, Iran	19.84 ± 0.23	2861	Camphene	12.62%	Karimi et al. (2020)
24.	Jandaq population	Jandaq, Isfahan province, Iran	21.46 ± 0.36	1262	Myrcene	10.93%	Karimi et al. (2020)
25.	Hotkan population	Reyhan Shahr-Kerman province, Iran	20.20 ± 0.52	2592	Limonene	0.14–17.79%	Karimi et al. (2020)
26.	Qaleasgar population	Qaleasgar village, Kerman province, Iran	12.17 ± 0.38	2652	Z-1-propenyl sec-butyl disulfide	0.62–15.86%	Karimi et al. (2020)
27.	Khajuei population	Khajuei mountain-Sirjan, Kerman Province, Iran	18.09 ± 0.38	2096	E-1-propenyl sec-butyl disulfide	0.18–49.25%	Karimi et al. (2020)

(continued)

Table 8.1 (continued)

Sr. no.	Population	Origin	Mean tem. (°C)	Average altitude (m)	Constituents	Concentration	Reference
28.	Ashkezar population	Zarband village, Yazd Province, Iran	21.06 ± 0.36	2113	7-epi-α-eudesmol	0.12–13.00%	Karimi et al. (2020)
29.	Taft population	Darreh-ye Gahan mountains, Yazd Province, Iran	20.26 ± 0.74	1844	5-epi-7-epi-α-eudesmol	0.07–16.47%	Karimi et al. (2020)
30.	Neyriz population	Gardaneh-ye Layzar, Fars Province, Iran	21.26 ± 0.48	1854	γ-Eudesmol	0.24–20.81%	Karimi et al. (2020)
31.	Tabas population	Neyaz village, South Khorasan Province, Iran	23.70 ± 0.37	1326	Dihydro-eudesmol	0.12–15.41%	Karimi et al. (2020)

Fig. 8.1 *Ferula assa-foetida* germplasm resource center at CSIR-IHBT, Palampur, India

assa-foetida populations (Haghi et al. 2017). Some threatened and endemics plant species (including *Ferula assa-foetida*) were identified and introduced to Markazi province based on their habitat, environmental factors, red-species flora database of that province, field survey, life span information and also on some topographic, and distribution factors. The mountain area of Markazi province contained 66% of the total red species (included *Ferula assa-foetida*) and out of these, *Ferula assa-foetida* is in endangered status, and poor geographical distribution and human interruption were the major reasons (Abdi 2008).

Principal component analysis (PCA) showed similarities among Azeran, Neyriz, Yasuj and Qaleasgar populations of *Ferula assa-foetida*. Also, Jandaq, Ashkezar, Taft, and Hotkan showed similarities and lied in group towards negative direction for eudesmane sesquiterpenoid at PC1 (Karimi 2022). Azeran population were separated far with PC2. Chemotype I (Azeran, Neyriz, and Yasuj populations) was rich in monoterpenes and lower for eudesmane sesquiterpenoids. While, Chemotype II (Jandaq, Ashkezar, Taft, Tabas, and Khajuei populations) were higher for eudesmane sesquiterpenoids and lowest for monoterpenes. Further, chemotype III (Hotkan and Qaleasgar populations) was characterized with the presence of high disulfide components such as Z- and E-1-propenylsec-butyl disulfide. These results would be used to select the appropriate population with unique fully characterized chemotypes for future breeding schemes (Karimi 2022).

8.3 Breeding Techniques

This species has several challenges in its cultivation as well as genetic improvement such as long juvenile phase, shorter vegetative period in a year, low yielding ability, difficult to harvest the resin, and mutually exclusive seed and economic yield. There are several studies reported to manage some of these challenges during cultivation, however studies on its genetic improvement for tackle these challenges are very scanty. Oleo-gum resin productivity were reported as lowest (6.3 g) in July and highest (13.2 g) in September month. Researcher suggested that these results could be helpful to produce much more quantity of oleo-gum resin at specific month (Karimian et al. 2020). Different methods for collection of oleo-gum resin, i.e., conventional cut, crown cut, concave cut and surface cut method were used with different number of cuts (5, 10, and 13 cuts) in different time intervals, to evaluate the oleo-gum resin yield from Mazar, Bajistan, Iran. The highest and lowest oleo-gum resin yields were 25.24 g/plant and 2.02 g/plant obtained from concave method with cutting ten times and superficial method with cutting five times, respectively. Conventional cut causes loss of crown buds, and leads to plant death. Therefore, it was not a good method for collection of oleo-gum resin. On the other hand, concave cutting provides highest resin yield and plant regeneration rates after ten number of cuttings (Gholami and Faravani 2014).

A total of nine population of *F. assa-foetida* were observed for environmental influence on oleo-gum resin yield, essential oil, and its components. The resin yield (50.76–69.64 g/plant) was positively correlated with precipitation rate, whereas maximum essential oil content and sulphur constituents in *Ferula assa-foetida* were positively correlated with increasing temperature and decreasing altitude (Moghaddam and Farhadi 2015). Eight to ten plants of *F. assa-foetida* natural habitats from Iran populations and their roots essential oil content and constituents were evaluated and studied for geographic and climatic conditions using metabolomics approach (GC-MS, Near-infrared (NIR) spectroscopy and Structural equation models). The results showed that essential oil content and their constituents were directly or indirectly effected by latitude, longitude, altitude, and soil profile such as soil texture, pH, temperature and Fe, K, and Al content. Therefore, the prediction of essential oil yield and its constituents based on geographical, weather and soil profile could be exploit for selection of populations with specific essential oil characteristics (Karimi et al. 2020).

DNA sequence of conserved regions of phenylalanine ammonia lyase 2 (pal2) and chalcone synthase (chs) genes were obtained using phenyl propanoid pathway (PPP) in Iranian native species, i.e., *Ferula assa-foetida* and others. Multiple alignments analysis of pal2 and chs genes successfully outline the similarity among *Ferula assa-foetida* and other native species of Iran, whereas neighbour-joining analysis (NJA) of chs gene revealed higher diversity among all species spread over all the groups (Akbarian et al. 2021). These genes have partial-length genomic DNA sequence, which may be helped to understand genetic and geographic kinship among the populations (Akbarian et al. 2021).

8.4 Tissue Culture Techniques

The challenges such as poor seed germination, plant multiplication and dormancy in vegetative phase associated with this species can be effectively tackled using tissue culture techniques. Alternative method for asafoetida production (in bioreactors), synthetic seed, rapid in vitro mass multiplication, and somaclonal variation would be some tissue culture-based techniques which can help a breeder in tackling the challenges associated with this species. A list of in vitro tissue culture protocols available in this species is given in Table 8.2. Plants were successfully regenerated through the arbitration of callus from various explants (roots, hypocotyl, and cotyledon) of mature seeds. The MS medium with 1 mg/L NAA (α-naphthaleneacetic acid) and 2 mg/L BA (N6-benzyladenine) was the best supplementation for multiplication of callus especially from the roots. The callus from hypocotyl explant in medium supplemented with 1 mg/L BA and 0.2 mg/L NAA given best shoot regeneration rate (81.1%). Moreover, medium supplemented with 1–3 mg/L of either BA or kinetin or both with 0.2–0.5 mg/L NAA resulted in indirect organogenesis. The shoot morphogenesis was better with BA as compared to kinetin. Therefore, NAA played the major role in the callus induction. Moreover, the supplementation of NAA and BA is important for callus induction from almost all types of explants in this species (Zare et al. 2010).

A higher embryogenesis rate was observed in a medium supplemented with 1.5 mg/L kinetin and 1 mg/L NAA, whereas the quantity of 1 mg/L kinetin and 0.5 mg/L NAA supplements were best for the highest callus induction. Callus induction frequency varied from 25% to 100%, depending on the ecotype and concentration of kinetin and NAA. Kinetin was an important supplement for somatic embryogenesis for both ecotypes (Hassani et al. 2008). The higher somatic embryogenesis rates were observed from hypocotyl explant in B5 supplemented with 2,4-D (0.5 mg/L) and kinetin (0.2 mg/L) (Otroshy et al. 2013). In another research, callus produced from hypocotyl explant in MS medium supplemented by 1 mg/L BA and 0.2 mg/L NAA regenerated the maximum shoots (81.1%) (Karizi 2008). Further, the somatic embryogenesis initiates directly from leaves in MS medium supplemented with 2,4-D (0.2 mg/L) and kinetin (0.2 mg/L). In contrast, best indirect somatic embryogenesis was obtained from root explant with B5 medium and 2,4-D (0.5 mg/L) and kinetin (0.2 mg/L). The MS with 2,4-D (0.2 mg/L) and B5 with

Table 8.2 Tissue culture protocols reported for in vitro cultures in *Heeng*

Sr. no.	In vitro culture	Explant	Medium	Hormone supplement	References
1.	Callus	Root	MS	1 mg/L NAA and 2 mg/L BA	Zare et al. (2010); Salehi et al. (2019)
2.	Callus	Hypocotyl	MS	1 mg/L BA and 0.2 mg/L NAA	Zare et al. (2010)
3.	Callus	Hypocotyl	MS	1.5 mg/L kinetin and 1 mg/L NAA	Hassani et al. (2008); Salehi et al. (2019)
4.	Callogenesis	Root	MS and B5	0.5 mg/L 2,4-D and 0.2 mg/L kinetin	Otroshy et al. (2013)
5.	Somatic embryos	Hypocotyl	B5	0.5 mg/L 2,4-D and 0.2 mg/L kinetin	Otroshy et al. (2013); Salehi et al. (2019)
6.	Somatic embryos	Cotyledon (leaf)	MS and B5	0.5 mg/L 2,4-D and 0.2 mg/L kinetin	Otroshy et al. (2013)
7.	Somatic embryos	Cotyledon	MS	1.0 or 1.5 mg/L ABA	Otroshy and Roozbeh (2015); Salehi et al. (2019)
8.	Somatic embryo	Root	MS and B5	2,4-D (0.5 mg/L) + kinetin (0.2 mg/L)	Roozbeh et al. (2012)
9.	Somatic embryo	Cotyledon (leaf)	B5	2,4-D (0.2 mg/L) + kinetin (0.2 mg/L)	Roozbeh et al. (2012)
10.	Callus	Root	MS	1 mg/L NAA and 2 mg/L BA	Karizi (2008)
11.	Callus	Hypocotyl	MS	1–3 mg/L BA +0.2 or 0.5 mg/L NAA	Karizi (2008)
12.	Callus	Cotyledon	MS	1 mg/L NAA and 2 mg/L BA	Karizi (2008)

2,4-D (0.5 mg/L) medium, both combined with kinetin (0.2 mg/L) are best for leaf and root explant, respectively for somatic embryogenesis (Roozbeh et al. 2012).

Moreover, effects of growth regulators on germination of *Ferula assa-foetida* seeds were also observed by various researchers. The Abscisic acid (1.0–1.5 mg/L) enhanced germination at temperature range of 15–25 °C whereas, sorbitol enhanced germination at 5 °C. In case of somatic embryogenesis, the good somatic embryogenesis rate (35.94%) were observed on MS medium with supplementation of ABA at 25 °C (Otroshy and Roozbeh 2015). However, somatic embryogenesis rate was reduced at 5 °C, when MS medium supplemented with abscisic acid, sorbitol/mannitol (Otroshy and Roozbeh 2015). Therefore MS medium supplemented with abscisic acid had a positive effect on the generation of somatic embryos, whereas, carbon source and mannitol reduced the generation of somatic embryos (Otroshy and Roozbeh 2015). The better generation of in vitro somatic embryos of *Ferula assa-foetida* were achieved from both MS medium supplemented by sucrose + PEG and activated charcoal (AC) with concentration of 30 g/L and 0.5–1%, respectively.

The MS medium with some media additives (Sorbitol and glutamine) was effective on germination and conversion of somatic embryos rather than AC, which reduced the secondary somatic embryogenesis (Roozbeh and Otroshy 2014).

8.5 Genomic Resources

DNA-based molecular markers are the most important source to representation the genetic diversity. There are many parameters to recognize the diversity through molecular marker approaches. These would be used in structural analysis and to study genetic variation. DNA-based marker approaches have improved time to time and also, they used to determine the population structure and their phylogenetic interactions. Recent studies of genomic analysis showed a diverse range of DNA-based markers such as wide range of new DNA-based marker systems or gene-targeted molecular markers, which could be used for future breeding programs (Poczai et al. 2013). Commonly, DNA-based molecular markers are used to measure genetic variation in *Ferula assa-foetida* species (Rahimmalek et al. 2009). A list of available molecular markers available in this species is given in Table 8.3.

A total of 127 polymorphic bands were scored using ten numbers of random amplified polymorphic DNA (RAPD) primers named as OPNO1, OPDO3, H, TIBMB19, OPC12, 398, TIBMBA14, TiBMBA11, UBC9, and OPK 10 in *Ferula assa-foetida* germplasms from Kerman and Yazd provinces (Sarhadipour et al. 2015). The RAPD markers explained high genetic variation in the studied germplasm, indicating that these markers would be an efficient tool for assessing genetic diversity in this plant. Moreover, 18 Start Codon Targeted (ScoT) and sequence-related amplified polymorphism (SRAP) molecular markers provide 211 and 194 polymorphic bands in Feragheh and Karaj population (Tajbakht et al. 2018). Both markers systems showed 97.07% and 92.87% polymorphism, respectively, and disclosed differences in allelic frequencies, higher genetic diversity and gene flow across the two populations (Tajbakht et al. 2018). Thirteen ISSR markers produced 186 polymorphic alleles (average number of 14.38 bands per locus) in a set of 90 *Ferula assa-foetida* genotypes (Sadeghi et al. 2021a). The high mean marker index (4.98) and the polymorphic information content (0.37) showed that ISSR primers

Table 8.3 Molecular markers available in *Ferula assa-foetida*

Sr.no.	Molecular marker	No. of primers	Polymorphism (%)	Reference
1	SCoT	9	96.07	Tajbakht et al. (2018)
2	SRAP	9	92.87	Tajbakht et al. (2018)
3	RAPD	10	–	Sarhadipour et al. (2015)
4	ISSR	13	–	Sadeghi et al. (2021a)
5	ISSR	22	83.33	Jafari et al. (2019)
6	ISSR	22	85.04	Jafari et al. (2019)
7	SCoT	12	98.37	Sadeghi et al. (2021b)
8	URP	12	97.91	Sadeghi et al. (2021b)

were highly efficient for identifying genetic diversity in *Ferula* populations (Sadeghi et al. 2021a).

Twenty-two ISSR primers, named as ISSR-12 to ISSR-26 and ISSR-53 to ISSR-59 were used for genetic analysis of *Ferula assa-foetida* and one other *Ferula* species (*Ferula pseudalliacea*) (Jafari et al. 2019). Out of 22, eight primers (ISSR-12 to ISSR-16, ISSR-53, ISSR-55, and ISSR-57) gave clear-cut DNA patterns with 234 DNA bands, whereas ISSR-16 and ISSR-55 markers were best for total polymorphic bands (218), their percent share (93.16%) and average alleles per loci (1.93 ± 0.25). Therefore, ISSR markers efficiently investigate the genetic diversity of *Ferula assa-foetida* and could be used for future breeding (Jafari et al. 2019). The molecular variation studies on six *Ferula assa-foetida* populations from Iran using rpl16 (F:5′-GATGGTTGGTTCTGAATTCC-3′ and R: 5′-GATAAACGGAAGGGTGAAAG-3′) found seven polymorphic sites with four distinct haplotypes. The transition to transversion proportions were k1 = 0.965 and k2 = 0.799 in purines and pyrimidines, respectively with an average bias (R) of 0.332. Therefore, chloroplast-encoded sequences and diversity among them could efficiently helpful in understanding the phylogenetic relationship of higher taxonomic classifications (Haghi et al. 2017).

Ninety accessions from 30 different *Ferula assa-foetida* populations produced polymorphic fragments using 24 SCoT and URP markers (Sadeghi et al. 2021b). Higher genetic variation was observed within populations than between the populations by SCoT (91%), URP (88%), and based on pooled data (90%). The SCoT markers were better as compared to URP markers in assessing genetic diversity within populations. The clustering using principal coordinate analysis (PCoA) showed high gene migration among the population (Sadeghi et al. 2021b). Root, leaf, stem, and flower of *Ferula assa-foetida* were put together for computational annotation to identify candidate genes cause biosynthesis of terpenoid and coumarin using de novo transcriptome analysis. The flowers have several genes involved in terpenoid and phenylpropanoid metabolic pathways, which facilitate genomic resources for future breeding programs (Amini et al. 2019; Wang et al. 2022). The phenylpropanoid pathway is responsible for several important secondary-metabolites, and the essential oil of *Ferula assa-foetida* belongs to secondary metabolites of plants. These secondary metabolites are actively involved directly or indirectly towards most of the plant defences mechanisms against both biotic and abiotic stresses. The essential oil used as a botanical pesticide has many advantages for several diseases and pest control (Karimi 2022). Further studies on accumulation of secondary metabolites, their pathways and genes involved will helpful to understand the medicinal properties of this plant (Akbarian et al. 2021). The molecular docking of galbanic acid and chloroplast DNA showed their interactions for partial intercalation via DNA groove binding and responsible for a complex by van der Waals and electrostatic interactions (Ahmadi et al. 2017).

8.6 Genetic Diversity for Phytochemical Markers

The 10-epi-ɣ-eudesmol, α-eudesmol, agarospirol, β-eudesmol, and guaiol are the main essential oil components from aerial parts of *Ferula assa-foetida* and its oleo-gum resin. In contrast, α-pinene, (Z)-1-propenyl sec-butyl disulfide, β-pinene, (E)-1-propenyl sec-butyl disulfide, and bicyclogermacrene were the major constituents of oleo-gum resin. Chlorogenic acid, ellagic acid, quercetin, and vanillin were major polyphenols, while, p-coumaric and 1,3-caffeoylquinic acids were main phenolic acids in plant extract (Tajbakht et al. 2018). Carvacrol, trans-ferulic acid, eugenol, hesperetin, hesperidin and vanillin are present in its oleo-gum-resin. Total phenolic contents assessed by folin-ciocalteu method showed that the oleo gum resin (76.12 ± 6.52 mg/g) has higher phenolic contents than aerial part (50.50 ± 0.74 mg/g) (Rowshandel et al. 2018).

A wide range of essential oil yield (2.53 to 20.85%) was observed from 14 accessions (Iranian and Afghans). Further, a total of six compounds E-1-propenyl-sec-butyl disulfide (13.66–49.35%), β pinene (1.06–21.18%), Z-1-propenyl-sec-butyl disulfide (2.02–15.29%), α pinene (2.04–17.61%), thiophene (0.03–36.81%), and thiourea (0.08–9.63%) were found copiously/abundantly in *Ferula assa-foetida* using GC–MS analysis. The variation in these compounds showed high variation among the accessions, which could be used in future breeding programme (Hassanabadi et al. 2019). Three distinct chemotypes (monoterpenes and Z-1-propenyl sec-butyl disulfide, eudesmane sesquiterpenoids and α-agarofuran and Z-and E-1-propenyl sec-butyl disulfide) were identified using NIRS and GCMS. These roots constituents were arbitrating directly or indirectly towards defence mechanism against biotic and abiotic stresses. Structural equation models (SEMs) found edaphic factors and temperature as the major influencers for essential constituents (Karimi 2022). *Ferula assa-foetida* had a main secondary metabolites (SMs) compound, i.e., terpenoid. Genes responsible for terpene biosynthesis were studied by transcript and metabolite analysis, which facilitate genomic resources for understanding metabolic engineering and terpene biosynthesis pathways. *Ferula* species roots are more active in biosynthesis of terpenes as compared to other organs (leaf and stem tissue) and also it is a most suitable host for endophytes as compared to other organs. The co-cultivation of endophytic microorganisms within the root zone would enhance the secondary metabolites production in this plant (Salehi et al. 2019). Two main constituents, i.e., E-propenyl-sec-butyl disulfide (37.30–53.99%) and Z-propenyl-sec-butyl disulfide (12.10–23.10%) were identified in nine populations (Moghaddam and Farhadi 2015), and these disulfides are majorly characterized in Hotkan and Qaleasgar populations (Karimi et al. 2020). A total of 28 endophytic fungi were extracted from roots, stems and leaves using hyphal tip technique. The internal-transcribed spacer sequencing (ITS) analysis identified SBUp1 in *Alternaria tenuissima*, where, further liquid chromatography with mass spectrometric characterized RA (6411 mg/g) as the major phenolic acid (Parvandi et al. 2021) proves it as a rich RA source. The markers Epithelial-mesenchymal transition (EMT) and surface glycoprotein (CD44) were mainly responsible for invasion and metastatic diffusion of cancer stem cells and gastric CSCs through

EMT, respectively. Results showed that hydroalcoholic extracts of *Ferula assa-foetida* could decrease the EMT markers, antiapoptotic factor (Bcl-2), surface marker (CD44 and CD54), leading to decrease the CSCs multiplication and their killing (Keyghobadi et al. 2022).

8.7 Conclusion

In conclusion, the *Ferula assa-foetida* species showed vast genetic and phytochemical diversity across its natural and wild habitats in Iran, Afghanistan, Uzbekistan, and Kazakhstan. However, the diversity is on extinct due to over and unsustainable destructive harvesting of oleo-gum resin from its roots before flowering and seed-set. Though the organized genetic improvement programs are lacking in *F. assa-foetida*, the divergence accrued through natural adaptability while dispersal across countries and their climatic conditions may aid in its improvement. The information compiled on diverse populations may be useful for developing future breeding programs for this species. There is a need to limit anthropogenic disruptions and the sites with natural habitats of *F. assa-foetida* should be identified as conservatories. However, the conservation of *Heeng* plants (*F. assa-foetida* L.) presents one of the biggest hurdles for scientific communities because this plant's seed-to-seed cycle takes almost 5 years. The scientific information on genetic and genomic resources aggregated here will unable us to develop long-term conservation plans, germplasm maintenance, population improvement, breeding for improved varieties and superior agrotechnologies for this priceless species.

References

Abd El-Razek MH, Ohta S, Ahmed AA, Hirata T (2001) Sesquiterpene coumarins from the roots of Ferula Assa-foetida. Phytochemistry 58:1289–1295. https://doi.org/10.1016/S0031-9422(01)00324-7

Abdi N (2008) Evalution of red plant biodiversity in Markazi Province. Iran J Rangelands For Plant Breed Genet Res 16:50–74

Abu-Zaiton AS (2010) Anti-diabetic activity of ferula assafoetida extract in Normal and Alloxan-induced diabetic rats. Pakistan J Biol Sci 13:97–100

Ahmadi F, Shokoohinia Y, Javaheri S, Azizian H (2017) Proposed binding mechanism of galbanic acid extracted from ferula Assa–foetida to DNA. J Photochem Photobiol B Biol 166:63–73. https://doi.org/10.1016/j.jphotobiol.2016.11.011

Akbarian A, Rahimmalek M, Sabzalian MR, Hodaei M (2021) Sequencing and phylogenetic analysis of phenylalanine ammonia lyase (pal) and chalcone synthase (chs) genes in some Iranian endemic species of Apiaceae. Sci Direct, Gene Reports 23:101147. https://doi.org/10.1016/j.genrep.2021.101147

Amalraj A, Gopi S (2017) Biological activities and medicinal properties of Asafoetida: a review. J Tradit Complement Med 7:347–359. https://doi.org/10.1016/j.jtcme.2016.11.004

Amini H, Naghavi MR, Shen T et al (2019) Tissue-specific transcriptome analysis reveals candidate genes for terpenoid and phenylpropanoid metabolism in the medicinal plant ferula assafoetida. G3 Genes Genom Genet 9:807–816. https://doi.org/10.1534/g3.118.200852

Bahrami G, Soltani R, Sajjadi S-E et al (2013) Essential oil composition of ferula Assa-Foetida L. fruits from Western Iran. J Reports Pharm Sci 2:90–97. https://doi.org/10.1080/10412905.1998.9701010

Bamoniri A, Mazoochi A (2009) Determination of bioactive and fragrant molecules from leaves and fruits of Ferula Assa-foetida L. growing in Central Iran by nano scale injection. Dig J Nanomater Biostruct 4:323–328

Evans WC (2009) 16th edition Trease and Evans Pharmacognosy. Elsevier, New York

Fatehi M, Farifteh F, Fatehi-Hassanabad Z (2004) Antispasmodic and hypotensive effects of ferula asafoetida gum extract. J Ethnopharmacol 91:321–324. https://doi.org/10.1016/j.jep.2004.01.002

Gholami BA, Faravani M (2014) Effects of different cutting methods and times of cutting on growth performance and gum resin production of ferula Assa-foetida. J Agric Sci 59:35–44. https://doi.org/10.2298/jas1401035g

Haghi H, Khederzadeh S, Samiei M et al (2017) Genetic comparison of Iranian asafetida (Ferula assa-foetida L.) populations based on cpDNA ribosomal protein L16 intron. Int J Agric Innov Res 5:2319–1473

Hassanabadi M, Ebrahimi M, Farajpour M, Dejahang A (2019) Variation in essential oil components among Iranian Ferula assa-foetida L. accessions. Ind Crops Prod 140:111598. https://doi.org/10.1016/j.indcrop.2019.111598

Hassani B, Saboora A, Radjabian T, Husseini HF (2008) Somatic embryogenesis of Ferula assa-foetida. J Sci (University Tehran) 33:15–23

Iranshahy M, Iranshahi M (2011) Traditional uses, phytochemistry and pharmacology of asafoetida (Ferula Assa-foetida oleo-gum-resin)—a review. J Ethnopharmacol 134:1–10. https://doi.org/10.1016/j.jep.2010.11.067

Jafari SH, Sepehry A, Soltanloo H, Karimian AA (2019) Genetic differentiation between bitter and sweet asafetida plants using ISSR markers. Mol Biol Rep 46:1069–1078. https://doi.org/10.1007/s11033-018-4565-1

Karimi A (2022) Environmental metabolomics approaches to identify and enhance secondary compounds in medicinal plants for bio-based plant protection. Refubium - Freie Univ Berlin Repos

Karimi A, Krahmer A, Herwig N et al (2020) Metabolomics approaches for analyzing effects of geographic and environmental factors on the variation of root essential oils of Ferula Assa-foetida L. J Agric Food Chem 68:9940–9952. https://doi.org/10.1021/acs.jafc.0c03681

Karimian V, Sepehry A, Barani H et al (2020) Productivity, essential oil variability and antioxidant activity of Ferula assa-foetida L. oleo-gum-resin during the plant exploitation period. J Essent Oil Res 32:545–555. https://doi.org/10.1080/10412905.2020.1794988

Karizi ARZ (2008) The effect of hormones and various explant on ferula assa foetida L. in vitro. Masters thesis, Univ Zabol

Keyghobadi N, Bagheri V, Rahnamaii MS, Sarab GA (2022) Evaluation of hydroalcoholic extract effects of Ferula Assa-foetida on expression change of EMT and CD44-related genes in gastric cancer stem cell. Gene Rep 27:101535. https://doi.org/10.1016/j.genrep.2022.101535

Khajeh M, Yamini Y, Bahramifar N et al (2005) Comparison of essential oils compositions of Ferula assa-foetida obtained by supercritical carbon dioxide extraction and hydrodistillation methods. Food Chem 91:639–644. https://doi.org/10.1016/j.foodchem.2004.06.033

Mahendra P, Bisht S (2012) Ferula asafoetida: traditional uses and pharmacological activity. Pharmacogn Rev 6:141–146. https://doi.org/10.4103/0973-7847.99948

Moghaddam M, Farhadi N (2015) Influence of environmental and genetic factors on resin yield, essential oil content and chemical composition of Ferula Assa-foetida L. populations. J Appl Res Med Aromat Plants 2:69–76. https://doi.org/10.1016/j.jarmap.2015.04.001

Mozaffarian V (1996) A dictionary of Iranian plant names. Farhang-e Moaser, Tehran, pp 228–230

Nasir E (1972) Apium in E. Nasir & SI Ali Flora of West Pakistan. Pakistan Agric Res Counc 20:32–34

Otroshy M, Roozbeh S (2015) Optimization of somatic embryogenesis in Ferula assa-foetida L. (a rare medicinal plant). Annu Res Rev Biol 5:439–448. https://doi.org/10.9734/arrb/2015/12110

Otroshy M, Edrisi S, Enteshary S (2013) Propagation of medicinal plant ferula assa foetida L. through indirect somatic embryogenesis. Int J Med Plant Res 2:179–186

Parvandi M, Rezadoost H, Farzaneh M (2021) Introducing Alternaria tenuissima SBUp1, as an endophytic fungus of Ferula assa-foetida from Iran, which is a rich source of rosmarinic acid. Lett Appl Microbiol 73:569–578. https://doi.org/10.1111/lam.13542

Poczai P, Varga I, Laos M et al (2013) Advances in plant gene-targeted and functional markers: a review. Plant Methods 9:1–31. https://doi.org/10.1186/1746-4811-9-6

Rahimmalek M, Bahreininejad B, Khorrami M, Sayed Tabatabaei BE (2009) Genetic variability and geographic differentiation in thymus daenensis subsp. daenensis, an endangered medicinal plant, as revealed by inter simple sequence repeat (ISSR) markers. Biochem Genet 47:831–842. https://doi.org/10.1007/s10528-009-9281-z

Rahman MU, Gul S, Odhano EA (2008) Antimicrobial activities of ferula assafoetida oil against gram positive and gram negative bacteria. Am J Agric Environ Sci 4:203–206

Roozbeh S, Otroshy M (2014) Improvement germination and conversion of somatic embryos and production of normal plantlets in Ferula assa-foetida L. (a rare medicinal plant). Eur J Med Plants 4:1447–1461. https://doi.org/10.2989/10220110509485863

Roozbeh S, Otroshy M, Bozorgipoor R, et al. (2012) Micropropagation of Ferula assa-foetida L. (a medicinal plant) via direct somatic embryogenesis. In: International symposium on medicinal and aromatic plants IMAPS2010 and history of Mayan Ethnopharmacology IMAPS2011. pp 143–152

Rowshandel L, Bahmanzadegan A, Zareiyan F, Rowshan V (2018) Essential oil composition, phenolic and total phenolic contents and antioxidant activity from aerial parts and oleo-gum-resin of ferula assa foetida L. In: 7th National Congress on medicinal plants, Shiraz, Iran. p 10

Sadeghi R, Omidi M, Azizinezhad R et al (2021a) Evaluation of structure and genetic diversity of wild populations of Ferula assa-foetida L. medicinal plant using ISSR molecular marker. Mod Genet Journal, Iran Soc Genet 16:309–319

Sadeghi R, Omidi M, Azizinezhad R et al (2021b) Universal Rice primer (URP) and start codon target (SCoT) markers in studying population structures and genetic variation in ferula assa foetida L. Accessions Res Sq:1–26

Sahranavard S, Ghafari S, Mosaddegh M (2014) Medicinal plants used in Iranian traditional medicine to treat epilepsy. Seizure 23:328–332. https://doi.org/10.1016/j.seizure.2014.01.013

Salehi M, Naghavi MR, Bahmankar M (2019) A review of ferula species: biochemical characteristics, pharmaceutical and industrial applications, and suggestions for biotechnologists. Ind Crop Prod 139:111511. https://doi.org/10.1016/j.indcrop.2019.111511

Sarhadipour S, Naghavi MR, Baghizadeh A, Hasani D (2015) Genetic diversity of Ferula assa-foetida L. germplasms using rapd analysis in southeast of Iran. Sci Inf Database 4:72–82

Tajbakht P, Talebi M, Rahimmalek M (2018) Genetic diversity among and within ferula asafoetida H. Karst. Populations using molecular and phytochemical markers. Phytochemistry 155:19–29. https://doi.org/10.1016/j.phytochem.2018.07.004

Upadhyay PK, Singh S, Agrawal G, Vishwakarma VK (2017) Pharmacological activities and therapeutic uses of resins obtained from ferula asafoetida Linn.: a review. Int J Green Pharm 11:S240–S247

Wang XJ, Luo Q, Li T et al (2022) Origin, evolution, breeding, and omics of Apiaceae: a family of vegetables and medicinal plants. Hortic Res 9:uhac076. https://doi.org/10.1093/hr/uhac076

Zare AR, Solouki M, Omidi M et al (2010) Callus induction and plant regeneration in ferula Assa Foetida L. (asafetida), an endangered medicinal plant. Trakia J Sci 8:11–18

Zare AR, Omidi M, Fallah Hoseini H et al (2011) A review on pharmacological effects of ferula assa - Foetida L.: a systematic review. J Med Plants 10:178424

Future Prospects: High-Altitude Crop Improvement via Genomics

9

Vikas Kumar Singh, Ronika Thakur, Jatin Sharma, Ashita Bisht, Kumar Sanu, Arushi Arora, Deepak Bhamare, Neeraj Pal, and Dinesh Kumar Saini

9.1 Introduction

High-altitude regions worldwide display unique climatic conditions, including lower temperatures, reduced oxygen levels, as well as diverse soil, geological, and vegetative characteristics, leading to variations in cropping practices, crop rotations, and precipitation patterns (Wolfe 2022; Das et al. 2018). In such demanding

V. K. Singh
Department of Genetics and Plant Breeding, Ch. Charan Singh University, Meerut, Uttar Pradesh, India

R. Thakur · K. Sanu
Department of Genetics and Plant Breeding, CSKHPKV, Palampur, Himachal Pradesh, India

J. Sharma · A. Arora
Department of Plant Breeding and Genetics, Punjab Agricultural University, Ludhiana, Punjab, India

A. Bisht
Highland Agricultural Research and Extension Centre, CSKHPKV, Kukumseri, Himachal Pradesh, India

D. Bhamare
School of Agricultural Biotechnology, Punjab Agricultural University, Ludhiana, Punjab, India

N. Pal
Department of Molecular Biology and Genetic Engineering, College of Basic Science & Humanities, Govind Ballabh Pant University of Agriculture and Technology, Pantnagar, Uttarakhand, India

D. K. Saini (✉)
Department of Plant Breeding and Genetics, Punjab Agricultural University, Ludhiana, Punjab, India

Department of Plant and Soil Science, Texas Tech University, Lubbock, TX, USA

V. Gahlaut, V. Jaiswal (eds.), *Genetics and Genomics of High-Altitude Crops*, https://doi.org/10.1007/978-981-99-9175-4_9

conditions, specific crops have acclimated and flourished, making substantial contributions to the economic prosperity of the local communities inhabiting these regions. The cultivation of high-altitude crops holds substantial economic significance on both local and global scales, influencing the well-being of residents, international trade, and food security (Merrey et al. 2018). Mountainous areas host a rich variety of crops specially adapted to thrive in challenging environments. These encompass essential staples such as potatoes, barley, quinoa, wheat, buckwheat, finger millet, barnyard millet, proso millet, foxtail millet, and maize (Singh et al. 2019; Li et al. 2019; Kumar et al. 2012; Giri et al. 2021; Babu et al. 2018; Jacobsen 2003), alongside distinctive specialty crops, including unique apple and berry varieties, as well as various medicinal herbs. Cultivating these crops not only plays a vital role in conserving biodiversity but also holds significance for ensuring food security. It brings attention to their distinct flavors and nutritional attributes, and additionally, it contributes to the boosts of the tourism sector (Gautam et al. 2022; Merrey et al. 2018; Liang et al. 2023; Bargali et al. 2022). For instance, potatoes are a vital source of carbohydrates and nutrients; quinoa, known for its high protein content; crops such as wheat, maize, and barley provide primary source of sustenance for mountainous areas (Bargali et al. 2022; Bazile et al. 2016). High-altitude crops demonstrate remarkable resilience in the face of climate change and extreme weather conditions, underscoring their vital role in ensuring food security as increasing worldwide temperatures (Fig. 9.1). Their ability to thrive in cooler climates serves as a valuable asset in mitigating the repercussions of climate change on agriculture. The distinct challenges inherent to high-altitude farming have catalyzed research and innovation in crop breeding and agricultural methods, yielding benefits not only for mountain communities but also fostering progress in the broader agricultural landscape. Traditional breeding methods have historically been utilized to enhance the resilience and tolerance of high-altitude crops to various challenges, resulting in the development of some resistant and tolerant varieties. Nevertheless, these conventional methods tend to be slow and time intensive. In recent years, there has been a surge in interest in adopting multi-omics approaches to improve resistance and tolerance in these crops. By integrating data from diverse omics disciplines, particularly genomics, researchers can uncover crucial genes and pathways associated with different forms of stress resistance and tolerance. This valuable knowledge can then be applied to formulate innovative breeding strategies aimed at improving high-altitude crops. Breakthroughs in genetic and molecular biology have led to the emergence of high-throughput sequencing technologies. As a result, plant scientists have identified different genomic regions linked to desirable traits. These genetic insights and information have been effectively harnessed for crop improvement, encompassing increased yield and resistance to both biotic and abiotic stress. Consequently, genomic techniques have significantly advanced our comprehension of high-altitude crop development.

Genomics, the study of an organism's complete genetic makeup, has revolutionized high-altitude crop cultivation. It enables the development of crops that thrive in extreme conditions, resist pests and diseases, and possess enhanced nutritional qualities, effectively addressing food security challenges in mountainous regions.

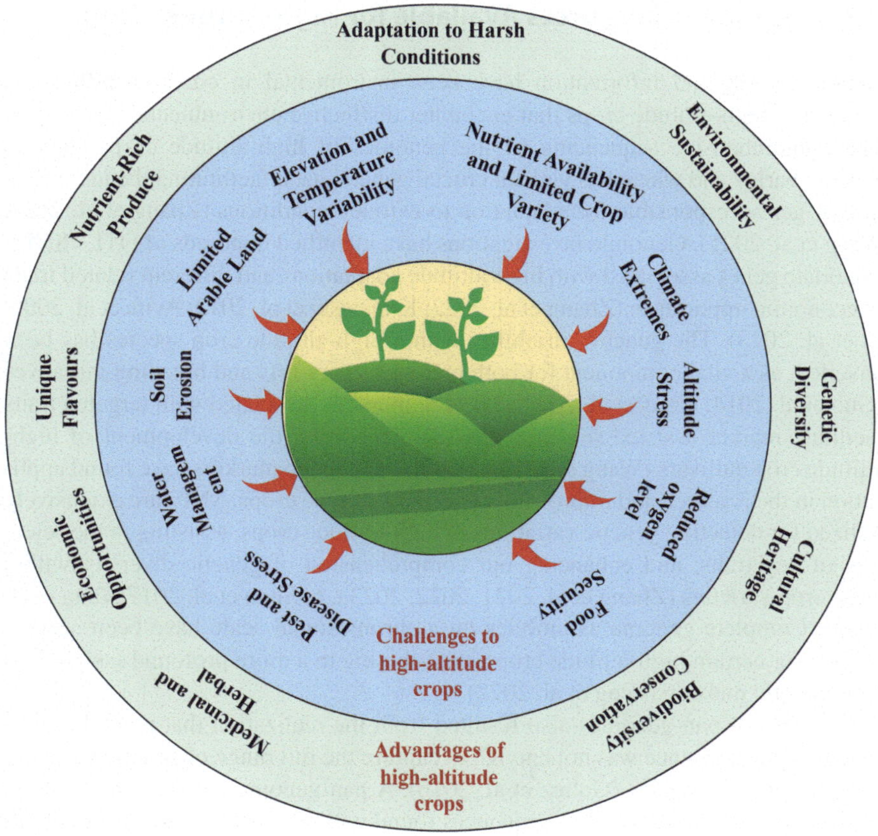

Fig. 9.1 An overview of different biotic and abiotic stress impacting high-altitude crops

Moreover, genomics contributes to the preservation of genetic diversity in high-altitude crops and holds promise for sustainable agriculture in the context of climate change (Guo et al. 2014; Arias et al. 2022). This chapter takes a comprehensive approach to examine the latest advancements in the enhancement of high-altitude crops. In this chapter, we explore the genomics resources available for high-altitude crops. We discuss the current status and future prospects of various genomics techniques, such as QTL mapping, genome-wide association mapping, marker-assisted selection, genomic selection, and genome editing, all aimed at improving crop yield, enhancing climate resilience and stress tolerance, and increasing the nutritional value of high-altitude crops. This chapter offers a comprehensive overview of the latest advancements in enhancing high-altitude crops, addressing genomics challenges, and providing insights into the future of agriculture in challenging landscapes.

9.2 Genomic Resources Available for High-Altitude Crops

Genomic tools and information have been instrumental in comprehending and enhancing high-altitude crops that encounter distinctive environmental challenges. The comprehensive sequencing of the genomes of high-altitude crops such as Tibetan barley and wheat has yielded crucial genetic data, facilitating the identification of genes responsible for adaptation to extreme conditions (Zhang et al. 2022; Wang et al. 2021). Genomic investigations have identified hundreds of QTLs/SNPs/candidate genes associated with high-altitude adaptations and different related traits of economic importance (Zhang et al. 2022; Krajewski et al. 2012; Avila et al. 2005; Li et al. 2023). The genetic variability within high-altitude crop species has been unveiled, as a vital component for both preservation efforts and breeding initiatives (Guo et al. 2014; Arias et al. 2022). Genetic markers associated with targeted traits facilitate marker assisted selection (MAS), expediting the development of high-altitude crop cultivars (Wang et al. 2021). Various genetic markers have found application in the research and improvement of high-altitude crops. SNPs are extensively utilized for detecting genetic variations in high-altitude crops, assisting in the selection of key traits, and enhancing our comprehension of genetic diversity within these crop varieties (Zhang et al. 2021, 2022, 2023a, b; Shao et al. 2017; Tang et al. 2022). Complete genome assemblies on a chromosomal scale have been accomplished for certain high-altitude crops, contributing to a more profound insight into their genetic makeup (Feng et al. 2022).

The idea of pan-genomes also resulted from the realization that a single reference genome sequence was not enough to capture the full range of genetic variation present within a species (Golicz et al. 2016). A pan-genome is made up of all the non-redundant genes and DNA sequences found within a species and includes both the core genome (containing genes present in all members) and the variable genome (containing genes that are shared partially or that are unique to specific members) (Saxena et al. 2014). Pangenome investigations have been conducted on maize (Hirsch et al. 2014) and soybean (Liu et al. 2020), Tartary buckwheat (Zhou et al. 2023), pearl millet (Raza et al. 2023) (Yan et al. 2023a, b), potato (Illayda 2017), and pea (Jha et al. 2022). In a study conducted by Lu et al. (2015), a high-resolution genetic mapping was utilized to correlate sequences to the maize pangenome. A recent study of the hexaploid bread wheat pangenome revealed the variability of this essential crop among 18 cultivars (Montenegro et al. 2017). Interestingly, variable gene annotations often reflect resemblance across plant species, with genes related to biotic and abiotic stress tolerance being recurrently enriched within variable gene clusters. The pan-genome studies (Table 9.1) have enabled us to understand the existence or lack of variation in a crop species is critical for developing superior crop varieties that are adapted to changing climates.

Further, mapping populations play a vital role in deciphering the genetic foundations of traits and developing crop varieties well-suited to demanding environments. Numerous research endeavours and resources are dedicated to mapping populations specifically tailored for high-altitude crops (Gage et al. 2020; Zhang et al. 2022; Mishra et al. 2023). However, there are many areas of research that remain largely

Table 9.1 Summary of pan-genomic studies of some high-altitude crops

Crop	Chrom.	Sequencing strategy (approaches of pan genome)	Accessions	Variable genes	Core genes	Pan genes (No.)	References
Soybean (*Glycine max*)	40	Illumina HiSeq, PacBio (graph-based de novo assembly)	27	49.9%	50.1%	57,492	Liu et al. (2020)
Wild soybean (*Glycine soja*)	40	Illumina HiSeq 2000 (de novo assembly and sequencing)	7	51.4%	48.6%	59,080	Jia et al. (2020)
Barley (*Hordeum vulgare*)	14	PacBio Iso-Seq, Illumina RNA-Seq (CSCS and assembly comparison)	20	45.26%	54.74%	40,176	Jayakodi et al. (2020)
Wheat (*Triticum aestivum*)	42	Illumina HiSeq (iterative mapping and assembly)	18	42.3%	57.7%	140,500	Montenegro et al. (2017)
Maize (*Zea mays*)	20	Illumina Hiseq (Pan-transcriptomics)	503	60.88%	39.12%	41,903	Hirsch et al. (2014)

unexplored. To enhance high-altitude crop improvement, it is imperative to delve into various avenues, such as conducting different omics studies and implementing advanced biotechnological approaches. In summary, the accessible genomics resources have fundamentally transformed our knowledge of high-altitude crops, enabling the cultivation of crop varieties that excel in demanding mountainous landscapes and play a pivotal role in enhancing food security in these regions.

9.3 Genomic Approaches for Improvement of High-Altitude Crops

In the following sections, we discuss the present status and future outlook of genomics approaches aimed at enhancing crop yield, improving climate resilience and stress tolerance, and elevating nutritional value.

9.3.1 Genomics Approaches for Yield Improvement: Current Status and Future Prospects

The green revolution led to a sharp increase in the yields during the late 1960s of major staple crops, viz., wheat, rice, and maize, specially grown in plain areas. With the ever-increasing population and climate fluctuations, there is a dire need to improve the yield of the crops grown in high-altitude regions as well. By 2050, crop yield improvement on the projected path may not be able to provide enough food for everyone on the planet. Increasing crop yields in high-altitude places can be difficult due to challenging environmental circumstances, which include lower temperatures, lower oxygen levels, and greater UV radiation. However, attempts have been made in the past using various breeding tools and strategies, viz., QTL mapping, genome-wide association studies, meta-QTL analysis and genome editing tools (CRISPR/Cas9) for the yield improvement in different high-altitude crops as discussed in the following section.

9.3.1.1 Genetic Dissection of Yield and Related Traits Using QTL Mapping, Meta-QTL Analysis, and GWAS

QTL mapping is a standard approach to unravelling genomic regions related to various crop traits. It has been used previously in various crops for discovering the genomic regions associated with yield, even in high-altitude crops (Krajewski et al. 2012; Avila et al. 2005; Li et al. 2023) (Table 9.2). Researchers can use the molecular markers flanking these QTL regions to transfer yield traits through marker-assisted breeding (MAB). In barley, 221 QTLs associated with yield-related traits were discovered in single environment analysis utilizing composite interval mapping (CIM) (Wang et al. 2016). In another study, using 188 doubled-haploid (DH) lines of barley (Chinese feed barley (TX9425) X Japanese malting barley (Naso Nijo), a total of 111 significant QTLs were identified from five years of

Table 9.2 List of yield and associated trait QTLs identified in different high-altitude crops

Crop	Mapping population	QTLs	Identified QTLs related to yield components	Linkage group	PVE (%)	Reference
Field pea	Two F_5 populations (Wt11238 × Wt3557) and (Wt10245 × Wt11238)	12	Peduncle number (2), pod number per peduncle (1), seed number per pod (2), thousand seed weight (4) and yield (3)	1, 2, 3, and 5	–	Krajewski et al. (2012)
Faba bean	$F_{2:3}$ (29H × Vf 136)	13	Yield distribution traits and yield characters	3, 4, 13	9.1–57.5%	Avila et al. (2005)
Tartary buckwheat	RILs (Xiaomiqiao × Jinqiaomai 2)	32	Grain yield, 1000 grain weight, grain length, grain width, and grain length to width ratio	1,3,4, 7 and 8	3.41–58.79%	Li et al. (2023)
Tartary buckwheat	RILs (Xiaomiqiao × Jinqiaomai 2)	9	1000 grain weight	1 and 4	2.9–47.5%	Shi et al. (2021)
Barley	DH (Huaai 11 × Huadamai)	111	Spike number per plant, main spike length, spike-let number per spike, spikelet number per plant, grain number per plant, grain number per spike, grain weight per plant, grain weight per spikes, thousand grain weight, and spike density	1, 2, 3, 4, 5, 6, 7	–	Wang et al. (2021)
	DH (Huaai 11 × Huadamai 6)	60	Grain size and weight-related traits	1, 2, 3, 4, 5, 6, 7	59–84.33%	Wang et al. (2016)
Oats	BC_2F_{2-6} IAH611× Iltis	13	Kernel weight, panicle number, test weight and grain yield	A1, A8, B1, B13 and B5	12.06–63.05%	Herrmann et al. (2014)
Potato	(12601ab1 × Stirling)	3	Size and yield	1, 2, 5, 6	5.4–13.3%	Bradshaw et al. (2008)
Carrot	BILs (Songzi × Amsterdam forcing)	5	Root length, root shoulder width, the dry material content of root, and the ratio of root shoulder width to root middle width	2	23.4–66.9%	Ou et al. (2022)
Radish	F_2 (835 × B2)	13	Root weight, root length and root diameter	1, 2, 3, 4, 6, 8 and 9	3–12.6%	Yu et al. (2016)

multi-environment. In both single and multiple environments, 71 (63.40%) of the 111 QTLs were discovered and can be the potential QTLs to be used in barley breeding (Wang et al. 2021). Seven significant QTLs for grain size were found on 1H, 2H, 3H, and 5H. Herrmann et al. (2014) discovered 13 QTLs related to kernel weight, panicle number, test weight and grain yield using two $BC_2F_{2\text{–}6}$ populations of the parents Iltis × IAH611 in oats. QTL "*QKw3.jki-B1*" explained the maximum phenotypic variation of 63.04%, followed by "*QKw1.jki-A1*," explaining 53.05% of phenotypic variation for kernel weight.

In Tartary buckwheat (*Fagopyrum tataricum*), four loci harboring nine QTLs were identified related to 1000-grain weight on chromosomes 1 and 4 in the three experimental trials conducted across different environments (Shi et al. 2021). A major locus, located in a 38.2–39.8 cM region on chromosome 1, with a LOD score ranging from 18.1 to 37.0, was found in all three experiments. This locus accounted for 23.6–47.5% of the phenotypic variation (Shi et al. 2021). In another study, Li et al. (2023) mapped the QTLs for grain yield and related traits in Tartary buckwheat, including 1000-grain weight, grain length, grain width, and grain length-width ratio (L/W), across four different environments. They used 221 RILs produced from a cross between "Xiaomiqiao and Jinqiaomai 2" and detected a total of 32 QTLs. On chromosome Ft1, they identified two QTL clusters, *qClu-1-3* and *qClu-1-5*, containing seven stable and significant QTLs for grain yield (*qGY1.2*), 1000-grain weight (*qTGW1.2*), grain length (*qGL1.1* and *qGL1.4*), grain width (*qGW1.7* and *qGW1.10*), and length-width ratio (*qL/W1.2*), all consistently observed across all three environments.

Krajewski et al. (2012) identified 12 QTLs for yield components in field pea (*Pisum sativum* L.) using two mapping populations derived from crosses "Wt11238 × Wt3557" and "Wt10245 × Wt11238." Of these, 2, 1, 2, 4, and 3 QTLs were identified for peduncle number, pod number per peduncle, seed number per pod, thousand seed weight and yield, respectively, spreading across chromosomes 1, 2, 3, and 5. Ávila et al. (2017) identified the 5 and 8 QTLs for plant architecture and yield related traits in faba bean. These QTLs were identified mainly in three regions of chromosomes 3, 4 and 13, forming QTL clusters/hotspots for yield contributing traits. Bradshaw et al. (2008) mapped 1 QTL for potato size and 2 QTLs for potato yield present on different potato chromosomes. Ou et al. (2022) identified five consistent QTLs on chromosome 2 that were found in both years of QTL mapping experiments and were related to carrot root length, root shoulder width, dry material content of the root, and the ratio of root shoulder width to root middle width. Together, these QTLs accounted for 23.4–66.9% of the total phenotypic variation. Yu et al. (2016) conducted an investigation into QTLs associated with radish yield, specifically focusing on root weight, root length, and root diameter. This study resulted in the identification of 13 QTLs located on chromosomes 1, 2, 3, 4, 6, 8, and 9, using (simple sequence repeapets) SSR markers.

QTLs identified for yield-related traits open up new possibilities for marker-assisted and genomic selection programs in the high-altitude crop, with a primary focus on improving yield. The QTL can be directly transferred into different cultivars of high-altitude crops using the markers flanking the QTL regions, or the underlying candidate genes can be further validated and cloned using a variety of biotechnological techniques for effective use in breeding programs for yield improvement. However, it has certain limitations, viz., phenotypic diversity is only limited to contrasting parents used for the development of mapping population, large confidence interval (CI) and low resolution of the mapped QTLs and limited stability, which varies with respect to the genetic background. By combining data from many QTL mapping studies carried out with bi-parental populations, a meta-analysis of QTLs is a potent technique for resolving variability among different studies. Using this approach, consensus and consistent QTLs or meta-QTL regions with reduced CI can be more easily identified (Saini et al. 2021). Among high-altitude crops, it has been used in the case of peas and barley for the identification of genomic regions associated with yield-associated traits. Klein et al. (2020) identified 27 meta-QTLs related to the yield traits, viz., seed weight per plant, seed number per plant, and thousand seed weight in pea. Of these, some potential meta-QTLs showed a CI of less than 2 cM. In another study, Khahani et al. (2019) carried out the meta-QTL analysis in barley using 639 yield-related QTLs reported in 26 studies published from both plains and high-altitude regions and identified 31 meta-QTLs with a reduced CI of less than 2 cM. However, this approach can be used only in those cases where there are enough QTL mapping studies for the yield-related traits in the crop of interest.

In addition, unravelling marker-trait associations (MTAs) has been successfully accomplished in different crops, including high-altitude crops, using an approach known as genome-wide association studies (GWAS). The GWAS enables the mapping of traits-linked markers into relatively narrower regions. The genes in these genomic areas can then be functionally annotated to identify potential candidate genes for the related traits. Here, we have discussed some of the most important GWA studies for yield and associated trait in high-altitude crops (Table 9.3). For instance, Zhang et al. (2022) identified 116 consistent single nucleotide polymorphisms (SNPs) spread across the genome of Siberian wild rye (confidence interval) related to the forage yield-related traits, viz., culm-related traits, leaf-related traits, and seed-related traits. In addition, Zhang et al. (2021) discovered two non-synonymous SNPs related to 1000-grain weight in Tartary buckwheat located on chromosome 4 at positions 46,350,596 (C/G) and 37,407,976 (T/G) co-localizing with two gene models (*FtPinG0404616900* and *FtPinG0280000714*) encoding AP2 transcription factors and an unidentified protein, respectively. Additionally, they discovered two potential genes linked to seed width, including the gene (*FtPinG0404616900*) linked to 1000-grain weight. According to previous studies from rice (Kakei et al. 2017) and wheat (Shao et al. 2017), the tryptophan aminotransferase-related protein encoded by *FtPinG0100980400* is essential for grain yield via auxin biosynthesis. As a result, *FtPinG0404616900* and *FtPinG0100980400* gene models can be chosen as the candidate gene linked to the 1000-grain weight trait for future functional investigation and ultimately for the developing high-yield Tartary wheat cultivars.

In a most recent study, Zhang et al. (2023a, b) identified loci (i.e., *C01*) for the weight of a single curd located on chromosome 1 of the cauliflower crop. They

Table 9.3 List of yield and associated trait MTAs identified in different high-altitude crops

Panel size	MTAs identified	Identified QTLs related to yield components	R^2-value	Chromosome	Crop	Reference
210 diverse accessions	116	Grass yield (culm-related trait and leaf-related traits)	7.62–31.85	1, 2, 3, 4, 5, 6, and 7	Siberian wild rye	Zhang et al. (2022)
510 diverse accessions	13	1000 grain weight (TGW)	–	2, 4, 5, 6, 7, and 8	Tartary buckwheat	Zhang et al. (2021)
220 diverse accessions	5	Weight of single curd	–	1	Cauliflower	Zhang et al. (2023a, b)
135 diverse accessions	9	Yield and seed weight	38–80	1, 4, 7	Field pea	Gali et al. (2018)
379 diverse accessions	360	Grain length, grain width, grain area, grain roundness, and thousand-grain weight	–	1, 2, 3, 4, 5, 6, and 7	Barley	Xu et al. (2018)
178 diverse accessions	82	Forage yield, viz., plant height and dry matter yield	–	–	Oat	Huang et al. (2020)
749 diverse accessions	12	Carrot root: Length, maximum width, and their quotient, aspect ratio	1–7	2, 3, 4, 5, 6, 7, and 9	Carrot	Brainard et al. (2021)

discovered the *BOB01G136670* gene with five SNPs encoding nonsynonymous mutations in this locus; these variants produced two haplotypes with notable variations in curd weight. Compared to cauliflower varieties with BOB01G136670^{Hap2}, the weight of a single curd was substantially higher in those with the BOB01G136670^{Hap} gene. Five SNPs were identified for seed yield, viz., Chr1LG6_57305683, Chr1LG6_366513463, Chr4LG4_373933955, Chr7LG7_488770913, SSc4454_324798 and four for seed weight, viz., Chr1LG6_176606388, Sc00398_17041, Sc01126_54371 and Sc01886_124838 in field pea explaining phenotypic variation ranging from 38 to 80% (Gali et al. 2018). Similarly, Xu et al. (2018) identified 360 MTAs related to yield-associated traits in the pan-European association mapping panel of 379 barley cultivars. These QTLs overlap with 12 of the loci for barley grain yield, 22 of the orthologs for other grain traits, and 2 of the loci each map to 9 QTL hotspots, which implies the consistent and robust nature of these QTLs and can be used in barley breeding program for yield improvement.

9.3.1.2 Transgenesis and Genome Editing for Yield Improvement: Current Status and Future Directions

Over the past century, transgenic breeding has evolved as a method for introducing genetic diversity into crops by incorporating foreign genes into plant genomes. This research has made substantial progress, enabling the transfer of genes from wild plant relatives into cultivated crops and surmounting reproductive barriers between different plant species. However, it is worth noting that there are only a few instances of such advancements in high-altitude crops. While progress in genomics and bioinformatics has advanced significantly across various high-altitude crops, leading to the identification and cloning of genes associated with yield and related traits and creating exciting opportunities for genetic engineering to enhance desired plant characteristics, the widespread acceptance and adoption of genetically modified (GM) crops face obstacles due to ethical and regulatory concerns related to environmental safety and human health in many countries worldwide.

In addition, yield improvement could also be facilitated quickly by new gene-editing technologies, the most popular of which being CRISPR/Cas9. The CRISPR/Cas9 system has quickly taken over as the most popular method for editing crop plant genomes due to its effectiveness and simplicity. For instance, Miao et al. (2018) produced a rice mutant with the *pyl1/4/6* triple knockout. When compared to the wild type, the mutant exhibited much more primary and secondary branches, longer panicles, and comparatively more yield. In another study, Xu et al. (2016) simultaneously edited 3-grain weight-related genes (*GW2*, *GW5*, and *TGW6*) in rice, which led to a substantial increase in rice grain weight, ultimately boosting the yield. In case of wheat, increased grain length and width resulted from deletion of the *TaGW2* gene, which encodes a RING E3 ligase (Wang et al. 2018). There are numerous examples of modifying yield-associated traits utilizing CRISPR in the case of wheat and rice. However, as per our knowledge, based on a survey of the literature, not much work has been done to improve yield in high-altitude crops. Nevertheless, this system can be utilized in the same manner in high-altitude crops as well to ameliorate yield-contributing traits. In conclusion, yield is a trait having complex architecture greatly influenced by plant genotypes and environmental conditions. To achieve the necessary increases in the yield of high-altitude crops, it is essential to have a deeper understanding of the mechanisms and processes that influence yield in varied environments.

9.3.2 Genomics Approaches for Improving Climate-Resilience and Stress Tolerance: Current Status and Future Directions

Ensuring food security to the ever-growing population is challenging in the era of climate change which has led to abrupt weather conditions consequently resulting in a decrease in yield due to various biotic and abiotic stresses. This poses even a huge problem primarily in the hilly regions where life sustains on a few crops. Climate change has favored the emergence of vast array of stronger strains and races of new pathogens, insects, and weeds in highland crops (Bajwa et al. 2020). So, there is a need to improve the genetic basis of high-altitude crops to tackle the shortage of food in the hilly areas. Advancement in the study at the genome level has come up with a great understanding that can contribute a lot towards climate resilience agriculture. A valuable combination of traditional breeding with advanced biotechnological tools such as GWAS, and genome editing with "omics" approaches has permitted us to explore and more precisely manipulate the genetic make-up of high-altitude crops to develop climate-resilient cultivars. Additionally, genomics-assisted breeding (GAB) has provided the weapon to fight climate change by permitting quick identification of genes concerned with the climate-associated agronomic traits, for the improvement of crops to accoutrement with climate change.

9.3.2.1 Genetic Dissection of Traits Contributing to Stress Tolerance

Several QTL mapping studies have been conducted in different high-altitude crops for the identification of the QTLs/genes governing resilience towards unfavorable environmental incidents (Tables 9.4 and 9.5). For instances, accumulation of water-soluble carbohydrates and their partitioning to storage organs (Rebetzke et al. 2008), root architecture (Uga et al. 2013; Lynch et al. 2014), stay-green character (Borrell et al. 2014), abscisic acid concentration (Rehman et al. 2011), carbon isotope discrimination and canopy temperature (Lopes et al. 2014) to name a few among many. However, the low allelic diversity and resolution of bi-parental populations places some restrictions on interval mapping. On the other hand, the GWAS method enables researchers to investigate the tremendous allelic variation present in the germplasm of various highland crops including wheat, barley, millets, quinoa, buckwheat, rajma, maize, potato, pea, soybean, and others. The resolution mapping of GWAS is more due to millions of crossing events that have been accumulated in due course of time in natural populations.

Subsequently, GWAS have been widely used for gene discovery of multiple traits of wheat, including stripe rust resistance (Jia et al. 2020), powdery mildew (Kang et al. 2020) and Karnal bunt (Gupta et al. 2019). Further, 32 MTAs for leaf

Table 9.4 Major QTLs identified for different biotic stresses in high-altitude crops

Crops	Traits	QTLs	Chr. No.	PVE %	References
Wheat	Powdery mildew, Karnal bunt, loose smut, aphids	4	1AS, 2BL, 6B, 3AS, 3A	11.19–74.9%	Simeone et al. (2020); Gupta et al. (2019); Kumar et al. (2018); Zhao et al. (2023)
Barley	Septoria speckled leaf blotch, leaf blight/spot blotch, powdery mildew	6	2 (2H), 6 (6H), 2 (2H), 1 (7H), 4 (4H), 5 (1H)	8.52–42.3%	Yun et al. (2005)
Maize	Banded leaf sheath, turcicum leaf blight	7	5, 2, 6, 2, 6, 4, 4	3.27–55%	Adhikari et al. (2022); Zhao et al. (2006)
Raj mash	Anthracnose, common blight, fusarium root rot, common mosaic virus	16	Pv01, Pv04, Pv04, Pv08, Pv08, Pv08. Pv10, Pv06, Pv06, 1 (B5), 8, 7, 7, Pv04, Pv07	2.17–53.3%	Zuiderveen et al. (2016); Choudhary et al. (2018); Banoo et al. (2020); Román-Avilés and Kelly (2005); Soler-Garzón et al. (2021)
Soybean	Charcoal rot, stem and root rot, soybean mosaic virus	3	15, 16, 20, 2	16.6–54.23%	da Silva et al. (2019); Lin et al. (2022); Chu et al. (2021)

Table 9.5 Major QTLs identified for different abiotic stresses in high altitude crops

Crops	Stress	Trait	QTL	Chr.	PVE (%)	References
Wheat	Drought	Coleoptile length, seedling height, root length, relative water content, plant height, days to anthesis, productive tillers/m2, root number, abscisic acid	11	1A, 1A, 1A, 2A, 2B, 5A, 5A, 7A, 4D, 3B, 5A	13.59–46.1%	Zhang et al. (2013a, b); Malik et al. (2015); Gahlaut et al. (2017); Barakat et al. (2015)
	Heat	Fv/Fm (maximum quantum efficiency of photosystem II), maximum fluorescence (Fm), thylakoid membrane damage, SPAD chlorophyll content	6	3B, 3B, 3B, 4A, 7A, 7A	11.9–34.8%	Sharma et al. (2017); Azam et al. (2015); Talukder et al. (2014)
	Salinity	Root Na exclusion, root zinc concentration, shoot Na exclusion, proportion of dead leaves	4	7A, 7A, 7A, 4B	11.23–18.79%	Hussain et al. (2017); Turki et al. (2015)
	Nitrogen	Root dry weight	1	4B	10.4%	Ren et al. (2017)
	Deficiency	Maximal root length	1	2B	21.6%	
Maize	Drought	ASI, plant height, senescence (6 weeks after flowering), ASI, senescence, normalized differential, vegetative index (NDVI), grain yield, ears per plant, plant height	13	1, 1, 2, 2, 4, 5, 5, 5, 6, 6, 7, 9, 19	10.23–23.65%	Trachsel et al. (2016); Almeida et al. (2013, 2014)
	Waterlogging	Plant height, chlorophyll content, Total dry weight, plant mortality (%), root fresh weight, root dry weight	7	1, 2, 4, 5, 5, 5, 5, 9	10.77–18.8%	Osman et al. (2013); Zaidi et al. (2015); Zhang et al. (2013a, 2013b)
	Heat	Heat susceptibility, index for dry grain yield, quantum efficiency of PSII, days to emergence	3	2, 1, 3	–	Frey et al. (2016); Revilla et al. (2016)
	Cold	Vigor, number of kernel rows per ear, 100 kernel weight, early vigor, number of leaves per plant, plant height, ear height	7	4, 5, 5, 5, 6, 8, 8	14–43.3%	Allam et al. (2016); Revilla et al. (2016)
	Salinity	Shoot fresh weight, root dry weight, root length	5	1, 1, 1, 1, 2	15.8–21.9%	Hoque and Avenue (2022)

rust (Amouzoune et al. 2022) and 23 MTAs for spot blotch (Visioni et al. 2020) in barley were identified in recent studies. Moreover, the identification of new resistance loci by GWAS contibutes to the enhancement of horizontal resistance with the exhisting array of resistance genes and improve durable resistance in diseases, viz., turcicum leaf blight (Rashid et al. 2020) and insects such as fall armyworm (Kamweru et al. 2022) in maize. Four *WRKY* gene family namely; *FtWRKY*6, 74, 31, and 7 responsible for abiotic stress including salt, drought, cold, and heat treatments (He et al. 2019) and 14 *FtHDAC* genes encoding histone deacetylases responsible for biotic and abiotic stress (Hou et al. 2022) were identified in Tartary buckwheat using GWAS. Tashi et al. (2018) identified 23 *Hsf* genes encoding heat shock transcription factors (*CqHsfs*) in *Chenopodium quinoa*. Six *CqMYB2R* genes were identified to be involved in salt stress response (Liu et al. 2023). Further 23 *SPL* genes playing essential roles in plant growth and development, signal transduction, and responses to biotic and abiotic stresses were also identified in *Chenopodium quinoa* through a GWAS (Ren et al. 2022). A study conducted on foxtail millet identified *CCT* gene family to be responsible for diurnal rhythm and abiotic stress response (Li et al. 2022). Moreover, GWAS are also helpful in detecting the genetic regions controlling stress-adaptive traits related to abiotic stress tolerance in pea (Tafesse et al. 2021). It also improved mapping resolution for the exact localization of resistance QTLs for anthracnose (Zuiderveen et al. 2016) and common blight (Zhu et al. 2016) in *Phaseolus vulgaris*. GWAS has proven to be extremely helpful in observing the variation in several physiological traits related to drought such as chlorophyll content and fluorescence, stomatal conductance, normalized difference vegetation index (NDVI), and leaf area and circumference in potato (Alvarez-Morezuelas et al. 2023). A summary of recent GWAS studies conducted for different biotic and abiotic stresses in high-altitude crops is provided in Table 9.6 and Table 9.7, respectively.

9.3.2.2 Genomic Selection for Climate Resilience and Stress Tolerance

The major issue in breeding for biotic and abiotic stress tolerance is their complex inheritance, low heritability, and high environmental effect on such characters. The genomic selection (GS) is a specialized form of MAS that provides us with a platform to efficiently select the superior germplasm for quantitative traits by assessment of an individual's genetic value based on genomic estimated breeding values (GEBVs) determined using genome-wide marker profiling data and robust phenotypic information. The GS improves the selection intensity, which is the prime goal of a plant breeder.

Extensive studies on GS have been reported in wheat for biotic stress which comprises an extensive spectrum of diseases such as three types of rusts, *Fusarium* head blight, powdery mildew, *septoria tritici* blotch, and *Stagonospora nodorum* blotch. For these diseases, the genomic prediction accuracies varied from 0.14 to 0.85 (Daetwyler et al. 2014; Sarinelli et al. 2019; Mirdita et al. 2015). GS has been effectively used in maize, to select lines from natural populations for resistance to *Stenocarpella maydis*, which causes ear rot (dos Santos et al. 2016), and from

Table 9.6 Marker trait associations and promising candidate genes identified for different biotic stresses in various highland crops

Crops	Traits	Panel size	Number of MTAs/QTNs	References
Wheat	Yellow rust	240, 230, 245, 139, 381	12, 12, 5, 30, 195	Jia et al. (2020); Shahinnia et al. (2022); Franco et al. (2022); Mehrabi et al. (2022); Page et al. (2023)
	Powdery mildew	329, 637	35, 7	Kang et al. (2020); Du et al. (2021)
	Karnal bunt	339, 179	18, 15	Gupta et al. (2019); Singh et al. (2020)
	Aphids	181, 163	27, 10	Kisten et al. (2020); Zhao et al. (2023)
Barley	Stem rust	431	1	Czembor and Czembor (2022)
	Leaf rust	316, 114	32, 12	Amouzoune et al. (2022); Ziems et al. (2023)
	Leaf blight/spot blotch	261	23	Visioni et al. (2020)
	Powdery mildew	431	5	Czembor and Czembor (2022)
	Loose smut	62	27	Abo-Elyousr et al. (2022)
Maize	Turcicum leaf blight	320, 324, 1487	14, 14, 4	Rizzardi et al. (2022); Rashid et al. (2020); Van Inghelandt et al. (2012)
	Stalk rot	165, 342	34, 7	Liu et al. (2021); Rashid et al. (2022)
	Fall armyworm	424, 289	56, 15	Kamweru et al. (2022); Warburton et al. (2023)
Raj mash	Anthracnose	192	–	Banoo et al. (2020)
	Fusarium root rot	205	11	Paulino et al. (2021)
	Halo blight	504	–	Tock et al. (2017)
Soybean	Frogeye leaf spot	234	69	Gu et al. (2021)
	Charcoal rot	459	19	Coser et al. (2017)
	Soybean mosaic virus	165	104	Che et al. (2020)
	Stem and root rot	185	18	Jing et al. (2021)
	Aphid	145	5	Natukunda et al. (2019)

Table 9.7 Marker trait associations and promising candidate genes identified for different abiotic stresses in various highland crops

Crops	Traits	No. of MTAs/QTNs	References
Wheat	Drought	12, 205, 62, 99, 75, 45	Ahmed et al. (2021); Abou-Elwafa and Shehzad (2021); Mwadzingeni et al. (2017); Sallam et al. (2022); Id et al. (2019); Bhatta et al. (2018)
	Cold	76, 53	Zhao et al. (2020); Soleimani et al. (2022)
	Salinity	238	Alotaibi et al. (2022)
Maize	Drought	123	Guo et al. (2020)
	Waterlogging	71	Guo et al. (2020)
Raj mash	Drought	68	Guo et al. (2020)
Barley	Cold	39	Elakhdar et al. (2023)
	Waterlogging	92	Borrego-Benjumea et al. (2021)
	Drought	519, 97, 36	Abdel-Ghani et al. (2019); Tarawneh et al. (2020); Xiong et al. (2023)

Table 9.8 Applications of CRISPR/Cas system to engineer abiotic/biotic stress tolerance varieties

Stress	Crop	Gene	Traits	Target result	Vector delivery	References
Abiotic stress	Potato	*StFLORE*	Drought	Knockout	*Agrobacterium tumefaciens*	Ramírez Gonzales et al. (2021)
	Maize	*ZmSRL5*	Drought	Knockout	*Agrobacterium tumefaciens*	Pan et al. (2020)
	Soybean	*GmNAC06*	Salinity	Knockout	*A. Rhizogenes*	Li et al. (2021)
Biotic stress	Potato	*Caffeoyl-CoA O-methyltransferase (StCCoAOMT)*	Late blight resistance	SNP mutation	*Agrobacterium tumefaciens*	Hegde et al. (2021)
		StDND1, StCHL1 and StDMR6–1 (S-genes: Susceptibility genes)	Late blight resistance	Knockout	*Agrobacterium* spp.	Kieu et al. (2021)
	Wheat	*TaMLO homologs*	Powdery mildew	Insertion and deletion (indels) mutations	Biolistic transformation	Wang et al. (2014)
		TaEDR1/cds region				Prihatna et al. (2018)
	Maize	*bW2 and bE1*	Corn smut	Deletion mutation	*Agrobacterium* spp.	Schuster et al. (2016)
	Barley	*MP, CP, rep/rep, IR*	No disease symptoms and virus presence	Insertion	*Agrobacterium* spp.	Kis et al. (2019)

biparental populations for superior yield under severe infestation of Striga (Badu-Apraku et al. 2019). In barley, Fusarium head blight severity was predicted using markers and prediction models, and the prediction accuracy was 0.72, which was significantly higher than that of conventional phenotyping (Lorenz et al. 2012).

The use of GS in wheat, maize, and other crops for various abiotic tolerance has been established. In eight biparental populations of maize under drought conditions, Beyene et al. (2015) found a gain of 0.086 t/ha for grain production using the rapid cycling GS approach, and an ultimate gain of 0.176 t/ha after three cycles of selection. Similarly in a study on maize, Das et al. (2020) reported that rapid cycling GS resulted in genetic gains of 0.110 and 0.135 t/ha/year for grain yield during drought and 0.038 and 0.113 t/ha/year under water-logging conditions namely in Maize Yellow Synthetic 1 and Maize Yellow Synthetic 2, respectively. Genomic selection has been an effective methodology for breeding value prediction in potato for late blight (*Phytophthora infestans*) and common scab (*Streptomyces scabies*; Enciso-Rodriguez et al. 2018). GS has the potential to be a valuable tool in pea breeding programs that aim to improve Ascochyta blight resistance (Carpenter et al. 2018). It can be used to predict breeding values for lines that have not yet undergone phenotyping and to enhance estimated breeding values for lines that already have trait data available.

9.3.2.3 Genome Editing for Climate Resilience and Stress Tolerance

The genome editing techniques such as transcription activator-like effector nucleases (TALENs) and zinc finger nucleases (ZFNs) have been in use for a decade, and the (CRISPR)/Cas system has become the most widely used approach in recent years. This system has enabled remarkable progress in plant genome engineering for crop improvement in various highland crops (Li et al. 2013). With this technology, it is possible to modify specific genes that control traits necessary for adaptation. Several studies have already been conducted on several high-altitude crops to enhance the agronomic traits of plants and their tolerance to environmental pressures, such as abiotic stresses (Table 9.8).

9.3.3 Genomic Approaches for Nutritional Value Improvement: Current Status and Future Directions

Biofortification is a proven and cost-effective strategy for enhancing the levels of essential micronutrients in crops, especially those that serve as staple foods in developing nations. Micronutrient deficiencies, commonly referred to as, hidden hunger, are widespread among populations due to imbalanced diets primarily reliant on starchy staple crops. Biofortification is an excellent strategy to improve the nutrient profiles of several crops utilizing various breeding techniques (Priya and Nayyar 2023). However, in high-altitude regions, due to several challenges as previously discussed, micronutrient deficiencies are highly prevalent. Moreover, due to lesser diversity of crop species in these high altitudes, the biofortification of available crops is essential to fight the hidden hunger. The objective of biofortification is to provide diverse nutrients to the local population from the minimum crops available in such regions (Dame and Nüsser 2011). With this context, several studies in crops such as oats, barley, several millets, mustard, carrot, peas, potato, carrot, and others aim to enhance the nutrient profile of such crops prominently grown in high-altitude regions. The conventional breeding approaches rely on selections solely based on phenotype which is often unreliable. However, with the advent of molecular markers and next-generation sequencing technologies, QTLs, and potential candidate genes along with selectable markers are tagged with nutritionally important traits to ease and offer reliability in the selection of traits. Genome editing techniques have also made significant progress over the last decade and have proven promising for knockdown or enhancing the expression of traits involved in nutrition.

9.3.3.1 Genetic Dissection of Biofortification-Related Traits in High-Altitude Crops

Most of the traits associated with nutrition such as protein content, high amylose, oil content, fatty acids, and several micronutrients such as zinc, iron, and boron are governed by multiple genes. The complex mechanism and genes governing these traits have been successfully identified by researchers through QTL mapping and GWAS approach. Identified QTLs and target genes can be utilized for enrichment of elite cultivated varieties. The soluble fiber β-glucan in oat is well known for the management of diabetes, blood pressure, weight, celiac disease and for lowering cholesterol levels (Gazal et al. 2014). GWA studies identified several SNPs in different linkage groups for β-glucan in oat crop with phenotypic variance ranging from 4.2 to 4.4% (Zimmer et al. 2020). The importance of protein is crucial for the nutritional enhancement of crops as protein energy malnutrition is categorized as one of the most important deficiencies in the world and is known to cause several health complications such as kwashiorkor, marasmus, impaired mental health, oedema, organ failure, wasting, shrinkage of muscle tissues and weak immune system among many others (Latham 1990; Khan et al. 2017). Protein-associated QTLs were tagged on chromosomes 2, 3, and 5, respectively, with total phenotypic variance ranging between 11 and 22% (Zhou et al. 2022). Similarly in oat crop two QTLs, viz., *QTL16* and *QTL11* were identified for groat protein and were linked to several SNPs explaining a variation between 10.57–16.81%. A significant MTA was established in potato plants for different amino acids (Pandey et al. 2023) which imparts the quality of protein.

Edible oils are high in fatty acids such as oleic, linoleic, linolenic, palmitic, erucic, and eicosenoic acids. Fatty acids regulate the thermostability of oil, particularly if the oil has a high amount of polyunsaturated fatty acids, it gets rancid faster. Erucic acid belongs to the group mono-unsaturated fatty acid (MUFA) is very high in Indian varieties of mustard and is known to cause several problems in children and an increase in blood cholesterol levels in adults. While the MUFA oleic acid is of nutritional importance and is more thermostable than others. Still there is an immense importance to identify genes associated with different fatty acids. Singh et al. (2013) identified several QTLs for oil content, erucic, eicosenoic, linoleic, linolenic acid and palmitic acid on several linkage groups. Apart from fatty acid, an

Table 9.9 List of QTLs associated with different biofortification traits identified in high-altitude crops

Crop	Trait	QTL	Chromosome/linkage group	PVE (%)	Mapping population	Reference
Oat	β glucan	2	6A, 7A	–	RIL population	Tinker et al. (2022)
Carrot	β carotene	3	1, 3, 6	8.25–30.38%	F2	Shibaya et al. (2022)
	Lutein	2	5, 6	6.78–20%	F2	Shibaya et al. (2022)
	Bitterness	5	1, 3, 4, 5, 8	5.57–9.93%	F_2	Dunemann et al. (2022)
Pea	Seed protein	3	2, 3, 5	11–22%	RIL	Zhou et al. (2022)
	Iron	7	LG (II), LG (II), LG (III), LG (V), LG (VI), LG (IIIb)	3.62–12.3%	RIL	Ma et al. (2017); Gali et al. (2018)
	Boron	3	LG (I), LG (VI), LG (VII)	10.3–15.1%	RIL	Ma et al. (2017)
	Potassium	4	LG (III), LG (V), LG (VI), LG (VI)	6.7–9.5%	RIL	Ma et al. (2017)
	Magnesium	4	LG (III), LG (VI), LG (VI), LG (VII)	7.7–15.9%	RIL	Ma et al. (2017)
	Molybdenum	1	LG (V)	24.60%	RIL	Ma et al. (2017)
	Phosphate	4	LG (III), LG (V), LG (VI), LG (VI)	6.40–20.5%	RIL	Ma et al. (2017)
	Selenium	3	LG (VI), LG (VII), LG (IV)	3.81–14.1%	RIL	Ma et al. (2017); Gali et al. (2018)
	Zinc	4	LG (II), LG (III), LG (V), LG (IIIb)	7.1–13.70%	RIL	Ma et al. (2017)
	Trypsin inhibitors	2	–	–	F_2, mutant TILLING	Domoney et al. (1994), Clemente et al. (2015)
Mustard	Oil content	2	LG (I), LG (II)	11–24.7%	RIL	Singh et al. (2013)
	Erucic acid	2	LG (II), LG (II)	14.8–20.4%	RIL	Singh et al. (2013)
	Eicosenoic acid	2	LG (II), LG (V)	18.2–22.5%	RIL	Singh et al. (2013)
	Linolenic acid	2	LG (I), LG (VI)	10.5–21.8%	RIL	Singh et al. (2013)
	Linoleic acid	3	LG (II), LG (III), LG (III)	6.9–24.8%	RIL	Singh et al. (2013)
	Palmitic acid	2	LG (II), LG (V)	12.8–16.2%	RIL	Singh et al. (2013)
	Total glucosinolates	1	LG (J3)	15.59–45.46%	DH population	Ramchiary et al. (2007)

Table 9.10 List of MTAs/QTNs associated with different biofortification traits identified in high-altitude crops

Crop	Trait	Chromosome	QTLs	PVE (%)	Mapping population	Reference
Oat	Groat protein	6C, 4D	2	6.97–22.24%	174 diverse oat accessions	Yan et al. (2023a, b)
	β—Glucan	20, 02, 19, 06, 02, 11, 11, 11, 12, 19	10	–	413 genotypes oat pannel	Zimmer et al. (2020)
Barley	Fe	2H	1	3.50%	336 spring barley	Gyawali et al. (2017)
	Zn	1H, 2H, 5H, 6H, 2H	5	3–4.4%	336 spring barley; 298 barley landraces; 150 DH lines	Gyawali et al. (2017); Mamo et al. (2014); Sadeghzadeh et al. (2010)
	Protein	2H	1	13%	F_6 derived recombinant inbred lines	Gous et al. (2012)
Pea	Amylose	–	4	4–21%	169 diverse pea accessions	Jha et al. (2015)
	Crude protein	–	1	17%	169 diverse pea accessions	Jha et al. (2015)
	Iron	LG (I), LG (III), LG (IV), LG (V), LG (VII)	5	–	94 diverse accessions	Diapari et al. (2015)
	Zinc	LG (III), LG (III)	2	–	94 diverse accessions	Diapari et al. (2015)
Foxtail millet	Boron	8	1	–	93 diverse accessions	Jaiswal et al. (2019)
	Iron	8	1	–	93 diverse accessions	
	Magnisium	2, 2, 7	3	–	93 diverse accessions	
	Zinc	2, 2, 7, 8, 9	5	–	93 diverse accessions	
Potato	Amino acid	6, 2, 2, 6, 2, 1, 1, 2, 2, 3, 2, 2	12	1.2–10.5%	217 accessions of potato	Pandey et al. (2023)
	Total starch content	5, 6, 2, 6, 9, 9	6	2.2.–5.1%	90 accessions	Schönhals et al. (2017)
	Protein content	3, 5, 7	3	0.1–0.15%	–	Klaassen et al. (2019)

antinutritional glucosinolate majorly found in the family Brassicaceae which gives pungency is higher in Indian varieties of mustard (Kumar et al. 2010). Ramchiary et al. (2007) identified two QTLs linked to total glucosinolate in mustard crop, viz., *ToGsl2* and *ToGsl4* on J3 and J9 linkage groups, respectively, thereby associating flanking markers with each QTL.

Starch is composed of two components, viz., amylose and amylopectin. Amylose is classified as a type of resistant starch beneficial for diabetics, obese and cardiovascular patients (Hossain Paulino et al. 2021). In pea crop, Jha et al. (2015) identified three QTLs, viz., *Agpl1*, *Gbsts1* and *Sbe2* associated with amylose content with variance ranging between 4 and 17% and linked them to various SNP markers in 169 diverse pea accessions. Several macro and micronutrients also play a key role in regulating the growth, metabolism, and functions of various body tissues. These include iron, zinc, boron, potassium, magnesium, molybdenum, phosphorus, and several others. Researchers have focused on identifying QTLs for these nutrients and tagging molecular markers with these nutrient traits. In barley, Gyawali et al. (2017) identified a QTL, *Fe-2H-84.74* with 21.9–91.0 mg/kg variation in iron content on chromosome 2 and two QTLs, viz., *Zn-1H-21.97*, *Zn-2H-87.34* on chromosome 1 and 2, respectively, with 10.4–54.5 mg/kg variation in zinc content. These QTLs were also associated with several SNP markers identified using the GWAS approach.

QTLs and MTAs were identified for a few more traits like carotenes, antinutritional and factors. The detailed information for nutritional-related traits in high-altitude crops are listed in Tables 9.9 and 9.10. However, there are several other crops grown in high altitudes such as buckwheat, and numerous nutrient deficiencies in other staple crops of these regions are often less emphasized for identification of potential QTLs. Nevertheless, there are tremendous opportunities to improve the nutrient profile of such crops (Luthar et al. 2020). Moreover, several QTLs for nutrient enhancement in these crops are not essentially validated, which creates a huge gap in the application of breeding techniques to transfer nutritional traits in such crops due to advancements in breeding techniques. Further research and extensive studies in breeding aspects of high-altitude crops need to be undertaken to fulfil the nutritional requirements of the local and tribal populations of these regions.

9.3.3.2 Transgenesis and Genome Editing for Nutritional Value Improvement

Molecular biology and variation in natural germplasm have fastened the process of enhancing genetic gain and ultimately changed the dynamics for successful varietal development for different traits. Although, the natural genetic variation is being challenged by several factors such as long-term domestication, accumulation of deleterious alleles, lower recombination frequencies, lack of understanding of complex genetic traits and adverse impact of climate change became limitations for breeding. The transgenic approach has been proven successful for the development of plants across the globe but it has to undergo strict regulations. However, plants developed through genome editing techniques have less regulation as well and genome edited varieties are being generated in less duration and hence considered as next-generation breeding (Gao 2021).

The food with high amylose starch is a promising solution for people with diabetes as it digests slowly in the gut and prevents the abrupt rise in the level of glucose also has several health benefits (Bojarczuk et al. 2022). The other component of starch, amylopectin also has a variety of applications such as making films, coatings and as an emulsifying agent in food industries, composite in polymers, filters, dye, and color in non-food industries (Šárka and Dvořáček 2017). Hence, the improvement of starch is primarily based on its application. High amylose potato was developed using CRISPR/Cas9 by targeting genes involved in the amylopectin biosynthesis pathway (Zhao et al. 2021). The starch branching enzyme coding gene, *SbeI* and *SbeII* were mutated singly and in combinations where the mutation frequency up to 72% was obtained using ribonucleoprotein mediated transformation of potato. Almost no detectable branching was observed in plants having a combination of mutated genes. Potato crop was also modified to produce high amylopectin using CRISPR/cas9 by targeting the granule-bound starch synthase (*GBSS*) gene at different regions and the use of different promoters drive expression of CRISPR cassette that resulted in a reduction in the level of amylose and ultimately enhanced the level of amylopectin (Andersson et al. 2017). In another study, the same gene was targeted to generate barley with enhanced levels of amylopectin (Zhong et al. 2019).

Antinutritional factors such as lectins, phytic acids, gossypols, protease, inhibitors, glycosides and saponins present in plants are responsible for adverse health effects in humans and animals (Thakur et al. 2019). Phytic acid majorly present in cereals and oilseeds chelates metabolic ions such as zinc and iron and makes them insoluble for absorption (Bloot et al. 2023). The most potent gene involved in the biosynthesis of phytic acid *ITPK* inositol 1,4,5-trisphosphate 3-kinase has been targeted by CRISPR/Cas9 in *Brassica napus* that reduced the amount of phytic acid up to 35% thereby enhancing free phosphorus (Sashidhar et al. 2020). Glucosinolate present in the *Brassica* family is the prime reason for lesser acceptance of Brassica oilseeds as it gives a pungent smell as well as causes ill effects in livestock when used as feed. However, glucosinolate plays a crucial role in plant defense against herbivores and pathogens (Lv et al. 2022). Hence, the glucosinolate transporter becomes a prime target for the level of glucosinolate in particular tissues in plants. Successful glucosinolate reduction from 146.09 μmoles/g DW to as low as 6.21 μmoles s/g DW in seeds of *Brassica* was achieved by simultaneously targeting two glucosinolate transporter genes *GTR1* and *GTR2* by CRISPR/Cas9 but the level of glucosinolate remains constant in other parts of plant like leaves that aids in defense mechanism of plants (Mann et al. 2023). Another anti-nutritional factor, steroidal glycosides was also reduced in potatoes by CRISPR/Cas9 mediated knockout of the *St16DOX* gene.

Global demand for edible oil boosted the cultivation of oilseed crops such as *Brassica napus* and *B. juncea*. The quality of oil with a particular level of MUFA plays an important role in acceptance and global demand. Several efforts have been made to improve different characteristics of oil, particularly in *Brassica napus*. Editing of gene *FAD2* using CRISPR/Cas9 reduced the desaturation of oleic acid and its conversion to polyunsaturated fatty acids (PUFA) which enhanced the thermal stability of oil (Okuzaki et al. 2018). In another study, mutation of gene *BnTT8* by CRISPR/Cas9 elevated the protein and seed oil content as well as altered fatty acid composition (Zhai et al. 2020). Genome-wide association studies have proved the association of seed fatty acid reducer (SFAR) with seed oil content in *Brassica napus*. In further experiments, *BnSFAR* and *BnSFAR5* genes were knocked out by CRISPR/Cas9 and chemical mutagenesis that resulting in a subsequent enhancement of both seed and protein content (Karunarathna et al. 2020).

9.4 Challenges and Future Directions

High altitudes pose severe and often life-threatening abiotic challenges that exhibit significant variations, not only throughout the yearly winter-summer cycles but also on a daily basis, particularly in high tropical regions. In these high-altitude regions, extreme temperature fluctuations are common, with cold nights and the ever-looming threat of frost posing a significant threat to crop. Additionally, the cold climate prevalent in high-altitude areas results in shorter growing seasons, limiting the types of crops that can be grown and reduces the overall yield potential. Furthermore, the reduced oxygen levels found at elevated altitudes can detrimentally impact crop respiration and metabolic processes, culminating in stunted growth and diminished crop yields. Moreover, the elevated altitudes receive elevated levels of ultraviolet (UV) radiation due to the thinness of the atmospheric layers, with this heightened UV exposure posing a direct threat to plant cells and, consequently, the growth and development of crops.

In addition to these challenges, high-altitude soils tend to be less fertile and exhibit diminished nutrient content compared to their lowland. Consequently, the successful cultivation of crops in high-altitude regions necessitates meticulous soil management practices and the supplementation of essential nutrients. Nonetheless, the crops at higher altitude have displayed remarkable adaptability, enabling them to thrive at elevations exceeding 5000 meters. This resilience is attributed to their specialized physiological processes, which exert a profound influence on chemical response of the crop plants. These processes encompass the synthesis of unique lipids that enhance cell membrane flexibility and water permeability, the production of antifreeze carbohydrates, the development of ultraviolet filters, the presence of potent antioxidants, and the existence of free radical quenchers, all of which are absent in crops at lower altitudes. These intricate biochemical mechanisms have enabled these high-altitude crop plants to overcome the harsh climatic forces that challenge the very essence of life. They achieve this by pioneering novel biosynthetic pathways, resulting in molecular structures not found in their lower elevation. Regrettably, the pristine environments in which these remarkable adaptations have evolved are increasingly being altered by the effects of global warming at a pace that we are only beginning to comprehend. This climatic shift is pushing the boundaries of vegetation higher up mountainsides and peaks.

Genomics has the potential to revolutionize high-altitude crop agriculture further by providing insights into the genetic makeup of plants and enabling the development of crops that are better adapted to the challenging conditions found at higher elevations. Here are some future directions for high-altitude crop improvement using genomics: (i) sequencing and annotation the genomes of high-altitude crops, to understand their genetic makeup (ii) identification of genes and genetic markers associated with traits that are important for high-altitude agriculture, such as cold tolerance, drought resistance, and UV radiation resistance (iii) development of marker-assisted breeding programs that utilize genetic markers linked to desirable traits to accelerate the breeding of high-altitude crop varieties (iv) use of haplotype-based breeding approaches for developing high-yielding climate resilient high-altitude crop cultivars, (v) applying genome editing techniques like CRISPR/Cas9 to introduce specific genetic modifications in crops to enhance their adaptation to high-altitude conditions (vi) using transcriptomics to analyze gene expression patterns in response to altitude-related stressors, helping to identify genes involved in stress responses and adaptation (vii) exploring the microbiome of high-altitude soils and plants to understand how microbial communities contribute to crop health and soil fertility (viii) combining genomics with advanced phenotyping technologies to assess how specific genetic variations impact plant growth, development, and stress responses. These future directions in high-altitude crop genomics hold the promise of developing crops that can thrive in challenging high-altitude environments, contribute to food security, and improve the livelihoods of communities living at higher elevations.

9.5 Conclusions

The accessibility of genomic resources, when combined with genotyping platforms, has become a crucial factor in the improvement of crops cultivated at high altitudes. Plant breeders now possess the means to harness the genetic potential of crop varieties by utilizing a wide array of genomic resources. Looking ahead, the ongoing advancement of high-throughput genotyping and phenotyping techniques holds the potential to furnish us with a comprehensive understanding of gene functionality and associations between genomic regions and traits. This, in turn, will strengthen our capacity to develop high-yield crops that are resilient to climate variations in high-altitude regions (Fig. 9.2).

To sum it up, the presence and utilization of genomic resources, along with inventive genotyping technologies, have inaugurated a new era in the field of crop breeding for high-altitude environments. As we dive deeper into the realm of genomics, the horizon for enhancing high-altitude crops continues to broaden, promising increased yield, resilience, and nutritional value. These developments are poised to address the challenges posed by a burgeoning global population and an ever-changing environment.

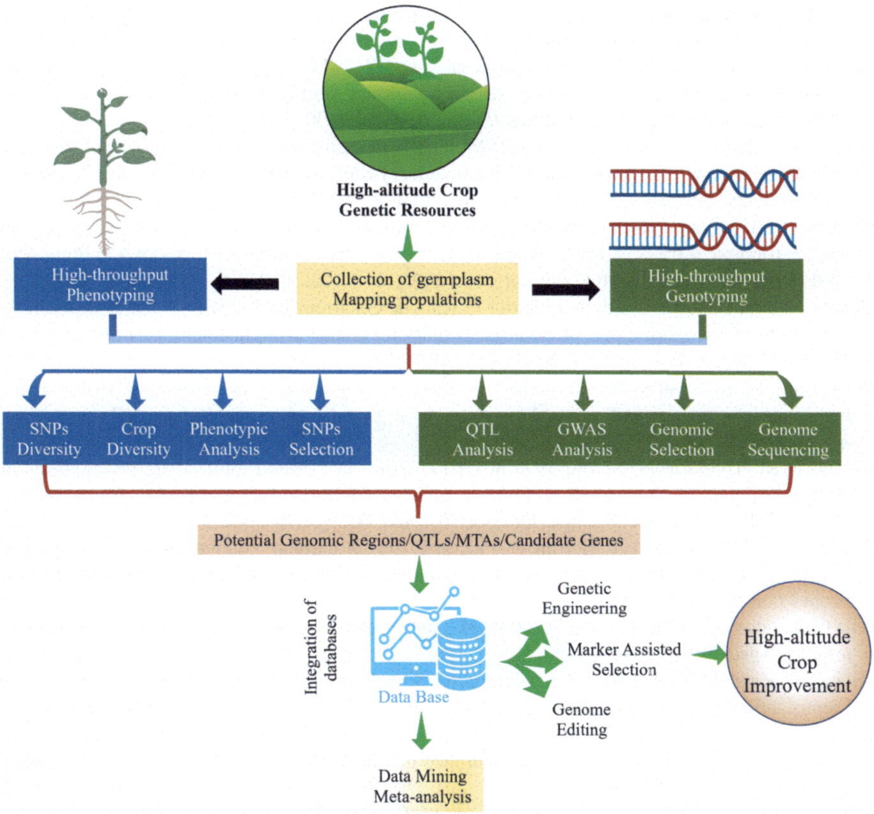

Fig. 9.2 Depiction of the integration of genotyping and phenotyping techniques for high-altitude crop improvement

References

Abdel-Ghani AH, Sharma R, Wabila C et al (2019) Genome-wide association mapping in a diverse spring barley collection reveals the presence of QTL hotspots and candidate genes for root and shoot architecture traits at seedling stage. BMC Plant Biol 19:1–19. https://doi.org/10.1186/s12870-019-1828-5

Abo-Elyousr KAM, Mourad AMI, Baenziger PS et al (2022) Identification of putative SNP markers associated with resistance to Egyptian loose smut race(s) in spring barley. Genes (Basel) 13:1–15. https://doi.org/10.3390/genes13061075

Abou-Elwafa SF, Shehzad T (2021) Genetic diversity, GWAS and prediction for drought and terminal heat stress tolerance in bread wheat (*Triticum aestivum* L.). Genet Resour Crop Evol 68:711–728. https://doi.org/10.1007/s10722-020-01018-y

Adhikari S, Joshi A, Kumar A et al (2022) Identification of QTL for banded leaf and sheath blight in teosinte-derived maize population. Agric Res 11:155–163. https://doi.org/10.1007/s40003-021-00567-7

Ahmed AAM, Mohamed EA, Hussein MY, Sallam A (2021) Genomic regions associated with leaf wilting traits under drought stress in spring wheat at the seedling stage revealed by GWAS. Environ Exp Bot 184:104393. https://doi.org/10.1016/j.envexpbot.2021.104393

Allam M, Revilla P, Djemel A et al (2016) Identification of QTLs involved in cold tolerance in sweet × field corn. Euphytica 208:353–365. https://doi.org/10.1007/s10681-015-1609-7

Almeida GD, Makumbi D, Magorokosho C et al (2013) QTL mapping in three tropical maize populations reveals a set of constitutive and adaptive genomic regions for drought tolerance. Theor Appl Genet 126:583–600. https://doi.org/10.1007/s00122-012-2003-7

Almeida GD, Nair S, Borém A et al (2014) Molecular mapping across three populations reveals a QTL hotspot region on chromosome 3 for secondary traits associated with drought tolerance in tropical maize. Mol Breed 34:701–715. https://doi.org/10.1007/s11032-014-0068-5

Alotaibi FS, Al-Qthanin RN, Aljabri M et al (2022) Identification of genomic regions associated with agronomical traits of bread Wheat under two levels of salinity using GWAS. Plant Mol Biol Report 40:595–609. https://doi.org/10.1007/s11105-022-01341-x

Alvarez-Morezuelas A, Barandalla L, Ritter E, Ruiz de Galarreta JI (2023) Genome-wide association study of agronomic and physiological traits related to drought tolerance in potato. Plan Theory 12. https://doi.org/10.3390/plants12040734

Amouzoune M, Rehman S, Benkirane R et al (2022) Genome-wide association study of leaf rust resistance at seedling and adult plant stages in a global barley panel. Agric 12. https://doi.org/10.3390/agriculture12111829

Andersson M, Turesson H, Nicolia A et al (2017) Efficient targeted multiallelic mutagenesis in tetraploid potato (*Solanum tuberosum*) by transient CRISPR/Cas9 expression in protoplasts. Plant Cell Rep 36:117–128. https://doi.org/10.1007/s00299-016-2062-3

Arias LA, Berli F, Fontana A, Bottini R, Piccoli P (2022) Climate change effects on grapevine physiology and biochemistry: benefits and challenges of high altitude as an adaptation strategy. Front Plant Sci 13:835425

Avila CM, Satovic Z, Sillero JC et al (2005) QTL detection for agronomic traits in faba bean (*Vicia faba* l.). Agric Conspec Sci 70:65–73

Ávila CM, Ruiz-Rodríguez MD, Cruz-Izquierdo S, Atienza SG, Cubero JI, Torres AM (2017) Identification of plant architecture and yield-related QTL in Vicia faba L. Mol Breed 37:1–3

Azam F, Chang X, Jing R (2015) Mapping QTL for chlorophyll fluorescence kinetics parameters at seedling stage as indicators of heat tolerance in wheat. Euphytica 202:245–258. https://doi.org/10.1007/s10681-014-1283-1

Babu S, Yadav GS, Singh R, Avasthe RK, Das A, Mohapatra KP, Tahashildar M, Kumar K, Prabha M, Devi MT, Rana DS (2018) Production technology and multifarious uses of buckwheat (Fagopyrum spp.): a review. Indian J Agron 63(4):415–427

Badu-Apraku B, Talabi AO, Fakorede MAB et al (2019) Yield gains and associated changes in an early yellow bi-parental maize population following genomic selection for Striga resistance and drought tolerance. BMC Plant Biol 19:1–17. https://doi.org/10.1186/s12870-019-1740-z

Bajwa AA, Farooq M, Al-Sadi AM et al (2020) Impact of climate change on biology and management of wheat pests. Crop Prot 137:105304. https://doi.org/10.1016/j.cropro.2020.105304

Banoo A, Nabi A, Rasool RS et al (2020) North-Western Himalayan common beans: population structure and mapping of quantitative anthracnose resistance through genome wide association study. Front Plant Sci 11:1–18. https://doi.org/10.3389/fpls.2020.571618

Barakat MN, Saleh MS, Al-Doss AA et al (2015) Mapping of QTLs associated with abscisic acid and water stress in wheat. Biol Plant 59:291–297. https://doi.org/10.1007/s10535-015-0499-9

Bargali SS, Shahi C, Bargali K, Negi B, Khatri K (2022) Energy and monetary efficiencies at the different altitudinal agroecosystems in central Himalaya, India. Heliyon. 8(11):e11500

Bazile D, Jacobsen SE, Verniau A (2016) The global expansion of quinoa: trends and limits. Front Plant Sci 7:622

Beyene Y, Semagn K, Mugo S et al (2015) Genetic gains in grain yield through genomic selection in eight bi-parental maize populations under drought stress. Crop Sci 55:154–163. https://doi.org/10.2135/cropsci2014.07.0460

Bhatta M, Morgounov A, Belamkar V, Baenziger PS (2018) Genome-wide association study reveals novel genomic regions for grain yield and yield-related traits in drought-stressed synthetic hexaploid wheat. Int J Mol Sci 19. https://doi.org/10.3390/ijms19103011

Bloot AP, Kalschne DL, Amaral JA, Baraldi IJ, Canan C (2023) A review of phytic acid sources, obtention, and applications. Food Rev Intl 39(1):73–92

Bojarczuk A, Skąpska S, Khaneghah AM, Marszałek K (2022) Health benefits of resistant starch: a review of the literature. J Funct Foods 93:105094. https://doi.org/10.1016/j.jff.2022.105094

Borrego-Benjumea A, Carter A, Zhu M et al (2021) Genome-wide association study of waterlogging tolerance in barley (*Hordeum vulgare* L.) under controlled field conditions. Front Plant Sci 12:10–16. https://doi.org/10.3389/fpls.2021.711654

Borrell AK, Mullet JE, George-Jaeggli B et al (2014) Drought adaptation of stay-green sorghum is associated with canopy development, leaf anatomy, root growth, and water uptake. J Exp Bot 65:6251–6263

Bradshaw JE, Hackett CA, Pande B, Waugh R et al (2008) QTL mapping of yield, agronomic and quality traits in tetraploid potato (Solanum tuberosum subsp. tuberosum). Theor Appl Genet 116(2):193–211. https://doi.org/10.1007/s00122-007-0659-1

Brainard SH, Ellison SL, Simon PW, Dawson JC, Goldman IL (2021) Genetic characterization of carrot root shape and size using genome-wide association analysis and genomic-estimated breeding values. Theor Appl Genet 135(2):605–622. https://doi.org/10.1007/s00122-021-03988-8

Carpenter MA, Goulden DS, Woods CJ et al (2018) Genomic selection for ascochyta blight resistance in pea. Front Plant Sci 871:1–13. https://doi.org/10.3389/fpls.2018.01878

Che Z, Yan H, Liu H et al (2020) Genome-wide association study for soybean mosaic virus SC3 resistance in soybean. Mol Breed 40. https://doi.org/10.1007/s11032-020-01149-1

Choudhary N, Bawa V, Paliwal R et al (2018) Gene/QTL discovery for anthracnose in common bean (Phaseolus vulgaris L.) from North-Western Himalayas. PLoS One 13:1–12. https://doi.org/10.1371/journal.pone.0191700

Chu J, Li W, Piao D et al (2021) Identification of a major QTL related to resistance to soybean mosaic virus in diverse soybean genetic populations. Euphytica 217:1–11. https://doi.org/10.1007/s10681-021-02907-8

Clemente A, Arques MC, Dalmais M, Le Signor C, Chinoy C et al (2015) Correction: eliminating anti-nutritional plant food proteins: the case of seed protease inhibitors in pea. PLoS One 10(9):e0138039. https://doi.org/10.1371/journal.pone.0138039

Coser SM, Reddy RVC, Zhang J et al (2017) Genetic architecture of charcoal rot (*Macrophomina phaseolina*) resistance in soybean revealed using a diverse panel. Front Plant Sci 8:1–12. https://doi.org/10.3389/fpls.2017.01626

Czembor JH, Czembor E (2022) Genome-wide association study of agronomic traits in European spring barley from polish Gene Bank. Agronomy 12. https://doi.org/10.3390/agronomy12092135

da Silva MP, Pereira A, Rupe JC et al (2019) Effectiveness of a seed plate assay for evaluating charcoal rot resistance in soybean and the relationship to field performance. Plant Dis 103:1947–1953. https://doi.org/10.1094/PDIS-10-18-1908-RE

Daetwyler HD, Bansal UK, Bariana HS et al (2014) Genomic prediction for rust resistance in diverse wheat landraces. Theor Appl Genet 127:1795–1803. https://doi.org/10.1007/s00122-014-2341-8

Dame J, Nüsser M (2011) Food security in high mountain regions: agricultural production and the impact of food subsidies in Ladakh, Northern India. Food Sec 3:179–194. https://doi.org/10.1007/s12571-011-0127-2

Das A, Thoithoi Devi M, Babu S, Ansari M, Layek J, Bhowmick SN, Yadav GS, Singh R (2018) Cereal-legume cropping system in indian himalayan region for food and environmental sustainability. Legumes Soil Health Sustain Manag:33–76

Das RR, Vinayan MT, Patel MB et al (2020) Genetic gains with rapid-cycle genomic selection for combined drought and waterlogging tolerance in tropical maize (Zea mays L.). Plant Genome 13:1–15. https://doi.org/10.1002/tpg2.20035

Diapari M, Sindhu A, Warkentin TD, Bett K, Tar'an B (2015) Population structure and marker-trait association studies of iron, zinc and selenium concentrations in seed of field pea (Pisum sativum L.). Mol Breed 35:1–14. https://doi.org/10.1007/s11032-015-0252-2

Domoney C, Welham T, Ellis N et al (1994) Inheritance of qualitative and quantitative trypsin inhibitor variants in *Pisum*. Theoret Appl Genet 89:387–391. https://doi.org/10.1007/BF00225370

dos Santos JPR, Pires LPM, de Castro Vasconcellos RC et al (2016) Genomic selection to resistance to Stenocarpella maydis in maize lines using DArTseq markers. BMC Genet 17:1–10. https://doi.org/10.1186/s12863-016-0392-3

Du X, Xu W, Peng C et al (2021) Identification and validation of a novel locus, Qpm-3BL, for adult plant resistance to powdery mildew in wheat using multilocus GWAS. BMC Plant Biol 21:1–13. https://doi.org/10.1186/s12870-021-03093-4

Dunemann F, He W, Böttcher C, Reichardt S, Nothnagel T, Heuvelmans P, Hermans F (2022) The genetic control of polyacetylenes involved in bitterness of carrots (Daucus carota L.): identification of QTLs and candidate genes from the plant fatty acid metabolism. BMC Plant Biol 22(1):92. https://doi.org/10.1186/s12870-022-03484-1

Elakhdar A, Slaski JJ, Kubo T et al (2023) Genome-wide association analysis provides insights into the genetic basis of photosynthetic responses to low-temperature stress in spring barley. Front Plant Sci 14:1–22. https://doi.org/10.3389/fpls.2023.1159016

Enciso-Rodriguez F, Douches D, Lopez-Cruz M et al (2018) Genomic selection for late blight and common scab resistance in tetraploid potato (Solanum tuberosum). G3 Genes Genom Genet 8:2471–2481. https://doi.org/10.1534/g3.118.200273

Feng L, Lin H, Kang M, Ren Y, Yu X, Xu Z, Wang S, Li T, Yang W, Hu Q (2022) A chromosome-level genome assembly of an alpine plant Crucihimalaya lasiocarpa provides insights into high-altitude adaptation. DNA Res 29(1):dsac004

Li X, El Solh M, Siddique KH (2019) Mountain agriculture: Opportunities for harnessing zero hunger in Asia. Food and Agriculture Organization of the United Nations, Regional Office for Asia and the Pacific

Franco MF, Polacco AN, Campos PE et al (2022) Genome-wide association study for resistance in bread wheat (*Triticum aestivum* L.) to stripe rust (Puccinia striiformis f. sp. Tritici) races in Argentina. BMC Plant Biol 22:1–17. https://doi.org/10.1186/s12870-022-03916-y

Frey FP, Presterl T, Lecoq P et al (2016) First steps to understand heat tolerance of temperate maize at adult stage: identification of QTL across multiple environments with connected segregating populations. Theor Appl Genet 129:945–961. https://doi.org/10.1007/s00122-016-2674-6

Gage JL, Monier B, Giri A, Buckler ES (2020) Ten years of the maize nested association mapping population: impact, limitations, and future directions. Plant Cell 32(7):2083–2093

Gahlaut V, Jaiswal V, Tyagi BS et al (2017) QTL mapping for nine drought-responsive agronomic traits in bread wheat under irrigated and rain-fed environments. PLoS One 12:1–27. https://doi.org/10.1371/journal.pone.0182857

Gali KK, Liu Y, Sindhu A, Diapari M, Shunmugam AS, Arganosa G, Daba K, Caron C, Lachagari RV, Tar'an B, Warkentin TD (2018) Construction of high-density linkage maps for mapping quantitative trait loci for multiple traits in field pea (Pisum sativum L.). BMC Plant Biol 18(1):1–25. https://doi.org/10.1186/s12870-018-1368-4

Gao C (2021) Genome engineering for crop improvement and future agriculture. Cell 184:1621–1635. https://doi.org/10.1016/j.cell.2021.01.005

Gautam R, Kandel BP, Chalaune S, Koirala B (2022) Importance of world high altitude Jumli Marshi rice with cultivation practices. Heliyon 8:e08885

Gazal A, Dar ZA, Zaffar G, Lone AA, Abidi I, Shabir A, Khan K, Yousuf N (2014) Trends in breeding oat for nutritional grain quality—an overview. J Appl Nat Sci 6(2):904–912. https://doi.org/10.31018/jans.v6i2.552

Giri A, Bharti VK, Kalia S, Acharya S, Kumar B, Chaurasia OP (2021 Nov) Health risk assessment of heavy metals due to wheat, cabbage, and spinach consumption at cold-arid high altitude region. Biol Trace Elem Res 8:1–3

Golicz AA, Batley J, Edwards D (2016) Towards plant pangenomics. Plant Biotechnol J 14:1099–1105. https://doi.org/10.1111/pbi.12499

Gous PW, Martin A, Lawson W, Kelly A, Fox GP, Sutherland MW (2012) QTL associated with barley (Hordeum vulgare) feed quality traits measured through in situ digestion. Euphytica 185:37–45. https://doi.org/10.1007/s10681-011-0608-6

Gu X, Huang S, Zhu Z et al (2021) Genome-wide association of single nucleotide polymorphism loci and candidate genes for frogeye leaf spot (Cercospora sojina) resistance in soybean. BMC Plant Biol:1–12. https://doi.org/10.1186/s12870-021-03366-y

Guo X, Long R, Kreuzer M, Ding L, Shang Z, Zhang Y, Yang Y, Cui G (2014) Importance of functional ingredients in yak milk-derived food on health of Tibetan nomads living under high-altitude stress: a review. Crit Rev Food Sci Nutr 54(3):292–302

Guo J, Li C, Zhang X et al (2020) Transcriptome and GWAS analyses reveal candidate gene for seminal root length of maize seedlings under drought stress. Plant Sci 292:110380. https://doi.org/10.1016/j.plantsci.2019.110380

Gupta V, He X, Kumar N et al (2019) Genome wide association study of Karnal bunt resistance in a wheat germplasm collection from Afghanistan. Int J Mol Sci 20. https://doi.org/10.3390/ijms20133124

Gyawali S, Otte ML, Chao S, Jilal A, Jacob DL, Amezrou R, Verma RPS (2017) Genome wide association studies (GWAS) of element contents in grain with a special focus on zinc and iron in a world collection of barley (Hordeum vulgare L.). J Cereal Sci 77:266–274. https://doi.org/10.1016/j.jcs.2017.08.019

He X, Li JJ, Chen Y et al (2019) Genome-wide analysis of the WRKY gene family and its response to abiotic stress in buckwheat (Fagopyrum tataricum). Open Life Sci 14:80–96. https://doi.org/10.1515/biol-2019-0010

Hegde N, Joshi S, Soni N, Kushalappa AC (2021) The caffeoyl-CoA O-methyltransferase gene SNP replacement in russet Burbank potato variety enhances late blight resistance through cell wall reinforcement. Plant Cell Rep 40:237–254. https://doi.org/10.1007/s00299-020-02629-6

Herrmann MH, Yu J, Beuch S, Weber WE (2014) Quantitative trait loci for quality and agronomic traits in two advanced backcross populations in oat (Avena sativa L.). Plant Breed 133:588–601

Hirsch CN, Foerster JM, Johnson JM et al (2014) Insights into the maize pan-genome and pan-transcriptome. Plant Cell 26:121–135. https://doi.org/10.1105/tpc.113.119982

Hoque ABMZ, Avenue MM (2022) Thesis (MS)-A-B-M-Zahidul-Hoque-2014

Hou Y, Lu Q, Su J et al (2022) Genome-wide analysis of the HDAC gene family and its functional characterization at low temperatures in Tartary buckwheat (Fagopyrum tataricum). Int J Mol Sci 23. https://doi.org/10.3390/ijms23147622

Huang CW, Liang WH, Esvelt Klos K, Chen CS, Huang YF (2020) Evaluation of agronomic performance and exploratory genome-wide association study of a diverse oat panel for forage use in Taiwan. Grassl Sci 66(4):249–260

Hussain B, Lucas SJ, Ozturk L, Budak H (2017) Mapping QTLs conferring salt tolerance and micronutrient concentrations at seedling stagein wheat. Sci Rep 7:1–14. https://doi.org/10.1038/s41598-017-15726-6

Id IM, Shimelis H, Isaac A, et al. (2019) Genome-wide association study of drought tolerance and biomass allocation in wheat. 1–21

Illayda B (2017) 2018_Pnas_Si_Spe. Proc Natl Acad Sci 120:2017. https://doi.org/10.1073/pnas

Jacobsen SE (2003) The worldwide potential for quinoa (Chenopodium quinoa Willd.). Food Rev Intl 19(1–2):167–177

Jaiswal V, Bandyopadhyay T, Gahlaut V, Gupta S, Dhaka A, Ramchiary N, Prasad M (2019) Genome-wide association study (GWAS) delineates genomic loci for ten nutritional elements in foxtail millet (Setaria italica L.). J Cereal Sci 85:48–55. https://doi.org/10.1016/j.jcs.2018.11.006

Jayakodi M, Padmarasu S, Haberer G et al (2020) The barley pan-genome reveals the hidden legacy of mutation breeding. Nature 588:284–289. https://doi.org/10.1038/s41586-020-2947-8

Jha AB, Tar'an B, Diapari M, Warkentin TD (2015) SNP variation within genes associated with amylose, total starch and crude protein concentration in field pea. Euphytica 206:459–471. https://doi.org/10.1007/s10681-015-1510-4

Jha UC, Nayyar H, von Wettberg EJB et al (2022) Legume Pangenome: status and scope for crop improvement. Plan Theory 11. https://doi.org/10.3390/plants11223041

Jia M, Yang L, Zhang W et al (2020) Genome-wide association analysis of stripe rust resistance in modern Chinese wheat. BMC Plant Biol 20:1–13. https://doi.org/10.1186/s12870-020-02693-w

Jing Y, Teng W, Qiu L et al (2021) Genetic dissection of soybean partial resistance to sclerotinia stem rot through genome wide association study and high throughout single nucleotide polymorphisms. Genomics 113:1262–1271. https://doi.org/10.1016/j.ygeno.2020.10.042

Kakei Y, Yamamoto M, Ishida Y, Yamazaki C, Sato A, Narukawa-Nara M et al (2017) Biochemical and chemical biology study of rice OsTAR1 revealed that tryptophan aminotransferase is involved in auxin biosynthesis: identification of a potent OsTAR1 inhibitor, pyruvamine2031. Plant Cell Physiol 58:598–606

Kamweru I, Anani BY, Beyene Y et al (2022) Genomic analysis of resistance to fall armyworm (Spodoptera frugiperda) in CIMMYT maize lines. Genes (Basel) 13. https://doi.org/10.3390/genes13020251

Kang Y, Barry K, Cao F, Zhou M (2020) Genome-wide association mapping for adult resistance to powdery mildew in common wheat. Mol Biol Rep 47:1241–1256. https://doi.org/10.1007/s11033-019-05225-4

Karunarathna NL, Wang H, Harloff HJ, Jiang L, Jung C (2020) Elevating SEED oil content in a polyploid crop by induced mutations in SEED FATTY ACID REDUCER genes. Plant Biotechnol J 18(11):2251–2266. https://doi.org/10.1111/pbi.13381

Khahani B, Tavakol E, Shariati J, V. (2019) Genome-wide meta-analysis on yield and yield-related QTLs in barley (Hordeum vulgare L.). Mol Breeding 39:56. https://doi.org/10.1007/s11032-019-0962-y

Khan A, Khan S, Jan A (2017) Health complication caused by protein deficiency. J Food Sci Nutr 1:645–647

Kieu NP, Lenman M, Wang ES et al (2021) Mutations introduced in susceptibility genes through CRISPR/Cas9 genome editing confer increased late blight resistance in potatoes. Sci Rep 11:1–12. https://doi.org/10.1038/s41598-021-83972-w

Kis A, Hamar É, Tholt G et al (2019) Creating highly efficient resistance against wheat dwarf virus in barley by employing CRISPR/Cas9 system. Plant Biotechnol J 17:1004–1006. https://doi.org/10.1111/pbi.13077

Kisten L, Tolmay VL, Mathew I et al (2020) Genome-wide association analysis of Russian wheat aphid (Diuraphis noxia) resistance in Dn4 derived wheat lines evaluated in South Africa. PLoS One 15:1–22. https://doi.org/10.1371/journal.pone.0244455

Klaassen MT, Willemsen JH, Vos PG, Visser RG, van Eck HJ, Maliepaard C, Trindade LM (2019) Genome-wide association analysis in tetraploid potato reveals four QTLs for protein content. Mol Breed 39:1–12. https://doi.org/10.1186/s12870-022-03484-1

Klein A, Houtin H, Rond-Coissieux C, Naudet-Huart M, Touratier M, Marget P, Burstin J (2020) Meta-analysis of QTL reveals the genetic control of yield-related traits and seed protein content in pea. Sci Rep 10(1):15925. https://doi.org/10.1038/s41598-020-72548-9

Krajewski P, Bocianowski J, Gawłowska M et al (2012) QTL for yield components and protein content: a multi-environment study of two pea (*Pisum sativum* L.) populations. Euphytica 183:323–336. https://doi.org/10.1007/s10681-011-0472-4

Kumar S, Chauhan JS, Kumar A (2010) Screening for erucic acid and glucosinolate content in rapeseed-mustard seeds using near infrared reflectance spectroscopy. J Food Sci Technol 47:690–692. https://doi.org/10.1007/s13197-010-0120-3

Kumar M, Baishaya LK, Ghosh DC, Gupta VK, Dubey SK, Das A, Patel DP (2012) Productivity and soil health of potato (Solanum tuberosum L.) field as influenced by organic manures, inorganic fertilizers and biofertilizers under high altitudes of eastern Himalayas. J Agric Sci 4(5):223

Kumar S, Knox RE, Singh AK et al (2018) High-density genetic mapping of a major QTL for resistance to multiple races of loose smut in a tetraploid wheat cross. PLoS One 13:1–17. https://doi.org/10.1371/journal.pone.0192261

Latham MC (1990) Protein-energy malnutrition—its epidemiology and control. J Environ Pathol Toxicol Oncol 10(4–5):168–180

Li W, Teng F, Li T, Zhou Q (2013) Simultaneous generation and germline transmission of multiple gene mutations in rat using CRISPR-Cas systems. Nat Biotechnol 31:684–686. https://doi.org/10.1038/nbt.2652

Li M, Chen R, Jiang Q et al (2021) GmNAC06, a NAC domain transcription factor enhances salt stress tolerance in soybean. Plant Mol Biol 105:333–345. https://doi.org/10.1007/s11103-020-01091-y

Li Y, Yu S, Zhang Q et al (2022) Genome-wide identification and characterization of the CCT gene family in Foxtail millet (Setaria italica) response to diurnal rhythm and abiotic stress. Genes (Basel) 13. https://doi.org/10.3390/genes13101829

Li R, Chen Z, Zheng R et al (2023) QTL mapping and candidate gene analysis for yield and grain weight/size in Tartary buckwheat. BMC Plant Biol 23:58. https://doi.org/10.1186/s12870-022-04004-x

Liang T, Tao W, Wang Y, Zhou N, Hu W, Zhang T, Liao D, Chen X, Wang X (2023) The extension of vegetable production to high altitudes increases the environmental cost and decreases economic benefits in subtropical regions. Land 12(3):662

Lin F, Chhapekar SS, Vieira CC et al (2022) Breeding for disease resistance in soybean: a global perspective, vol 135. Springer, Berlin Heidelberg, p 3773

Liu Y, Du H, Li P et al (2020) Pan-genome of wild and cultivated soybeans. Cell 182:162–176.e13. https://doi.org/10.1016/j.cell.2020.05.023

Liu S, Fu J, Shang Z et al (2021) Combination of genome-wide association study and QTL mapping reveals the genetic architecture of fusarium stalk rot in maize. Front Agron 2:1–12. https://doi.org/10.3389/fagro.2020.590374

Liu Y, Wang M, Huang Y, et al (2023) Genome-Wide Identification and Analysis of R2R3-MYB Genes Response to Saline–Alkali Stress in Quinoa. Int J Mol Sci 24. https://doi.org/10.3390/ijms24119132

Lopes MS, Rebetzke GJ, Reynolds M (2014) Integration of phenotyping and genetic platforms for a better understanding of wheat performance under drought. J Exp Bot 65:6167–6177. https://doi.org/10.1093/jxb/eru384

Lorenz AJ, Smith KP, Jannink JL (2012) Potential and optimization of genomic selection for fusarium head blight resistance in six-row barley. Crop Sci 52:1609–1621. https://doi.org/10.2135/cropsci2011.09.0503

Lu F, Romay MC, Glaubitz JC et al (2015) High-resolution genetic mapping of maize pan-genome sequence anchors. Nat Commun 6:6914. https://doi.org/10.1038/ncomms7914

Luthar Z, Zhou M, Golob A, Germ M (2020) Breeding buckwheat for increased levels and improved quality of protein. Plan Theory 10(1):14. https://doi.org/10.3390/plants10010014

Lv Q, Li X, Fan B, Zhu C, Chen Z (2022) The cellular and subcellular organization of the glucosinolate–myrosinase system against herbivores and pathogens. Int J Mol Sci 23(3):1577. https://doi.org/10.3390/ijms23031577

Lynch JP, Chimungu JG, Brown KM (2014) Root anatomical phenes associated with water acquisition from drying soil: targets for crop improvement. J Exp Bot 65:6155–6166. https://doi.org/10.1093/jxb/eru162

Ma Y, Coyne CJ, Grusak MA, Mazourek M, Cheng P, Main D, McGee RJ (2017) Genome-wide SNP identification, linkage map construction and QTL mapping for seed mineral concentrations and contents in pea (*Pisum sativum* L.). BMC Plant Biol 17(1):1–17. https://doi.org/10.1186/s12870-016-0956-4

Malik S, Mehboob-Ur-Rahman, Malik TA (2015) Genetic mapping of potential QTLs associated with drought tolerance in wheat. J Anim Plant Sci 25:1032–1040

Mamo BE, Barber BL, Steffenson BJ (2014) Genome-wide association mapping of zinc and iron concentration in barley landraces from Ethiopia and Eritrea. J Cereal Sci 60(3):497–506. https://doi.org/10.1016/j.jcs.2014.08.007

Mann A, Kumari J, Kumar R, Kumar P, Pradhan AK, Pental D, Bisht NC (2023) Targeted editing of multiple homologues of GTR1 and GTR2 genes provides the ideal low-seed, high-leaf glucosinolate oilseed mustard with uncompromised defence and yield. Plant Biotechnol J 21(11):2182–2195. https://doi.org/10.1111/pbi.14121

Mehrabi AA, Steffenson BJ, Pour-Aboughadareh A et al (2022) Genome-wide association study identifies two loci for stripe rust resistance in a durum Wheat panel from Iran. Appl Sci 12. https://doi.org/10.3390/app12104963

Merrey DJ, Hussain A, Tamang DD, Thapa B, Prakash A (2018) Evolving high altitude livelihoods and climate change: a study from Rasuwa District, Nepal. Food Security 10:1055–1071

Miao C, Xiao L, Hua K, Zou C, Zhao Y, Bressan RA, Zhu JK (2018) Mutations in a subfamily of abscisic acid receptor genes promote rice growth and productivity. Proc Natl Acad Sci U S A 4:774

Mirdita V, He S, Zhao Y et al (2015) Potential and limits of whole genome prediction of resistance to fusarium head blight and Septoria tritici blotch in a vast central European elite winter wheat population. Theor Appl Genet 128:2471–2481. https://doi.org/10.1007/s00122-015-2602-1

Mishra B, Bhandari R, Bhandari KP, Bhandari DM, Luintel N, Dahal A, Poudel S (2023) High-resolution mapping of seasonal crop pattern using sentinel imagery in mountainous region of Nepal: a semi-automatic approach. Geomatics 3(2):312–327

Montenegro JD, Golicz AA, Bayer PE et al (2017) The pangenome of hexaploid bread wheat. Plant J 90:1007–1013. https://doi.org/10.1111/tpj.13515

Mwadzingeni L, Shimelis H, Rees DJG, Tsilo TJ (2017) Genome-wide association analysis of agronomic traits in wheat under drought-stressed and non-stressed conditions. 1–13. https://doi.org/10.1371/journal.pone.0171692

Natukunda MI, Parmley KA, Hohenstein JD et al (2019) Identification and genetic characterization of soybean accessions exhibiting antibiosis and Antixenosis resistance to Aphis glycines (Hemiptera: Aphididae). J Econ Entomol 112:1428–1438. https://doi.org/10.1093/jee/toz017

Okuzaki A, Ogawa T, Koizuka C, Kaneko K, Inaba M, Imamura J et al (2018) CRISPR/Cas9-mediated genome editing of the fatty acid desaturase 2 gene in brassica Napus. Plant Phys Biochem 131:63–69. https://doi.org/10.1016/j.plaphy.2018.04.025

Osman KA, Tang B, Wang Y et al (2013) Dynamic QTL analysis and candidate gene mapping for waterlogging tolerance at maize seedling stage. PLoS One 8:e79305. https://doi.org/10.1371/journal.pone.0079305

Ou C, Sun T, Liu X, Li C, Li M, Wang X, Ren H, Zhao Z, Zhuang F (2022) Detection of chromosomal segments Introgressed from wild species of carrot into cultivars: quantitative trait loci mapping for morphological features in backcross inbred lines. Plants (Basel) 11(3):391. https://doi.org/10.3390/plants11030391

Page R, Huang S, Ronen M, et al (2023) Genome-wide association mapping of rust resistance in Aegilops longissima. 1–18. https://doi.org/10.3389/fpls.2023.1196486

Pan Z, Liu M, Zhao H et al (2020) ZmSRL5 is involved in drought tolerance by maintaining cuticular wax structure in maize. J Integr Plant Biol 62:1895–1909. https://doi.org/10.1111/jipb.12982

Pandey J, Thompson D, Joshi M, Scheuring DC, Koym JW, Joshi V, Vales MI (2023) Genetic architecture of tuber-bound free amino acids in potato and effect of growing environment on the amino acid content. Sci Rep 13(1):13940. https://doi.org/10.1038/s41598-023-40880-5

Paulino JFC, de Almeida CP, Bueno CJ et al (2021) Genome-wide association study reveals genomic regions associated with fusarium wilt resistance in common bean. Genes (Basel) 12. https://doi.org/10.3390/genes12050765

Prihatna C, Barbetti MJ, Barker SJ (2018) A novel tomato fusarium wilt tolerance gene. Front Microbiol 9:1–11. https://doi.org/10.3389/fmicb.2018.01226

Priya M, Nayyar H (2023) Towards developing biofortified food crops for enhancing nutritional aspects and human health. In: Biomedical research, medicine, and disease. CRC Press, pp 571–588

Ramchiary N, Bisht NC, Gupta V, Mukhopadhyay A, Arumugam N, Sodhi YS, Pental D, Pradhan AK (2007) QTL analysis reveals context-dependent loci for seed glucosinolate trait in the oilseed Brassica juncea: importance of recurrent selection backcross scheme for the identification of 'true' QTL. Theor Appl Genet 116:77–85. https://doi.org/10.1007/s00122-007-0648-4

Ramírez Gonzales L, Shi L, Bergonzi SB et al (2021) Potato CYCLING DOF FACTOR 1 and its lncRNA counterpart StFLORE link tuber development and drought response. Plant J 105:855–869. https://doi.org/10.1111/tpj.15093

Rashid Z, Sofi M, Harlapur SI et al (2020) Genome-wide association studies in tropical maize germplasm reveal novel and known genomic regions for resistance to northern corn leaf blight. Sci Rep 10:1–16. https://doi.org/10.1038/s41598-020-78928-5

Rashid Z, Babu V, Sharma SS et al (2022) Identification and validation of a key genomic region on chromosome 6 for resistance to fusarium stalk rot in tropical maize. Theor Appl Genet 135:4549–4563. https://doi.org/10.1007/s00122-022-04239-0

Raza A, Bohra A, Varshney RK (2023) Pan-genome for pearl millet that beats the heat. Trends Plant Sci 28:857–860. https://doi.org/10.1016/j.tplants.2023.04.016

Rebetzke GJ, Van Herwaarden AF, Jenkins C et al (2008) Quantitative trait loci for water-soluble carbohydrates and associations with agronomic traits in wheat. Aust J Agric Res 59:891–905. https://doi.org/10.1071/AR08067

Rehman AU, Malhotra RS, Bett K et al (2011) Mapping QTL associated with traits affecting grain yield in chickpea (Cicer arietinum L.) under terminal drought stress. Crop Sci 51:450–463. https://doi.org/10.2135/cropsci2010.03.0129

Ren Y, Qian Y, Xu Y et al (2017) Characterization of QTLS for root traits of wheat grown under different nitrogen and phosphorus supply levels. Front Plant Sci 8:1–8. https://doi.org/10.3389/fpls.2017.02096

Ren Y, Ma R, Fan Y et al (2022) Genome-wide identification and expression analysis of the SPL transcription factor family and its response to abiotic stress in quinoa (Chenopodium quinoa). BMC Genomics 23:1–17. https://doi.org/10.1186/s12864-022-08977-9

Revilla P, Rodríguez VM, Ordás A et al (2016) Association mapping for cold tolerance in two large maize inbred panels. BMC Plant Biol 16:1–10. https://doi.org/10.1186/s12870-016-0816-2

Rizzardi DA, Peterlini E, Scapim CA et al (2022) Genome wide association study identifies SNPs associated with northern corn leaf blight caused by Exserohilum turcicum in tropical maize germplasm (Zea mays L.). Euphytica 218:1–12. https://doi.org/10.1007/s10681-022-02986-1

Román-Avilés B, Kelly JD (2005) Identification of quantitative trait loci conditioning resistance to fusarium root rot in common bean. Crop Sci 45:1881–1890. https://doi.org/10.2135/cropsci2005.0028

Sadeghzadeh B, Rengel Z, Li C, Yang HA (2010) Molecular marker linked to a chromosome region regulating seed Zn accumulation in barley. Mol Breed 25:167–177. https://doi.org/10.1007/s11032-009-9317-4

Saini DK, Chopra Y, Pal N, Chahal A, Srivastava P, Gupta PK (2021) Meta-QTLs, ortho-MQTLs and candidate genes for nitrogen use efficiency and root system architecture in bread wheat (Triticum aestivum L.). Physiol Mol Biol Plants 27(10):2245–2267. https://doi.org/10.1007/s12298-021-01085-0

Sallam A, Eltaher S, Alqudah AM et al (2022) Genomics combined GWAS and QTL mapping revealed candidate genes and SNP network controlling recovery and tolerance traits associated with drought tolerance in seedling winter wheat. Genomics 114:110358. https://doi.org/10.1016/j.ygeno.2022.110358

Sarinelli JM, Murphy JP, Tyagi P et al (2019) Training population selection and use of fixed effects to optimize genomic predictions in a historical USA winter wheat panel. Theor Appl Genet 132:1247–1261. https://doi.org/10.1007/s00122-019-03276-6

Šárka E, Dvořáček V (2017) Waxy starch as a perspective raw material (a review). Food Hydrocoll 69:402–409. https://doi.org/10.1016/j.foodhyd.2017.03.001

Sashidhar N, Harloff HJ, Potgieter L, Jung C (2020) Gene editing of three BnITPK genes in tetraploid oilseed rape leads to significant reduction of phytic acid in seeds. Plant Biotechnol J 18(11):2241–2250. https://doi.org/10.1111/pbi.13380

Saxena RK, Edwards D, Varshney RK (2014) Structural variations in plant genomes. Briefings Funct Genomics Proteomics 13:296. https://doi.org/10.1093/bfgp/elu016

Schönhals EM, Ding J, Ritter E, Paulo MJ, Cara N, Tacke E, Hofferbert HR, Lübeck J, Strahwald J, Gebhardt C (2017) Physical mapping of QTL for tuber yield, starch content and starch yield in tetraploid potato (*Solanum tuberosum* L.) by means of genome wide genotyping by sequencing and the 8.3 K SolCAP SNP array. BMC Genomics 18(1):1–20. https://doi.org/10.1186/s12864-017-3979-9

Schuster M, Schweizer G, Reissmann S, Kahmann R (2016) Genome editing in Ustilago maydis using the CRISPR-Cas system. Fungal Genet Biol 89:3–9. https://doi.org/10.1016/j.fgb.2015.09.001

Shahinnia F, Geyer M, Schürmann F et al (2022) Genome-wide association study and genomic prediction of resistance to stripe rust in current central and northern European winter wheat germplasm. Theor Appl Genet 135:3583–3595. https://doi.org/10.1007/s00122-022-04202-z

Shao A, Ma W, Zhao X, Hu M, He X, Teng W et al (2017) The auxin biosynthetic TRYPTOPHAN AMINOTRANSFERASE RELATED TaTAR2.1-3A increases grain yield of wheat. Plant Physiol 174:2274–2288

Sharma DK, Torp AM, Rosenqvist E et al (2017) Qtls and potential candidate genes for heat stress tolerance identified from the mapping populations specifically segregating for Fv/Fm in wheat. Front Plant Sci 8. https://doi.org/10.3389/fpls.2017.01668

Shi T, Li R, Zheng R, Chen Q, Li H, Huang J et al (2021) Mapping QTLs for 1000-grain weight and genes controlling hull type using SNP marker in Tartary buckwheat (Fagopyrum tataricum). BMC Genomics 22(1):142

Shibaya T, Kuroda C, Tsuruoka H et al (2022) Identification of QTLs for root color and carotenoid contents in Japanese orange carrot F_2 populations. Sci Rep 12:8063. https://doi.org/10.1038/s41598-022-11544-7

Simeone R, Piarulli L, Nigro D et al (2020) Mapping powdery mildew (Blumeria graminis f. sp. tritici) resistance inwild and cultivated tetraploid wheats. Int J Mol Sci 21:1–19. https://doi.org/10.3390/ijms21217910

Singh S, Mohapatra T, Singh R, Hussain Z (2013) Mapping of QTLs for oil content and fatty acid composition in Indian mustard [Brassica juncea (L.) Czern. And Coss.]. J Plant Biochem Biotechnol 22:80–89. https://doi.org/10.1007/s13562-012-0113-6

Singh RB, Khan S, Chauhan AK, Singh M, Jaglan P, Yadav P, Takahashi T, Juneja LR (2019) Millets as functional food, a gift from Asia to western world. In: The role of functional food security in global health. Academic Press, pp 457–468

Singh S, Sehgal D, Kumar S et al (2020) GWAS revealed a novel resistance locus on chromosome 4D for the quarantine disease Karnal bunt in diverse wheat pre-breeding germplasm. Sci Rep 10:1–11. https://doi.org/10.1038/s41598-020-62711-7

Soleimani B, Lehnert H, Babben S et al (2022) Genome wide association study of frost tolerance in wheat. Sci Rep:1–11. https://doi.org/10.1038/s41598-022-08706-y

Soler-Garzón A, Oladzad A, Beaver J et al (2021) NAC candidate gene marker for bgm-1 and interaction with QTL for resistance to bean Golden yellow mosaic virus in common bean. Front Plant Sci 12:1–18. https://doi.org/10.3389/fpls.2021.628443

Tafesse EG, Gali KK, Reddy Lachagari VB et al (2021) Genome-wide association mapping for heat and drought adaptive traits in pea. Genes (Basel) 12. https://doi.org/10.3390/genes12121897

Talukder SK, Babar MA, Vijayalakshmi K et al (2014) Mapping QTL for the traits associated with heat tolerance in wheat (Triticum aestivum L.). BMC Genet 15:1–13. https://doi.org/10.1186/s12863-014-0097-4

Tang X, Li J, Liu L, Jing H, Zuo W, Zeng Y (2022) Transcriptome analysis provides insights into Potentilla bifurca adaptation to high altitude. Life 12(9):1337

Tarawneh RA, Alqudah AM, Nagel M, Börner A (2020) Genome-wide association mapping reveals putative candidate genes for drought tolerance in barley. Environ Exp Bot 180:104237. https://doi.org/10.1016/j.envexpbot.2020.104237

Tashi G, Zhan H, Xing G et al (2018) Genome-wide identification and expression analysis of heat shock transcription factor family in Chenopodium quinoa Willd. Agronomy 8. https://doi.org/10.3390/agronomy8070103

Thakur A, Sharma V, Thakur A (2019) An overview of anti-nutritional factors in food. Int J Chem Stud 7(1):2472–2479

Tinker NA, Wight CP, Bekele WA et al (2022) Genome analysis in *Avena sativa* reveals hidden breeding barriers and opportunities for oat improvement. Commun Biol 5:474. https://doi.org/10.1038/s42003-022-03256-5

Tock AJ, Fourie D, Walley PG et al (2017) Genome-wide linkage and association mapping of halo blight resistance in common bean to race 6 of the globally important bacterial pathogen. Front Plant Sci 8:1–17. https://doi.org/10.3389/fpls.2017.01170

Trachsel S, Sun D, Sanvicente FM et al (2016) Identification of QTL for early vigor and stay-green conferring tolerance to drought in two connected advanced backcross populations in tropical maize (Zea mays L.). PLoS One 11:1–22. https://doi.org/10.1371/journal.pone.0149636

Turki N, Shehzad T, Harrabi M, Okuno K (2015) Detection of QTLs associated with salinity tolerance in durum wheat based on association analysis. Euphytica 201:29–41. https://doi.org/10.1007/s10681-014-1164-7

Uga Y, Sugimoto K, Ogawa S et al (2013) Control of root system architecture by DEEPER ROOTING 1 increases rice yield under drought conditions. Nat Genet 45:1097–1102. https://doi.org/10.1038/ng.2725

Van Inghelandt D, Melchinger AE, Martinant JP, Stich B (2012) Genome-wide association mapping of flowering time and northern corn leaf blight (Setosphaeria turcica) resistance in a vast commercial maize germplasm set. BMC Plant Biol 12. https://doi.org/10.1186/1471-2229-12-56

Visioni A, Rehman S, Viash SS et al (2020) Genome wide association mapping of spot blotch resistance at seedling and adult plant stages in barley. Front Plant Sci 11:1–13. https://doi.org/10.3389/fpls.2020.00642

Wang Y, Cheng X, Shan Q et al (2014) Simultaneous editing of three homoeoalleles in hexaploid bread wheat confers heritable resistance to powdery mildew. Nat Biotechnol 32:947–951. https://doi.org/10.1038/nbt.2969

Wang J, Sun G, Ren X et al (2016) QTL underlying some agronomic traits in barley detected by SNP markers. BMC Genet 17:103. https://doi.org/10.1186/s12863-016-0409-y

Wang W, James S, Pan Q et al (2018) Gene editing and mutagenesis reveal inter-cultivar differences and additivity in the contribution of TaGW2 homoeologues to grain size and weight in wheat. Theor Appl Genet 131:2463–2475

Wang X, Liu S, Zuo H, Zheng W, Zhang S, Huang Y, Pingcuo G, Ying H, Zhao F, Li Y, Liu J (2021) Genomic basis of high-altitude adaptation in Tibetan Prunus fruit trees. Curr Biol 31(17):3848–3860

Warburton ML, Woolfolk SW, Smith JS et al (2023) Genes and genetic mechanisms contributing to fall armyworm resistance in maize. Plant Genome 16:1–13. https://doi.org/10.1002/tpg2.20311

Wolfe, S.A., 2022. Cold-climate aeolian environments

Xiong J, Chen D, Chen Y et al (2023) Current plant biology genome-wide association mapping and transcriptomic analysis reveal key drought-responding genes in barley seedlings. Curr Plant Biol 33:100277. https://doi.org/10.1016/j.cpb.2023.100277

Xu R, Yang Y, Qin R et al (2016) Rapid improvement of grain weight via highly efficient CRISPR/Cas9-mediated multiplex genome editing in rice. J Genet Genom 43:529–532

Xu X, Sharma R, Tondelli A et al (2018) Genome-wide association analysis of grain yield-associated traits in a Pan-European barley cultivar collection. Plant Genome 11(1). https://doi.org/10.3835/plantgenome2017.08.0073

Yan H, Sun M, Zhang Z et al (2023a) Pangenomic analysis identifies structural variation associated with heat tolerance in pearl millet. Nat Genet 55:507–518. https://doi.org/10.1038/s41588-023-01302-4

Yan H, Zhang H, Zhou P, Ren C, Peng Y (2023b) Genome-wide association mapping of QTL underlying Groat protein content of a diverse panel of oat accessions. Int J Mol Sci 24(6):5581. https://doi.org/10.3390/ijms24065581

Yu X, Choi SR, Dhandapani V et al (2016) Quantitative trait loci for morphological traits and their association with functional genes in Raphanus sativus. Front Plant Sci 7:255. https://doi.org/10.3389/fpls.2016.00255

Yun SJ, Gyenis L, Hayes PM et al (2005) Quantitative trait loci for multiple disease resistance in wild barley. Crop Sci 45:2563–2572. https://doi.org/10.2135/cropsci2005.0236

Zaidi PH, Rashid Z, Vinayan MT et al (2015) QTL mapping of agronomic waterlogging tolerance using recombinant inbred lines derived from tropical maize (Zea mays L) germplasm. PLoS One 10:1–14. https://doi.org/10.1371/journal.pone.0124350

Zhai Y, Yu K, Cai S et al (2020) Targeted mutagenesis of BnTT8 homologs controls yellow seed coat development for effective oil production in Brassica napus L. Plant Biotechnol J 18(5):1153–1168. https://doi.org/10.1111/pbi.13281

Zhang H, Cui F, Wang L et al (2013a) Conditional and unconditional QTL mapping of drought-tolerance-related traits of wheat seedling using two related RIL populations. J Genet 92:213–231. https://doi.org/10.1007/s12041-013-0253-z

Zhang X, Tang B, Yu F et al (2013b) Identification of major QTL for waterlogging tolerance using genome-wide association and linkage mapping of maize seedlings. Plant Mol Biol Report 31:594–606. https://doi.org/10.1007/s11105-012-0526-3

Zhang K, He M, Fan Y et al (2021) Resequencing of global Tartary buckwheat accessions reveals multiple domestication events and key loci associated with agronomic traits. Genome Biol 22(1):23. https://doi.org/10.1186/s13059-020-02217-7

Zhang Z, Zheng Y, Zhang J, Wang N, Wang Y, Liu W, Bai S, Xie W (2022) High-altitude genetic selection and genome-wide association analysis of yield-related traits in Elymus sibiricus L. using SLAF sequencing. Front Plant Sci 13:874409

Zhang X, Kuang T, Dong W, Qian Z, Zhang H, Landis JB, Feng T, Li L, Sun Y, Huang J, Deng T (2023a) Genomic convergence underlying high-altitude adaptation in alpine plants. J Integr Plant Biol 65(7):1620–1635

Zhang X, Wen Z, Jiang H et al (2023b) Identification of loci for four important agronomic traits in loose-curd cauliflower based on genome-wide association studies. Horticulturae 9:970. https://doi.org/10.3390/horticulturae9090970

Zhao M, Zhang Z, Zhang S, et al (2006) Quantitative trait loci for resistance to banded leaf and sheath blight in maize. pp 1039–1045. https://doi.org/10.2135/cropsci2005.0166

Zhao Y, Li J, Zhao R, Xu K (2020) Genome-wide association study reveals the genetic basis of cold tolerance in wheat

Zhao X, Jayarathna S, Turesson H et al (2021) Amylose starch with no detectable branching developed through DNA-free CRISPR/Cas9 mediated mutagenesis of two starch branching enzymes in potato. Sci Rep 11(1):4311. https://doi.org/10.1038/s41598-021-83462-z

Zhao Y, Yan X, Zeng Z et al (2023) Integrated genome-wide association study and QTL mapping reveals qSa-3A associated with English grain aphid, Sitobion avenae (Fabricius) resistance in wheat. Pest Manag Sci 79(10):3970–3978. https://doi.org/10.1002/ps.7598

Zhong Y, Blennow A, Kofoed-Enevoldsen O et al (2019) Protein targeting to starch 1 is essential for starchy endosperm development in barley. J Exp Bot 70(2):485–496. https://doi.org/10.1093/jxb/ery398

Zhou J, Gali KK, Jha AB, Tar'an B, Warkentin TD (2022) Identification of quantitative trait loci associated with seed protein concentration in a pea recombinant inbred line population. Genes 13:1531. https://doi.org/10.3390/genes13091531

Zhou J, He W, Wang J et al (2023) The pan-plastome of tartary buckwheat (*Fagopyrum tataricum*): key insights into genetic diversity and the history of lineage divergence. BMC Plant Biol 23:1–13. https://doi.org/10.1186/s12870-023-04218-7

Zhu J, Wu J, Wang L et al (2016) QTL and candidate genes associated with common bacterial blight resistance in the common bean cultivar Longyundou 5 from China. Crop J 4:344–352. https://doi.org/10.1016/j.cj.2016.06.009

Ziems LA, Singh L, Dracatos PM, et al (2023) Characterization of leaf rust resistance in international barley germplasm using genome-wide association studies

Zimmer CM, McNish IG, Klos KE, Oro T, Arruda KM, Gutkoski LC, Pacheco MT, Smith KP, Federizzi LC (2020) Genome-wide association for β-glucan content, population structure, and linkage disequilibrium in elite oat germplasm adapted to subtropical environments. Mol Breed 40(11):103. https://doi.org/10.1007/s11032-020-01182-0

Zuiderveen GH, Padder BA, Kamfwa K et al (2016) Genome-wide association study of anthracnose resistance in andean beans (*Phaseolus vulgaris*). PLoS One 11:1–17. https://doi.org/10.1371/journal.pone.0156391

The manufacturer's authorised representative in the EU is Springer Nature Customer Service Centre GmbH, Europaplatz 3, 69115 Heidelberg, Germany. If you have any concerns regarding our products, please contact ProductSafety@springernature.com

Printed and bound by CPI Group (UK) Ltd, Croydon, CR0 4YY
01/10/2026
02225899-0007